Lecture Notes in Statistics

Volume 221

Lecture Notes in Statistics (LNS) includes research work on topics that are more specialized than volumes in Springer Series in Statistics (SSS). The series editors are currently Peter Bickel, Peter Diggle, Stephen Fienberg, Ursula Gather, and Scott Zeger. Ingram Olkin was an editor of the series for many years.

More information about this series at http://www.springer.com/series/694

Kai-Tai Fang · Min-Qian Liu
Hong Qin · Yong-Dao Zhou

Theory and Application of Uniform Experimental Designs

Science Press
Beijing

Springer

Kai-Tai Fang
Beijing Normal University-Hong Kong Baptist University United International College
Zhuhai, Guangdong, China

and

Institute of Applied Mathematics
Chinese Academy of Sciences
Beijing, China

Min-Qian Liu
School of Statistics and Data Science
Nankai University
Tianjin, China

Hong Qin
Faculty of Mathematics and Statistics
Central China Normal University
Wuhan, Hubei, China

Yong-Dao Zhou
School of Statistics and Data Science
Nankai University
Tianjin, China

ISSN 0930-0325 ISSN 2197-7186 (electronic)
Lecture Notes in Statistics
ISBN 978-981-13-2040-8 ISBN 978-981-13-2041-5 (eBook)
https://doi.org/10.1007/978-981-13-2041-5

Jointly published with Science Press, Beijing, China

The print edition is not for sale in Mainland China. Customers from Mainland China please order the print book from: Science Press.

Library of Congress Control Number: 2018950939

This Springer imprint is published by the registered company Springer Nature Singapore Pte Ltd.
The registered company address is: 152 Beach Road, #21-01/04 Gateway East, Singapore 189721, Singapore

Foreword

Experiment is essential to scientific and industrial areas. How do we conduct experiments so as to lessen the number of trials while still achieving effective results? In order to solve this frequently encountered problem, there exists a special technique called experimental design. The better the design, the more effective the results.

In the 1960s, Prof. Loo-Keng Hua introduced J. Kiefer's method, the "golden ratio optimization method," in China, also known as the Fibonacci method. This method and orthogonal design which were popularly used in industry promoted by Chinese mathematical statisticians are the two types of experimental designs. After these methods became popular, many technicians and scientists used them and made a series of achievements, resulting in huge social and economic benefits. With the development of science and technology, these two methods were not enough. The golden ratio optimization method is the best method to deal with a single variable, assuming the real problem has only one interesting factor. However, this situation is almost impossible. This is why we only consider one most important factor and fix the others. Therefore, the golden ratio optimization method is not a very accurate approximation method. Orthogonal design is based on Latin square theory and group theory and can be used to do multifactor experiments. Consequently, the number of trials is greatly reduced for all combinations of different levels of factors. However, for some industrial or expensive scientific experiments, the number of trials is still too high and cannot be facilitated.

In 1978, due to the need for missile designs, a military unit proposed a five-factor experiment, where the level of every factor should be higher than 18 and the total number of trials should be not larger than 50. Neither the golden ratio optimization method nor orthogonal design could be applied. Several years before 1978, Prof. Kai-Tai Fang asked me about an approximate calculation of a multiple integration problem. I introduced him to use the number-theoretical methods for solving that problem, which inspired him to think of using number-theoretical methods in the design of the problem. After a few months of research, we put forward a new type of experimental designs that is known as uniform design. This method had been successfully applied to the design of missiles. After our article was published in the early 1980s, uniform design has been widely applied in China and has resulted in a series of gratifying achievements.

Uniform design belongs to the quasi-Monte Carlo methods or number-theoretical methods, developed for over 60 years. When the calculation of a single variable problem (the original problem) is generalized to a multivariable problem, the calculation complexity is often related to the number of variables. Even if computational technology advances greatly, this method is still impossible in application. Ulam and Von Neumann proposed the Monte Carlo method (i.e., statistical stimulation) in the 1950s. The general idea of this method is to put an analysis problem into a probability problem with the same solution and then use a statistical simulation to deal with the latter. This solves some difficult analysis, including the approximate calculation of multiple definite integrals. The key to the Monte Carlo method is to find a set of random numbers to serve as a statistical simulation sample. Thus, the accuracy of this method lies in the uniformity and independence of random numbers.

In the late 1950s, some mathematicians tried to use deterministic methods to find evenly distributed points in space in order to replace the random numbers used in the Monte Carlo method. The set of points that had been found was by using number theory. According to the measure defined by Weyl, the uniformity (of a uniform design) is good, but the independence is relatively poor. By using these points to replace the random numbers used in the Monte Carlo method, we usually get more precise results. This kind of method is called a quasi-Monte Carlo method, or the number-theoretical method. Mathematicians successfully applied this method into approximate numerical calculations for multiple integrals.

In statistics, pseudo-random numbers can be regarded as representative points of the uniform distribution (in cubed units). Numerical integration requires a large sample, but uniform design just uses small samples. Since the sample is more uniform than orthogonal designs, it is preferred for settling the experiment. Of course, when seeking a small sample, the method of seeking a large sample can be used as a reference.

Uniform design is only one of the applications of the number-theoretical method, which is also widely used in other areas, such as the establishment of multiple interpolation formulas, the approximate solutions of systems of some integrals or differential equations, the global extremes of the functions, the approximate representation points for some multivariate distributions, and some problems for statistical inference, such as multivariate normality test and the sphericity test.

When the Monte Carlo method was first discovered in the late 1950s, Prof. Loo-Keng Hua initiated and led a study of this method in China. Loo-Keng Hua and his pioneering results were summarized in our monograph titled "Applications of Number Theory to Numerical Analysis" published in Springer-Verlag Science Press in 1981. These results are one of the important backgrounds and reference materials for my work with Prof. Kai-Tai Fang.

I have worked with Prof. Kai-Tai Fang for nearly 40 years. As a mathematician and a statistician with long-term valuable experience in popularizing mathematical statistics in Chinese industrial sector, he has excellent insight and experience in applied mathematics. He always provided valuable research questions and possible ways to solve the problem in a timely manner. Our cooperation has been pleasant and fruitful, and the results were summarized in our monograph "Number-Theoretic Methods in Statistics" published by Chapman and Hall in 1994.

This book focuses on the theory and application of uniform designs, but also includes many latest results in the past 20 years. I strongly believe that this book will be important for further development and application of uniform designs. I would like to take this opportunity to wish the book success.

Beijing, China

Yuan Wang
Academician of Chinese Academy
of Sciences

Preface

The purpose of this book is to introduce theory, methodology, and applications of the *Uniform experimental design*. The uniform experimental design can be regarded as a fractional factorial design with model uncertainty, a space-filling design for computer experiments, a robust design against the model specification, a supersaturated design and can be applied to experiments with mixtures. The book provides necessary knowledge for the reader who is interested in developing theory of the uniform experimental design.

The *experimental design* is extremely useful in multifactor experiments and has played an important role in industry, high tech, sciences and various fields. Experimental design is a branch of statistics with a long history. It involves rich methodologies and various designs. Comprehensive reviews for various kinds of designs can be found in *Handbook of Statistics, Vol. 13*, edited by S. Ghosh and C. R. Rao.

Most of the traditional experimental designs, like fractional factorial designs and optimum designs, have their own statistical models. The model for a factorial plan wants to estimate the main effects of the factors and some interactions among the factors. The optimum design considers a regression model with some unknown parameters to be estimated. However, the experimenter may not know the underlying model in many case studies. How to choose experimental points on the domain when the underlying model is unknown is a challenging problem. The natural idea is to spread experimental points uniformly distributed on the domain. A design that chooses experimental points uniformly scattered on the domain is called *uniform experimental design* or *uniform design* for simplicity. The uniform design was proposed in 1980 by Fang and Wang (Fang 1980; Wang and Fang 1981) and has been widely used for thousands of industrial experiments with model unknown.

Computer experiments are for simulations of physical phenomena which are governed by a set of equations including linear, nonlinear, ordinary, and partial differential equations or by several softwares. There is no analytic formula to describe the phenomena. The so-called space-filling design becomes a key part of computer simulation. In fact, the uniform design is one of the space-filling designs.

Computer experiment is a hot topic in the past decades. It involves two parts: design and modeling. The book focuses on the theory of construction of uniform designs and connections among the uniform design, orthogonal array, combinatorial design, supersaturated design, and experiments with mixtures. There are many useful techniques in the literature, such as polynomial regression models, Kriging models, wavelets, Bayesian approaches, neural networks as well as various methods for variable selection. This book gives a brief introduction to some of these methods; the reader can refer to Fang et al. (2006) for details of these methods.

There are many other space-filling designs among which the Latin hypercube sampling has been widely used. Santner et al. (2003) and Fang et al. (2006) give the details of the Latin hypercube sampling.

The book involves eight chapters. Chapter 1 gives an introduction to various experiments and their models. The reader can easily understand the key idea and method of the uniform experimental design from a demo experiment. Many basic concepts are also reviewed. Chapter 2 concerns with various measures of uniformity and introduces their definitions, computational formula, and properties. Many useful lower bounds are derived. There are two chapters for the construction of uniform designs. Chapter 3 focuses on the deterministic approach while Chap. 4 on numerical optimization approach. Various useful modeling techniques are briefly recommended in Chap. 5. The uniformity has played an important role not only for construction of uniform designs, but also for many other designs such as factorial plans, block designs, and supersaturated designs. Chapters 6 and 7 present a detailed description on the usefulness of the uniformity. Chapter 8 introduces design and modeling for experiments with mixtures.

The book can be used as a textbook for postgraduate level and as a reference book for scientists and engineers who have been implementing experiments often. We have taught partial contents of the book for our undergraduate students and our postgraduate students.

We sincerely thank our coauthors for their significant contribution to the development of the uniform design, who are Profs. Yuan Wang in the Chinese Academy of Science, Fred Hickernell in the Illinois Institute of Technology, Dennis K. J. Lin in the Pennsylvania State University, R. Mukerjee in Indian Institute of Management Calcatta, P. Winker in Justus-Liebig-Universität Giessen, C. X. Ma in the State University of New York at Buffalo, H. Xu in University of California, Los Angeles, and K. Chatterjee in Visva-Bharati University. Many thanks to Profs. Z. H. Yang, R. C. Zhang, J. X. Yin, R. Z. Li, L. Y. Chan, J. X. Pan, R. X. Yue, M. Y. Xie, Y. Tang, G. N. Ge, Y. Z. Liang, E. Liski, G. L. Tian, J. H. Ning, J. F. Yang, F. S. Sun, A. J. Zhang, Z. J. Ou, and A. M. Elsawah for successful collaboration and their encouragement. We particularly thank Prof. K. Chatterjee who spent so much time to read our manuscript and to give valuable useful comments.

The first author would thank several Hong Kong UGC research grants, BNU-HKBU UIC grant R201409, and the Zhuhai Premier Discipline Grant for partial support. The second author would thank the National Natural Science Foundation of China (Grant Nos. 11431006 and 11771220), National Ten

Thousand Talents Program, Tianjin Development Program for Innovation and Entrepreneurship, and Tianjin "131" Talents Program. The third author would thank the National Natural Science Foundation of China (Grant Nos. 11271147 and 11471136) and the self-determined research funds of CCNU from the college's basic research and operation of MOE (CCNU16A02012 and CCNU16JYKX013). The last author would thank the National Natural Science Foundation of China (Grant Nos. 11471229 and 11871288) and Fundamental Research Funds for the Central Universities (2013SCU04A43). The authorship is listed in alphabetic order.

Zhuhai/Beijing, China Kai-Tai Fang
Tianjin, China Min-Qian Liu
Wuhan, China Hong Qin
Tianjin, China Yong-Dao Zhou

References

Fang, K.T.: The uniform design: application of number-theoretic methods in experimental design. Acta Math. Appl. Sin. **3**, 363–372 (1980)

Fang, K.T., Li, R., Sudjianto, A.: Design and Modeling for Computer Experiments. Chapman and Hall/CRC, New York (2006)

Santner, T.J., Williams, B.J., Notz, W.I.: The Design and Analysis of Computer Experiments. Springer, New York (2003)

Wang, Y., Fang, K.T.: A note on uniform distribution and experimental design. Chin. Sci. Bull. **26**, 485–489 (1981)

Contents

Chapter 1
Introduction

Experimental design is an important branch of statistics. This chapter concerns with experiments in various fields and indicates their importance, purpose, type of experiments, statistical models, and related designs. Section 1.1 demonstrates several experiments for different purposes and characteristics. This section also presents discussion on two popular types of experiments: (1) physical experiments and (2) computer experiments. Basic terminologies used in experimental design are introduced in Sect. 1.2. Various kinds of experimental designs based on different kinds of statistical models are introduced in Sect. 1.3. They involve the factorial design under ANOVA model, the optimum design under linear regression model, and the uniform design under model uncertainty (or nonparametric regression model). There are many criteria for assessing fractional factorial designs, among which the minimum aberration criterion has been widely used. Section 1.4 gives a brief introduction to this concept and its extensions. Section 1.5 shows the implementation of the uniform design for a multifactor experiment. Readers are recommended to read this chapter carefully so that they can understand the methodology of uniform design and will easily follow the remaining contents of the book.

1.1 Experiments

Scientific experiments are of essential importance for exploring nature. Experiments are performed almost everywhere, usually the purpose of discovering something about a particular process/system. Experiments are often implemented in agriculture, industry, natural sciences, and high-tech. The purpose of an experiment in industrial and chemical engineering is

K.-T. Fang et al., *Theory and Application of Uniform Experimental Designs*, Lecture Notes in Statistics 221,
https://doi.org/10.1007/978-981-13-2041-5_1

- To increase process yields;
- To improve the quality of the products, such as to reduce variability and increase reliability;
- To reduce the development time; or/and
- To reduce the overall costs.

In natural sciences and high-tech, the purpose of an experiment would be accommodated in different tasks:
- To evaluate the material alternatives;
- To screen and select the design parameters;
- To determine the values of the key product parameters which have an impact on product performance;
- To evaluate the effects and the interactions of the factors; or/and
- To explore the relationships between factors and responses.

How to find a good design for a specific experiment is an important research area in statistics as most experiments involve random errors. Design and modeling for experiments are a branch of statistics and have been playing an important role in the development of sciences and new techniques.

1.1.1 Examples

Let us present some motivating examples of experiments. We omit the details in the following experiments so that readers may concentrate on the problems and related methodology we are going to introduce.

Example 1.1.1 In a chemical experiment, the experimenter wishes to explore the relationship between the composition of a chemical material (x) and its strength (y) by an experiment. Suppose that the underlying relationship shown in Fig. 1.1 is unknown. How does one design an experiment to find an approximate model (or metamodel) to describe the desired relationship. In this experiment, the chemical composition is called a *factor* and the strength is called a *response*. A natural idea for this experiment is to choose several values of the composition, $x_1, \ldots, x_n$ say, to conduct experiments at these values, and measure the corresponding strengths, denoted by $y_1, \ldots, y_n$. Various modeling techniques applied to the data $(x_1, y_1), \ldots, (x_n, y_n)$ can result in different metamodels, among which the experimenter can choose a suitable one. The choice of experimental points, $x_1, \ldots, x_n$, and modeling techniques are important issues.

Example 1.1.2 This experiment is a typical situation in chemical engineering encountered by the first author in Nanjing in 1972. For increasing the yield (y), three controllable variables varied for study. They are

Fig. 1.1 Underlying nonlinear model

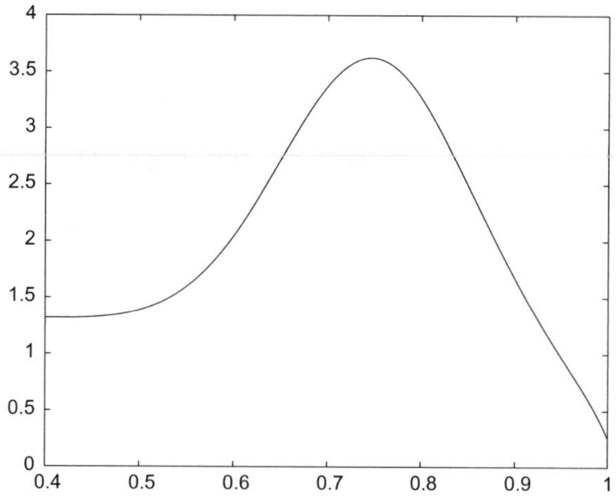

A: The temperature of the reaction;
B: The time allowed for the reaction; and
C: The alkali percentage.

The varied intervals of these variables are chosen to be

$$A: 80\,^{\circ}\mathrm{C}–90\,^{\circ}\mathrm{C};\; B: 90\,\mathrm{min}–150\,\mathrm{min};\; C: 5\%–7\%,$$

respectively. From experience, it is known that there are no interactions among the three factors. There may have a nonlinear relationship between the yield y and the factors. How do we find a design for this experiment? In the literature, a factorial experiment for multiple factors is recommended. Sections 1.3.1 and 1.3.2 will introduce this kind of designs.

Example 1.1.3 This is a real case study introduced by Fang and Chan (2006). Accelerated stress testing is an important method in studying the lifetime of systems. As a result of advancement in technology, the lifetimes of products are increasing, and as new products emerge quickly, their life cycles are decreasing. Manufacturers need to quickly determine the lifetimes of new products and launch them into the market before another new generation of products emerges. In many cases, it is not viable to determine the lifetimes of products by testing them under normal operating conditions. Instead, accelerated stress testing is commonly used, in which products are tested under high-stress physical conditions. The median times to failure of the products are extrapolated from the data obtained using lifetime models. Many different models, such as the Arrhenius model, inverse power rule model, the proportional hazards model, have been proposed based on physical or statistical considerations. Readers may refer to Elsayed (1996) for an introduction of accelerated stress testing. An experimental design is needed to choose the environmental parameters of the accelerated stress test. Three factors are considered as voltage V (Volts), temperature T (Kelvin), and relative humidity H (%). The response is its median time to failure t that is given by

$$t = aV^{-b}e^{c/T}e^{-dH},$$

where a, b, c, d are known constants to be determined. The median time to failure t_0 of an electronic device under the normal operating condition has to be determined under accelerated stress testing. The experimenter wants to explore the model via an experiment. The uniform design was used for this case study.

Example 1.1.4 In an environmental study, an experimenter wishes to conduct a quantitative risk assessment of toxic chemicals present and their interactions. Six chemicals are considered: cadmium (Cd), copper (Cu), zinc (Zn), nickel (Ni), chromium (Cr), and lead (Pb). The experimenter varies the concentration of each chemical in the experiment in order to determine how the concentration affects toxicity. Unfortunately, the underlying model between the response and the six chemical concentrations is unknown. One wants to find a metamodel to the true one by an experiment. Clearly, the range for each chemical concentration should be substantial. The experimenter might choose the following concentrations for each chemical

0.01, 0.05, 0.1, 0.2, 0.4, 0.8, 1, 2, 4, 5, 8, 10, 12, 14, 16, 18, and 20.

Given these 17 levels, there are $17^6 = 24, 137, 569$, almost 24 million, concentration-combinations! It is impossible to conduct an experiment for each concentration-combination. A good experimental design can choose a small number of representative concentration-combinations that still yield a reliable result. Fang and Wang (1994) and Fang et al. (2006) discussed the issues of design and modeling for this experiment in details.

Example 1.1.5 The reversible chemical reaction is a class of important basic reactions in chemistry and chemical engineering. Traditionally, chemists used the deterministic methods to obtain the kinetic rate constants, according to the characteristic of a chemical reaction. Chemists often used the techniques that make a large excess of some reactants involved in reaction, so that their concentration changes can be negligible. Thus, simple relation among the reactants and products is obtained. The chemical kinetics is modeled by a linear system of 11 differential equations:

$$h_j(x,t) = g_j(\eta, x, t), \ \ j = 1, \ldots, 11, \tag{1.1.1}$$

where x is a set of rate constants, the inputs to the system. A solution to (1.1.1) can be obtained numerically for any input x by the use of 11 differential equations solver, yielding concentrations of five chemical species at a reaction time of 7×10^{-4} seconds. One might be interested in finding a closed-form approximate model that is much simpler than the original one. Atkinson et al. (1998) discussed the possibility of applying D-optimal designs to the kinetics of reversible chemical reaction.

Example 1.1.6 Many products are formed by mixing two or more ingredients together. For making a coffee cake, the ingredients are: X_1 (flour), X_2 (water), X_3 (sugar), X_4 (vegetable shortening), X_5 (flaked coconut), X_6 (salt), X_7 (yeast), X_8 (emulsifiers), X_9 (calcium propionate), X_{10} (coffee powder), and X_{11} (liquid flavoring). Choosing a suitable percentage for each ingredient requires much rich experience. However, if the cooker does not have any experience, a design of experiment with mixtures is helpful. Note that the percentages of the above 11 ingredients must satisfy $x_1 + \cdots + x_{11} = 1$ and $x_i \geqslant 0, i = 1, \ldots, 11$. These restrictions imply that the related design of experiments is quite different from the above experiments and is called *experiment with mixtures*. Besides the constraints already mentioned, there may be other upper and lower bounds on the amounts of certain ingredients, i.e., $a_i \leqslant x_i \leqslant b_i, i = 1, \ldots, 11$. For instance, the water and flour in the coffee cake making experiment are major ingredients, and the amount of salt should be less than 1%, and the quantities of other ingredients should be very small. Such an experiment is called an *experiment with restricted mixtures* which has been useful in developing new materials.

A design of n runs for mixtures of s ingredients is a set of n points in the domain

$$T^s = \{(x_1, \ldots, x_s) : x_j \geqslant 0,\ j = 1, \ldots, s,\ x_1 + \cdots + x_s = 1\}. \quad (1.1.2)$$

Due to the constrain $x_1 + \cdots + x_s = 1$, to find a design for experiments with mixtures is quite different from experiments without any constrains on the factors.

Example 1.1.7 (*Robot arm*) A robot is an electromechanical machine that is guided by a computer program. A new branch of robot technology involves design, construction, operation, and computer system.

The movement trajectory of a robot arm is frequently used as an illustrative example for computer experiments (see Fang et al. 2006). Consider a robot arm with m segments. The shoulder of the arm is fixed at the origin in the (u, v)-plane. The segments of this arm have lengths $L_j, j = 1, \ldots, m$. The first segment is at angle θ_1 with respect to the horizontal coordinate axis of the plane. For $k = 2, \ldots, m$, segment k makes angle θ_k with segment $k - 1$. The end of the robot arm is at

$$\begin{cases} u = \sum_{j=1}^{m} L_j \cos(\sum_{k=1}^{j} \theta_k), \\ v = \sum_{j=1}^{m} L_j \sin(\sum_{k=1}^{j} \theta_k), \end{cases} \quad (1.1.3)$$

and the response y is the distance $y = \sqrt{u^2 + v^2}$ from the end of the arm to the origin expressed as a function of $2m$ variables $\theta_j \in [0, 2\pi]$ and $L_j \in [0, 1]$. Ho (2001) gave an approximation model $y = g(\theta_1, \theta_2, \theta_3, L_1, L_2, L_3)$ for the robot arm with three segments.

1.1.2 Experimental Characteristics

In the previous subsection, we list several experiments. Each experiment has its own purpose and characteristic. Example 1.1.1 wants to explore a nonlinear relationship between the composition of a chemical material and its strength. It is a one-factor experiment with model uncertainty. Example 1.1.2 is a multifactor experiment and needs a efficient design to explore the relationship between the response and the factors. Some prior knowledge provided in Example 1.1.2 (no interactions among the factors) is very useful. Example 1.1.3 wants to explore the nonlinear relationship between the response t and three factors V, T, H, and to choose the environmental parameters of the accelerated stress test. The goal of Example 1.1.4 is crucial. The relationships between the response and the six chemicals are complicated, and the underlying model is unknown. Example 1.1.5 is from chemistry and chemical engineering. Atkinson et al. (1998) discussed the possibility of applying D-optimal designs to the kinetics of reversible chemical reaction when the underlying model is known. However, the D-optimal design is not robust against the model change. Xu et al. (2000) considered three different kinds of designs for this chemical reaction and compared their performance. They concluded that the uniform design is robust for this experiment. Many products are formed by mixtures of several ingredients. Experiments with mixtures have played an important role in various fields such as chemical engineering, rubber industry, material and pharmaceutical engineering. Example 1.1.6 is such an experiment. Example 1.1.7 is related to computer experiments, where the underlying model is known, but too complicated. One wishes to use a metamodel to approximate the true model. Design and modeling for computer experiments are a rapidly growing area and have been widely used in system engineering.

A good experimental design should minimize the number of runs to acquire as much information as possible. Experimental design is a branch of statistics and has a long history. The experimental design has been playing an important role in development of sciences and new techniques, especially in development of high-tech. It involves rich methodologies and various designs. Comprehensive reviews for various kinds of designs can be found in *Handbook of Statistics, Vol. 13*, edited by S. Ghosh and C.R. Rao, *Handbook of Statistics, Vol. 22: Statistics in Industry*, edited by R. Khattree and C.R. Rao, as well as in *Handbook of Design and Analysis of Experiments*, edited by A. Dean, M. Morris, J. Stufken and D. Bingham. Historically, in the natural sciences, some fundamental laws of nature have only been discovered or verified empirically through carefully designed experiments. The information and testing conjectures through these experiments may lead to the birth of a new branch of science. Developments in life sciences have frequently arisen in this manner. A classical example is Mendel's genetic model now well known to the high school students.

The experimental design is extremely useful in many multifactor experiments. From the above examples, we find that experiments, especially in high-tech applications, have one or more of the following characteristics:

- Multiple factors;
- Nonlinear relationship between the factors and the response;
- Experimental region is large or very large so that the number of levels for the factors cannot be too small;
- The underlying model is unknown;
- Experiments can be simulated in computer.

We need some new designs that can treat experiments with the above complexities. There are many designs that can meet the above requirements, and the *uniform experimental design* or the *uniform design* is one of such designs.

1.1.3 Type of Experiments

Experiments can be classified under various considerations. For example,
I. The number of factors

- Single-factor experiments;
- Multifactor experiments.

II. The number of responses

- Single response;
- Multiple responses;
- Functional response.

III. Constraint condition

- No constraint experiment: The choice of the levels of every factor is independent;
- Experiments with constraints: The choice of the levels of every factor depends on other factors, for example, experiments with mixtures.

IV. The underlying model

- Known;
- Unknown.

V. Operating environment

- Physical experiments; and
- Computer experiments.

Let us give more words on the last classification.

A. Physical Experiments

Traditionally, an experiment is implemented in a laboratory, in a factory, or in an agricultural field. This is called a *physical experiment* or an *actual experiment*, where the experimenter physically carried out the experiment. There always exist random errors in physical experiments so that we might obtain different outputs under the identical experimental environment. Existence of random errors creates complexity in data analysis and modeling. Therefore, the experimenter may choose one or few factors in the experiment so that it is easy to explore the relationship between the output and inputs or propose some powerful statistical experimental designs.

Statistical approach to design experiments is usually based on a statistical model. A good design is an optimal one to the underlying statistical model. There are many designs for physical experiments, among which the *fractional factorial design* that is based on an ANOVA model and the *optimum regression design* (optimum design for short) that is based on a regression model are most popular in practice. The corresponding models involve some unknown parameters such as main effects, interactions, regression coefficients, and variance of random errors. A good design can provide unbiased estimators of the parameters with smaller or even the smallest variance–covariance matrix under a certain sense. Orthogonal arrays and various optimal designs are such popularly used designs. Useful concepts in physical experiments and basic knowledge on the orthogonal array and the optimal design are given in the next section.

The orthogonal design and optimum regression design assume the underlying model is known except for some parameters in the model. However, the underlying model is unknown in most of the experiments. One wishes to find a metamodel based on experiments. The uniform design is one of the most popular designs.

B. Computer Experiments

Computer experiments and/or computer simulations have been widely used for studying physical phenomena in various fields of industry, system engineering, and others because many physical processes/phenomena are difficult or even impossible to study by conventional experimental methods. Computer models are often used to describe complicated physical phenomena encountered in science and engineering. These phenomena are often governed by a set of equations, including linear, nonlinear, ordinary, and partial differential equations. The equations are often too difficult to be solved simultaneously in a short time, but can be by a computer modeling program. These programs, due to the number and complexity of the equations, may have long running times, making their use difficult for comprehensive scientific investigation. Santner et al. (2003) indicated "Many physical processes are difficult or even impossible to study by conventional experimental methods. As computing power has increasing, it has become possible to model some of these processes by sophisticated computer code."

In the past decades, *computer experiments* or *simulation experiments* become a hot topic in statistics and engineering. The underlying model in a computer experiment is deterministic and given, but is too complicated to be managed and analyzed. One of the goals of computer experiments is first to find an approximate model (metamodel) that is much simpler than the true one. Simulation experiments study the underlying process by simulating the behavior of the process on a computer. The underlying model in a simulation experiment is also deterministic and given, but errors on the inputs are considered.

The computer experiment has played its role as an artificial means for simulating a physical environment so that experiments can be implemented virtually, if such experiments are not performed physically for some reasons. Many scientific researches involve modeling complicated physical phenomena using the mathematical model

$$y = f(x_1, \ldots, x_s) = f(\boldsymbol{x}), \quad \boldsymbol{x} = (x_1, \ldots, x_s)' \in T, \tag{1.1.4}$$

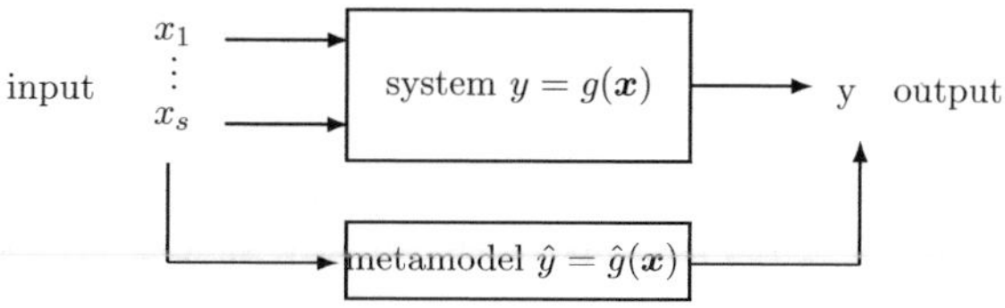

Fig. 1.2 Computer experiments

where $\boldsymbol{x}$ consists of the input variables, $\boldsymbol{x}'$ is the transpose of $\boldsymbol{x}$, y is the output variable, the function f may not have an analytic formula, and T is the input variable space. Model (1.1.4) may be regarded as a solution of a set of equations, including linear, nonlinear, ordinary, and/or partial differential equations, and it is often impossible to obtain an analytic solution for the equations or even impossible to study by conventional experimental methods. Engineers or scientists make use of the models to perform various tasks and decisions making by interrogating the model to predict behavior of systems under different input variable settings. Hence, the model plays a crucial role in scientific investigation and engineering design. However, in many situations, it is very expensive or time-consuming to conduct physical experiments in order to fully understand the relationship between response y and inputs x_j's. Alternatively, scientists and engineers use computer simulation to explore the relationship. Thus, metamodels become very important to investigate complicated physical phenomena. One of the goals of computer experiments is first to find a good metamodel that is much simpler than the true one (Cf. Fig. 1.2) by simulating the behavior of the device/process on a computer. Due to the number and complexity of the equations, these programs require special designs.

1.2 Basic Terminologies Used

This section reviews some basic terminologies used in experimental designs. Let us see a typical example.

Example 1.2.1 In a chemical experiment for increasing the yield, three controllable variables are varied for study. They are:
 A (or x_1): The type of raw materials;
 B (or x_2): The amount of pyridine;
 C (or x_3): The duration of the reaction.
Ranges of these variables are chosen to be $\{m_1, m_2, m_3\}$, $[10, 28]$, and $[0.5, 3.5]$, respectively. The following values for each variable are selected for experiments:
 A: The type of raw materials: m_1, m_2, m_3;
 B: The amount of pyridine (*in ml*): 10, 19, 28;
 C: The time length of reaction (*in hours*): 0.5, 1.5, 2.5, 3.5.

Factor: A controllable variable that is of interest in the experiment. Factors are denoted by $A, B, \ldots$ or $x_1, x_2, \ldots$ according to convenience. In Example 1.2.1, the type of raw materials, the amount of pyridine, and the duration of reaction are factors. A factor may be *quantitative* or *qualitative*. A *quantitative factor* is one whose values can be measured on a numerical scale and fall in an interval, e.g., temperature, pressure, ratio of two raw materials, reaction time. A *qualitative factor* is one which is measured categorically, such as education levels, types of material, gender. A qualitative factor is also called a *categorical factor* or *indicator factor*. In this example, the factor A is qualitative while the factors B and C are quantitative. In computer experiments, the factors are sometimes called *input variables*.

Environmental and nuisance variables: The variables that are not studied in the experiment and are generally controlled at their pre-decided values are called *environmental variables*. Environmental variables are not regarded as factors. With reference to Example 1.2.1, the amounts of other chemicals, the pressure, etc., are examples of environment variables.

The variables that cannot be controlled at all are treated as *nuisance variables*. In most agriculture experiments, the weather cannot be controlled and may create some noise variables. The season is considered as a nuisance variable in many physical experiments.

Experimental domain: The space where the factors take their values. In Example 1.2.1, the experimental domain is the *Cartesian product* $\{m_1, m_2, m_3\} \times [10, 28] \times [0.5, 3.5]$. We shall use $\mathcal{X}$ to denote the experimental domain that is a subset of R^s, where s is the number of the factors. In the development of new products or procedures, it is better to choose a larger domain so that the experimenter may have a better chance of finding the desired result. For experiments in a production environment, smaller experimental domain is recommended for safety.

Levels: The specific values at which a factor is tested. Denote by $A_1, A_2, \ldots$ the levels of the factor A. In Example 1.2.1, the levels of the reaction time are $C_1 = 0.5$, $C_2 = 1.5$, $C_3 = 2.5$, and $C_4 = 3.5$, and the levels of the type of raw materials are $A_1 = m_1$, $A_2 = m_2$, and $A_3 = m_3$. Usually, the experimenter chooses an experimental domain first and then he chooses levels for each factor.

Level-combination: A possible combination of levels of the factors. With reference to Example 1.2.1, $(m_2, 10, 2.5)$ and $(m_1, 28, 0.5)$ are two level-combinations. A level-combination is also called a *treatment combination*. A level-combination can be considered as a point in the experimental domain $\mathcal{X}$.

Run or **trial**: The implementation of a level-combination in the experimental environment. Throughout the book, we use the symbol "n" to denote the number of runs in an experiment.

Design: A set of level-combinations or a set of points in the experimental domain $\mathcal{X}$. A design of an experiment with n runs and s factors is often denoted by $\mathcal{P} = \{\boldsymbol{x}_1, \ldots, \boldsymbol{x}_n\}$ where $\boldsymbol{x}_j \in \mathcal{X}$ or $\mathsf{U} = (u_{ij}) : n \times s$ where the rows of U are transpose of $\boldsymbol{x}_1, \ldots, \boldsymbol{x}_n$, respectively. For simplicity, sometimes a design $\mathcal{P} = \{\boldsymbol{x}_1, \ldots, \boldsymbol{x}_n\}$ can be regarded as a matrix U.

Response: The result of a run, depending on the purpose of the experiment. The response or output of an experiment could be qualitative or quantitative. The response of the experiment considered in Example 1.2.1 is the yield. We always use y for the response. It is very common that there are several responses corresponding to one single run. Such experiments are called *experiments with multiple responses*. The response, for example, in chemical, biology, and medical studies, can be a curve, such as a chemical fingerprint or DNA fingerprint. A curve response is called *functional response*.

Random error: The variability of the response due to uncontrollable (nuisance) variables in a run of the experiment. In general, it is not necessary to obtain the same result in two runs for the identical level-combination due to a random error. Industrial experiments always involve random errors. When we repeat the experiment under the same experimental environment, called *repeated experiments* or *duplicated experiments*, we may get different results due to random errors. The random error can often be assumed to be distributed as a *normal distribution*, $N(0, \sigma^2)$, in most experiments. The variance σ^2 measures magnitude of the random error. The existence of random error necessitates more complex designs and modeling.

Interaction of the factors: Suppose two factors A and B are considered in an experiment. If the change in response due to a change in the levels of any of the factors remains invariant at all levels of the other factor, we say that there is no interaction between A and B. Otherwise, we say there is an interaction between A and B. In a similar way, we can define interaction among more factors. How to define and estimate the interaction of the factors is an important issue. Some discussion on this concept is given in the next section.

The following gives some useful concepts in algebra that will be used in our book.

Hadamard matrix: A Hadamard matrix $\boldsymbol{H}$ of order n is an $n \times n$ matrix with entries 1 and -1, which satisfies $\boldsymbol{H}\boldsymbol{H}^T = n\boldsymbol{I}$. It is known that n is necessarily 1, 2, or a multiple of four. The Hadamard matrix plays an important role in the coding theory and experimental designs.

Hadamard product or Dot product: The Hadamard product of two $n \times m$ matrices $A = (a_{ij})$ and $B = (b_{ij})$ is an $n \times m$ matrix whose (i, j)th entry is $a_{ij}b_{ij}$ for each ordered pair (i, j).

$$\boldsymbol{A} \cdot \boldsymbol{B} = \begin{pmatrix} a_{11}b_{11} & \cdots & a_{1m}b_{1m} \\ \vdots & & \vdots \\ a_{n1}b_{n1} & \cdots & a_{nm}b_{nm} \end{pmatrix}. \tag{1.2.1}$$

The Hadamard product is also called *dot product* in the literature.

Hamming distance: The Hamming distance between two strings with the same length is the number of positions at which the corresponding symbols are different. For example, two strings (a, f, t, e, r) and (f, f, b, r, r) have Hamming distance 3; two runs (2, 1, 2, 3) and (2, 2, 3, 1) in $L_9(3^4)$ have Hamming distance 3.

Kronecker product: The Kronecker product of two matrices $\boldsymbol{A} = (a_{ij})$ of $n \times p$ and $\boldsymbol{B} = (b_{kl})$ of $m \times q$ is an $nm \times pq$ matrix defined as

$$A \otimes B = (a_{ij}B) = \begin{pmatrix} a_{11}B & \cdots & a_{1m}B \\ \vdots & & \vdots \\ a_{n1}B & \cdots & a_{nm}B \end{pmatrix}. \tag{1.2.2}$$

Cartesian product: In mathematics, a Cartesian product is a mathematical operation which returns a set (or product set or simply product) from multiple sets. For example, the Cartesian product of $\mathcal{A} = \{a_1, \ldots, a_m\}$ and $\mathcal{B} = \{b_1, \ldots, b_n\}$ is the set

$$\mathcal{A} \times \mathcal{B} = \{(a, b) : a \in \mathcal{A}, b \in \mathcal{B}\}. \tag{1.2.3}$$

More generally, a Cartesian product of n sets, also known as an n-fold Cartesian product, can be represented by an array of n dimensions, where each element is an n-tuple.

Modulo operation: In this book, the concept of the modulo operation is in a special sense. Let m and n are two positive integers. It is easy to see that m can be expressed as

$$m = qn + r, \tag{1.2.4}$$

where q is the quotient and r is the remainder of m divided by n. When $m \leqslant n$, we have $q = 0$ and $r = m$; when $m > n$, q is a positive integer and $0 \leqslant r < n$. We define $m (\text{mod}\, n) = r$. Clearly, the range of $m(\text{mod}\, n)$ is $\{0, 1, \ldots, n-1\}$. In experimental design, the modulo operation is modified as $m\widetilde{(\text{mod}\ n)} = r$, if $r > 0$; otherwise, $m\widetilde{(\text{mod}\, n)} = n$ if $r = 0$.

1.3 Statistical Models

Each experiment has its own goal. A specific statistical model with some unknown parameters may be used for fitting the relationship between the response and the factors for the experiment. As mentioned earlier, there are two types of experiments: (1) The underline model involves a number of unknown parameters even though the model is known, and (2) The true model is not known.

When the underlying model is known, some optimality (or criterion) is raised for measuring estimation of the parameters. We wish to find an experimental design such that its data can give the best estimation of the parameters under the optimality. There are various statistical designs each based on its specific goal model. A comprehensive introduction to these designs can be referred to *Handbook of Statistics, Vol. 13* edited by Ghosh and Rao (1996) and *Handbook of Design and Analysis of Experiments* edited by Dean et al. (2015).

We use the following example to demonstrate the application of statistical models with respect to different types of designs.

Example 1.1.1 (*Continuity*) Suppose that the experimenter chooses the experimental domain as $\mathcal{X} = [0.4, 1]$ for his professional knowledge and the number of runs to be 12 from their financial budget. There are several statistical approaches to make a design for this experiment. Each approach focuses on its own statistical model. The following subsections introduce ANOVA models that are based on factorial designs; regression models based on optimum designs; nonparametric regression models that motivated uniform designs.

1.3.1 Factorial Designs and ANOVA Models

A *factorial design* is a set of level-combinations, and one of the main purposes of the experiment is to estimate main effects and some interactions of the factors. Factorial designs have been widely used in various kind of experiments. Let us give a brief introduction to the factorial design.

A. One-Factor Experiments

For Example 1.1.1, a *factorial design* suggests to observe the response at several compositions of the chemical material, $x_1, \ldots, x_q$, that are called *levels* as indicated in the previous section. For each x_i, we repeat experiment n_i times and related responses are denoted by $y_{i1}, \ldots, y_{in_i}$. Figure 1.3 shows a case of $q = 4$, and $n_1 = n_2 = n_3 = n_4 = 3$. A statistical model

$$y_{ij} = \mu_i + \varepsilon_{ij}, \quad i = 1, \ldots, q, \, j = 1, \ldots, n_i, \tag{1.3.1}$$

is considered, where μ_i is the true value y at $x = x_i$, denoted by $y(x_i)$, and ε_{ij} are random errors that are independently identically distributed according to a normal distribution $N(0, \sigma^2)$ with unknown $\sigma^2 > 0$. Let μ be the overall mean of y over $x_1, \ldots, x_q$. Then, the mean μ_i can be decomposed into $\mu_i = \alpha_i + \mu$, where α_i is called the *main effect* of y at x_i and they satisfy $\alpha_1 + \cdots + \alpha_q = 0$. Model (1.3.1) now can be expressed as

$$y_{ij} = \mu + \alpha_i + \varepsilon_{ij}, \quad i = 1, \ldots, q, \, j = 1, \ldots, n_i. \tag{1.3.2}$$

The main effects measure the influence of factor A to the response. If some $\alpha_i \neq 0$, the factor A is said to have significant influence to the response y. The least squares estimation has been used for $\mu, \alpha_1, \ldots, \alpha_{q-1}$, and the analysis of variances gives unbiased estimate of σ^2 and F test. Therefore, the model (1.3.2) is called ANOVA model.

B. Two-Factor Experiments

For a two-factor experiment, its ANOVA model may involve the overall mean, main effects, and interactions between the factors. Suppose that Factor A chooses q_1 levels

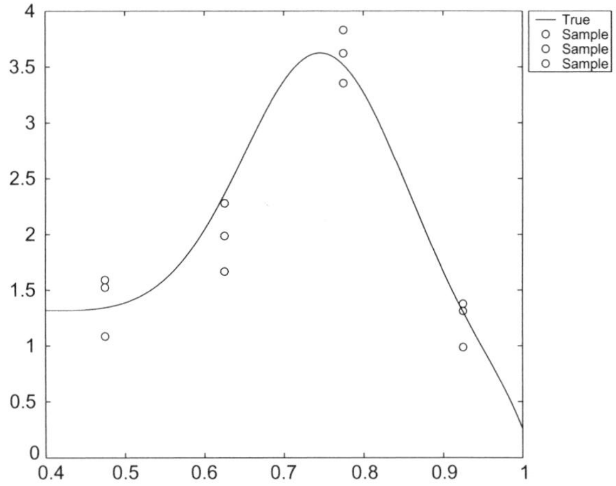

Fig. 1.3 A factorial design

$A_1, \ldots, A_{q_1}$ and Factor B chooses q_2 levels $B_1, \ldots, B_{q_2}$ in a factorial experiment, respectively. For each level-combination, we repeat the experiment r ($r \geqslant 1$) times. The statistical model of this experiment is expressed as

$$y_{ijk} = \mu + \alpha_i + \beta_j + (\alpha\beta)_{ij} + \varepsilon_{ijk}, \qquad (1.3.3)$$
$$i = 1, \ldots, q_1; \;\; j = 1, \ldots, q_2; k = 1, \ldots, r,$$

where
y_{ijk} is the kth response at the level-combination $A_i B_j$,
μ is the overall mean,
α_i is the main effect of factor A at A_i,
β_j is the main effect of factor B at B_j,
$(\alpha\beta)_{ij}$ is the interaction of factors A and B at the level-combination $A_i B_j$,
ε_{ijk} is random error in the kth experiment at $A_i B_j$. Usually, we assume that ε_{ijk} are *i.i.d.*, $\varepsilon_{ijk} \sim N(0, \sigma^2)$ with unknown σ^2.

The main effects and interactions in the model satisfy

$$\alpha_1 + \cdots + \alpha_{q_1} = 0; \beta_1 + \cdots + \beta_{q_2} = 0;$$
$$\sum_{i=1}^{q_1} (\alpha\beta)_{ij} = 0, \;\; j = 1, \ldots, q_2; \sum_{j=1}^{q_2} (\alpha\beta)_{ij} = 0, i = 1, \ldots, q_1.$$

So the number of linearly independent main effects of A is $q_1 - 1$, the number of linearly independent main effects of B is $q_2 - 1$, and the number of linearly independent interactions of $A \times B$ is $(q_1 - 1)(q_2 - 1)$. Here, $q_1 - 1$, $q_2 - 1$, and $(q_1 - 1)(q_2 - 1)$ are called *degrees of freedom* of A, B and $A \times B$, respectively, in the ANOVA. This model also can be expressed as

$$y_{ijk} = \mu_{ij} + \varepsilon_{ijk}$$

where μ_{ij} is the mean of y at $A_i B_j$. Rewrite

$$\mu_{ij} = \mu + (\mu_{i.} - \mu) + (\mu_{.j} - \mu) + (\mu_{ij} - \mu_{i.} - \mu_{.j} + \mu), \tag{1.3.4}$$

where $\mu_{i.}$ is the mean of y at A_i and $\mu_{.j}$ the mean of y at B_j. The above decomposition implies that

$$\alpha_i = \mu_{i.} - \mu,\ \beta_j = \mu_{.j} - \mu,\ (\alpha\beta)_{ij} = \mu_{ij} - \mu_{i.} - \mu_{.j} + \mu. \tag{1.3.5}$$

The least squares estimators of these parameters are given by

$$\begin{aligned}
\hat{\mu} &= \bar{y}, \\
\hat{\alpha}_i &= \bar{y}_{i.} - \bar{y}, \quad \hat{\mu}_{i.} = \bar{y}_{i.}, \\
\hat{\beta}_j &= \bar{y}_{.j} - \bar{y}, \quad \hat{\mu}_{.j} = \bar{y}_{.j}, \\
\widehat{(\alpha\beta)}_{ij} &= \bar{y}_{ij} - \bar{y}_{i.} - \bar{y}_{.j} + \bar{y}, \quad \hat{\mu}_{ij} = \bar{y}_{ij},
\end{aligned}$$

where

$\bar{y}_{ij}$ is the average response at the combined level $A_i B_j$,
$\bar{y}_{i.}$ is the average response at the level A_i,
$\bar{y}_{.j}$ is the average response at the level B_j,
$\bar{y}$ is the total average response,

and $\hat{\varepsilon}_{ijk} = y_{ijk} - \bar{y}_{ij}$ are *residuals*.

C. Experiments with More Factors

An experiment with three factors A, B, and C may involve main effects of A, B, and C, two-factor interactions $A \times B$, $A \times C$, and $B \times C$, and three-factor interaction $A \times B \times C$. Suppose that factor A has q_1 levels, factor B has q_2 levels, and factor c has q_3 levels. Suppose the underlying model is

$$\begin{aligned}
y_{ijkt} &= \mu + \alpha_i + \beta_j + \gamma_k + (\alpha\beta)_{ij} + (\alpha\gamma)_{ik} + (\beta\gamma)_{jk} \\
&\quad + (\alpha\beta\gamma)_{ijk} + \varepsilon_{ijkt}, \qquad (1.3.6) \\
&\quad i = 1, \ldots, q_1;\ j = 1, \ldots, q_2; k = 1, \ldots, q_3, t = 1, \ldots, r,
\end{aligned}$$

in which the main effects of A, B, and C have $q_1 - 1$, $q_2 - 1$, $q_3 - 1$ degrees of freedom, respectively; their two-factor effects have $(q_1 - 1)(q_2 - 1)$, $(q_1 - 1)(q_3 - 1)$, and $(q_2 - 1)(q_3 - 1)$ degrees of freedom, respectively; the interaction $A \times B \times C$ has $(q_1 - 1)(q_2 - 1)(q_3 - 1)$ degrees of freedom. If we need to estimate all the above main effects and interactions, the number of runs is at least $q_1 q_2 q_3 + 1$ as we also need to estimate the variance of error, σ^2. In general, for an experiment of s factors each having $q_1, \ldots, q_s$ levels, respectively, if you want to consider all of main effects,

Table 1.1 Number of unknown parameters

Number of factors, s	1	2	3	4	5	6	7	8
Linear ($q = 2$)	2	4	8	16	32	64	128	256
Quadratic ($q = 3$)	3	9	27	81	243	729	2187	6561
Cubic ($q = 4$)	4	16	64	256	1024	4096	16384	65536

two-factor interactions, three-factor interactions, ..., s-factor interactions, the total number of parameters to be estimated is $\prod_{i=1}^{s} q_i + 1$ that increases exponentially when $(s, q_1, \ldots, q_s)$ increases. In this case, a full factorial design is recommended.

Definition 1.3.1 A *full factorial design* or *complete design* is a factorial design that requires all the level-combinations of the factors to appear equally often.

One advantage of the full factorial design is that we can estimate all the main effects and interactions among the factors. However, the number of parameters (main effects and interactions) in a full factorial design increases exponentially as the number of factors increases. For example, there are $15626 = 5^6 + 1$ unknown parameters to be estimated in an experiment of six factors each having five levels. If a full factorial design is employed, the minimum number of runs is at least 15626 including to estimate σ^2. On the other hand, the number of runs should be a multiple of 15625 that is too much for a physical experiment. Table 1.1 shows the number of unknown parameters in various ANOVA models.

1.3.2 Fractional Factorial Designs

Do we need a model that involves so many unknown parameters in an experiment with multiple factors? By the experience, experimenters believe the following two principles:

Sparsity principle: The number of relatively important effects/interactions in a factorial design is small.

Hierarchical ordering principle: Lower-order effects are more likely to be important than higher-order effects; main effects are more likely to be important than interactions; and effects of the same order are likely to be equally important.

Considering the above two principles, the number of unknown parameters can be significantly reduced. For example, for an experiment of s factors each having q levels, there are $s(q-1)$ main effects and $\frac{1}{2}s(s-1)(q-1)^2$ two-factor interactions. If the high-order interactions can be ignored, the total number of unknown parameters reduces from q^s to $s(q-1)[1+\frac{1}{2}(s-1)(q-1)]+1$. When some two-factor interactions lack significance, the number of unknown parameters can be further reduced. Based on this consideration, a subset of the full factorial design is employed, and

Table 1.2 Orthogonal design $L_8(2^7)$

No	1	2	3	4	5	6	7
1	1	1	1	1	1	1	1
2	1	1	1	2	2	2	2
3	1	2	2	1	1	2	2
4	1	2	2	2	2	1	1
5	2	1	2	1	2	1	2
6	2	1	2	2	1	2	1
7	2	2	1	1	2	2	1
8	2	2	1	2	1	1	2

it is called a *fractional factorial design* (FFD for simplicity). The reader can refer to Dey and Mukerjee (1999) for the details and references therein. Some notations often appear in the literature, such as 2^{s-k} denotes an FFD of s two-level factors with the number of runs of $n = 2^{s-k}$. This design is called a $1/2^k$ fraction of the 2^s design. Similarly, notation 3^{s-k} denotes for an FFD of s factors each having three levels with $n = 3^{s-k}$ runs.

For the use of FFD, the experimenter needs to have some prior knowledge about which main effects and interactions are significant. A design that can estimate all the unknown parameters of the model is called *estimable*, otherwise called *inestimable*. For the latter one, some main effects and interactions will be *confounded*. This phenomenon is called *aliasing*.

A good fractional factorial design with a limited number of runs can allow us to estimate all of the main effects and interactions of interest and can allow non-interesting parameters to be confounded. The most popular fractional factorial design is the so-called *orthogonal array*.

Definition 1.3.2 An orthogonal array (OA) of strength t with n runs and s factors, denoted by $OA(n, s, q, r)$, is an FFD where any subdesign of n runs and m $(m \leqslant r)$ factors each having q levels is a full design.

Orthogonal arrays of strength two have been extensively used for planning experiments in various fields (cf. Hedayat et al. 1999). Orthogonal arrays can be expressed as a table. A general definition is given by:

Definition 1.3.3 An $n \times s$ matrix, denoted by $L_n(q_1 \times \cdots \times q_s)$ with entries $1, 2, \ldots, q_j$ at the jth column, is called an orthogonal design (OD) table, if it satisfies:
(1) Each entry in each column appears equally often.
(2) Each entry-combination in any two columns appears equally often.

Obviously, the condition (2) implies the condition (1). When some q_i are the same, the table $L_n(q_1 \times \cdots \times q_s)$ is denoted by $L_n(q_1^{r_1} \times \cdots \times q_m^{r_m})$ where $r_1 + \cdots + r_m = s$ and by $L_n(q^s)$ if all $q_i = q$. Two examples are given in Tables 1.2 and 1.3.

Table 1.3 Orthogonal design $L_8(4 \times 2^4)$

No	1	2	3	4	5
1	1	1	1	1	1
2	1	2	2	2	2
3	2	1	1	2	2
4	2	2	2	1	1
5	3	1	2	1	2
6	3	2	1	2	1
7	4	1	2	2	1
8	4	2	1	1	2

Definition 1.3.4 Saturated design. An orthogonal design $L_n(q^s)$ is called saturated if n, q, and s satisfy $n-1 = s(q-1)$. More general, an orthogonal design $L_n(q_1 \dots, q_s)$ is called saturated if $n - 1 = \sum_{j=1}^{s}(q_j - 1)$.

The designs $L_9(3^4)$, $L_8(2^7)$, and $L_8(4 \times 2^4)$ are saturated. The design $L_{18}(3^7)$ is not saturated as $(3-1) \times 7 = 14 < 17 = (18-1)$, but the design $L_{18}(2 \times 3^7)$ is still unsaturated. A saturated design cannot be inserted by any column such that the new design is still orthogonal.

The uniform design can be regarded as a kind of fractional factorial designs with model unknown.

Definition 1.3.5 Uniform design table. An $n \times s$ matrix, denoted by $U_n(q_1 \times \cdots \times q_s)$ with entries $1, 2, \dots, q_j$ at the jth column, is called a uniform design (UD) table, if it satisfies:

(1) Each entry in each column appears equally often.

(2) The n experimental points decided by $U_n(q_1 \times \cdots \times q_s)$ are uniformly scattered on the experimental domain in a certain sense.

When some q_i are the same, the design table $U_n(q_1 \times \cdots \times q_s)$ is denoted by $U_n(q_1^{r_1} \times \cdots \times q_m^{r_m})$ where $r_1 + \cdots + r_m = s$ and by $U_n(q^s)$ if all $q_i = q$.

The uniform design is based on a nonparametric regression model

$$y = f(x_1, \dots, x_s) + \varepsilon, \tag{1.3.7}$$

where the true model $f(\cdot)$ is unknown, $(x_1, \dots, x_s)$ are an experimental point on the domain, and ε is random error. The main purpose of the uniform design is to find a metamodel to approximate the true model $f(\cdot)$. It is easy to see that there is essential difference between the orthogonal design and the uniform design. But these two kinds of design tables are based on so-called U-type designs.

Definition 1.3.6 A U-type design, denoted by $U(n; q_1 \times \cdots \times q_s)$, is an $n \times s$ matrix with entries $\{1, \dots, q_j\}$ at the jth columns such that $\{1, \dots, q_j\}$ appear in this column equally often. When some q_j are equal, we denote it by $U(n; q_1^{r_1} \times \cdots \times q_m^{r_m})$ with $r_1 + \cdots + r_m = s$. When all q_j's are equal to q, we write $U(n; q^s)$ and the corresponding design is called symmetric, otherwise asymmetric or U-type design with mixed levels. Let $\mathcal{U}(n; q_1 \times \cdots \times q_s)$ be the set of all U-type designs $U(n; q_1 \times \cdots \times q_s)$. Similarly, we have notations $\mathcal{U}(q_1^{r_1} \times \cdots \times q_m^{r_m})$ and $\mathcal{U}(n; q^s)$.

From the definition of U-type design, it immediately follows that each q_j will be divisible by n. Let $\boldsymbol{U} = (u_{ij})$ be a U-type design in $\mathcal{U}(n; q_1 \times \cdots \times q_s)$. Make the transformation

$$x_{ij} = \frac{u_{ij} - 0.5}{q_j}, \ i = 1, \ldots, n, \ j = 1, \ldots, s, \tag{1.3.8}$$

and denote $\boldsymbol{X}_u = (x_{ij})$. $\boldsymbol{X}_u$ is then called the *induced matrix* of $\boldsymbol{U}$. The n rows of the matrix $\boldsymbol{X}_u$ are n points on $[0, 1]^s$.

The second condition in Definition 1.3.5 requires that the chosen n experimental points are uniformly scattered on the experimental domain in a certain sense. To achieve this, we need some uniformity measures. Most measures of uniformity are defined on $[0, 1]^s$ in the literature. Chapter 2 will introduce various discrepancies as uniformity measures. Each discrepancy defines uniformity of a U-type design $\boldsymbol{U}$ through its induced matrix by

$$D(\boldsymbol{U}) = D(\boldsymbol{X}_u). \tag{1.3.9}$$

Definition 1.3.7 A design $\boldsymbol{U} \in \mathcal{U}(n; q_1 \times \cdots \times q_s)$ is called a **Uniform design** under the measure D if

$$D(\boldsymbol{U}) = \min_{\boldsymbol{V} \in \mathcal{U}(n; q_1 \times \cdots \times q_s)} D(\boldsymbol{V}),$$

and $\boldsymbol{U}$ is denoted by $U_n(q_1 \times \cdots \times q_s)$.

Tables 1.4 and 1.5 give UD tables $U_{12}(12^4)$ and $U_6(3^2 \times 2)$, respectively.

For a given (n, q, s), the corresponding uniform design is not unique as each discrepancy is invariant by permuting rows and columns of the design table (Fig. 1.4).

Definition 1.3.8 Two U-type designs are called *equivalent* if one can be obtained from another by permuting rows and/or columns.

We shall not distinguish equivalent uniform designs. For a given uniform design table $U_n(q^s)$, by permuting rows and columns of the design we can obtain $n!s!$ equivalent designs. Chapters 3 and 4 will discuss the construction of uniform designs, where we search only one of the equivalent uniform designs.

1.3.3 Linear Regression Models

Kiefer (1959) initiated research on statistical design under a specific regression model. Suppose the experimenter has prior knowledge about the underlying model

Table 1.4 $U_{12}(12^4)$

No	1	2	3	4
1	1	10	4	7
2	2	5	11	3
3	3	1	7	9
4	4	6	1	5
5	5	11	10	11
6	6	9	8	1
7	7	4	5	12
8	8	2	3	2
9	9	7	12	8
10	10	12	6	4
11	11	8	2	10
12	12	3	9	6

Table 1.5 $U_6(3^2 \times 2)$

No	1	2	3
1	1	1	1
2	2	1	2
3	3	2	1
4	1	2	2
5	2	3	1
6	3	3	2

between the response and the factors $\boldsymbol{x} = (x_1, \ldots, x_s)$ which is stated below:

$$y(\boldsymbol{x}) = \beta_1 g_1(\boldsymbol{x}) + \cdots + \beta_m g_m(\boldsymbol{x}) + \varepsilon, \tag{1.3.10}$$

where $\boldsymbol{x} = (x_1, \ldots, x_s) \in \mathcal{X}$, $\mathcal{X}$ is the experimental domain in R^s, functions $g_1, \ldots, g_m$ are known, $\beta_1, \ldots, \beta_m$ are unknown parameters, and ε is the random error with $E(\varepsilon) = 0$ and $\text{Var}(\varepsilon) = \sigma^2$. The model (1.3.10) is linear in β's and involves many useful models, such as a simple linear model

$$y = \beta_0 + \beta_1 x_1 + \cdots + \beta_s x_s + \varepsilon, \tag{1.3.11}$$

a quadratic model

$$y = \beta_0 + \sum_{i=1}^{s} \beta_i x_i + \sum_{1 \leqslant i \leqslant j \leqslant s} \beta_{ij} x_i x_j + \varepsilon, \tag{1.3.12}$$

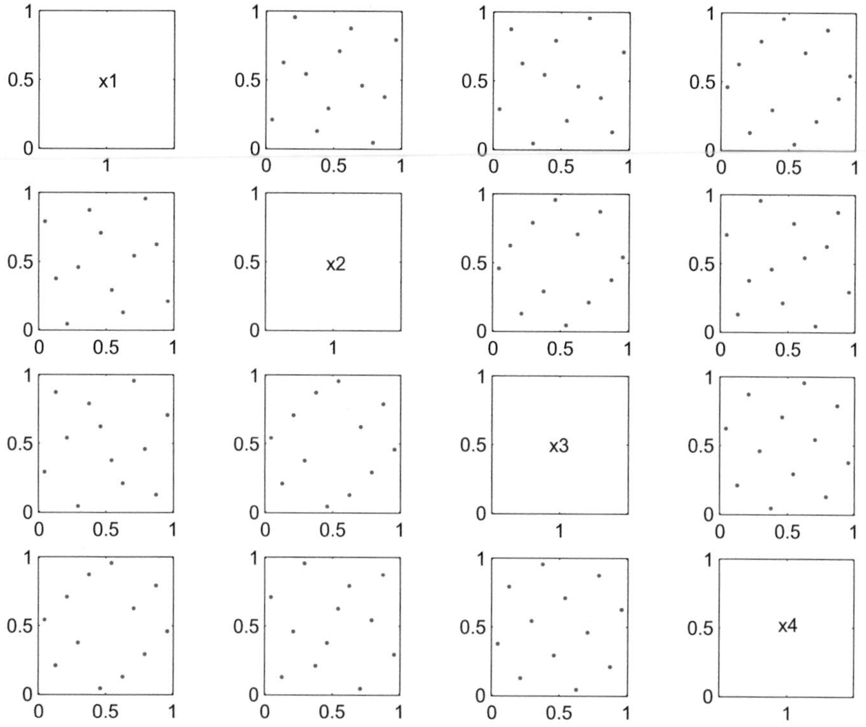

Fig. 1.4 Scatter plots of any two columns for $U_{12}(12^4)$

and a centered quadratic model

$$y = \beta_0 + \sum_{i=1}^{s} \beta_i (x_i - \bar{x}_i) + \sum_{1 \leqslant i \leqslant j \leqslant s} \beta_{ij} (x_i - \bar{x}_i)(x_j - \bar{x}_j) + \varepsilon, \quad (1.3.13)$$

where $\bar{x}_i$ is the sample mean of x_i. Note that functions g_j can be nonlinear in $\boldsymbol{x}$, such as $\exp(-\frac{1}{2}x_i)$, $\log(x_j)$, $1/(10 + x_i x_j)$. Suppose that one wants to employ model (1.3.10) to fit a data set $\mathcal{P} = \{\boldsymbol{x}_1, \ldots, \boldsymbol{x}_n\} \in \mathcal{X}$, the model can be expressed as

$$\boldsymbol{y} = \boldsymbol{G}\boldsymbol{\beta} + \boldsymbol{\varepsilon}, \quad (1.3.14)$$

where

$$\boldsymbol{G} = \begin{bmatrix} g_1(\boldsymbol{x}_1) & \cdots & g_m(\boldsymbol{x}_1) \\ \vdots & & \vdots \\ g_1(\boldsymbol{x}_n) & \cdots & g_m(\boldsymbol{x}_n) \end{bmatrix}, \quad \boldsymbol{\beta} = \begin{bmatrix} \beta_1 \\ \vdots \\ \beta_m \end{bmatrix}, \quad \boldsymbol{\varepsilon} = \begin{bmatrix} \varepsilon_1 \\ \vdots \\ \varepsilon_n \end{bmatrix}, \quad (1.3.15)$$

where $\varepsilon_1, \ldots, \varepsilon_n$ are random errors. The data can be expressed as an $n \times s$ matrix

$$\boldsymbol{X} = \begin{bmatrix} x_{11} & \cdots & x_{1s} \\ \vdots & & \vdots \\ x_{n1} & \cdots & x_{ns} \end{bmatrix} = \begin{bmatrix} \boldsymbol{x}'_1 \\ \vdots \\ \boldsymbol{x}'_n \end{bmatrix}.$$

The matrix $\boldsymbol{X}$ is called *design matrix*. The matrix $\boldsymbol{G}$, combining model and data information, is called *structure matrix*, and $\boldsymbol{M} = \boldsymbol{M}(\mathcal{P}) = \frac{1}{n}\boldsymbol{G}'\boldsymbol{G}$ *information matrix*. The least squares estimator of $\boldsymbol{\beta}$ and its covariance matrix are given by

$$\hat{\boldsymbol{\beta}} = (\boldsymbol{G}^T\boldsymbol{G})^{-1}\boldsymbol{G}^T\boldsymbol{y}, \mathrm{Cov}(\hat{\boldsymbol{\beta}}) = \frac{\sigma^2}{n}\boldsymbol{M}^{-1}. \tag{1.3.16}$$

Clearly, we wish $\mathrm{Cov}(\hat{\boldsymbol{\beta}})$ to be as small as possible in a certain sense that implies to maximize $\boldsymbol{M}(\mathcal{P})$ with respect to $\mathcal{P}$. As $\boldsymbol{M}$ is an $m \times m$ matrix, a more convenient way is to find a scale function of $\boldsymbol{M}$ as a criterion, denoted by $\phi(\boldsymbol{M}(\mathcal{P}))$, and a corresponding *$\phi$-optimal design* that maximizes $\phi(\boldsymbol{M})$ over the design space. Many criteria have been proposed, such as

1. *D-optimality*: Maximize the determinant of $\boldsymbol{M}$. In the multivariate analysis, the determinant of the covariance matrix is called the *generalized variance*. The D-optimality is equivalent to minimize the volume of the confidence ellipsoid $(\boldsymbol{\beta} - \hat{\boldsymbol{\beta}})'\boldsymbol{M}(\boldsymbol{\beta} - \hat{\boldsymbol{\beta}}) \leqslant a^2$ for any $a^2 > 0$.
2. *A-optimality*: Minimize the trace of $\boldsymbol{M}^{-1}$, which is equivalent to minimize the sum of variances of $\hat{\beta}_1, \ldots, \hat{\beta}_m$, where $\hat{\boldsymbol{\beta}} = (\hat{\beta}_1, \ldots, \hat{\beta}_m)'$.
3. *E-optimality*: Minimize the largest eigenvalue of $\boldsymbol{M}^{-1}$, or minimize the value $\max_{||\boldsymbol{\alpha}||=1} \mathrm{Var}(\boldsymbol{\alpha}'\hat{\boldsymbol{\beta}})$.
4. *G-optimality*: Minimize the maximum variance of the predicted response over the domain. Let $\hat{y}(\boldsymbol{x})$ be the prediction of the response at $\boldsymbol{x} \in \mathcal{X}$, i.e., $\hat{y}(\boldsymbol{x}) = \sum_{j=1}^m \hat{\beta}_j g_j(\boldsymbol{x})$. Then, the variance of $\hat{y}(\boldsymbol{x})$ is given by

$$\mathrm{Var}(\hat{y}(\boldsymbol{x})) = \frac{\sigma^2}{n} g(\boldsymbol{x})'\boldsymbol{M}^{-1}g(\boldsymbol{x}), \tag{1.3.17}$$

where

$$g(\boldsymbol{x}) = (g_1(\boldsymbol{x}), \ldots, g_m(\boldsymbol{x}))'. \tag{1.3.18}$$

The G-optimality minimizes the value $\max_{\boldsymbol{x}\in\mathcal{X}} g(\boldsymbol{x})'\boldsymbol{M}^{-1}g(\boldsymbol{x})$.

The concepts of "determinant," "trace," and "eigenvalue" of a matrix can be found in textbooks on linear algebra. The reader can find more optimalities and related theory in Atkinson and Donev (1992) and Pukelsheim (1993).

An optimal design is the best design if the underlying model is known. Optimal designs have many attractive properties, but they have lack of robustness against the model specification. When the true model is known, it may have some difficulty to find the corresponding optimal designs very often. When the underlying model is unknown, optimal designs may have a poor performance. Therefore, the optimum regression design is not robust.

The D-optimum design can improve the accuracy of the estimates since it minimizes the content of the confidence region of the estimated parameters. However, as the authors pointed out, the D-optimum design has the following problem. Because of nonlinearity of the kinetic model, the D-optimum design is only locally optimal in this case. In other words, it depends on the prior chosen parameters. If the initial values of the parameters to be estimated are not located close to the true one, the results may be not good. But in practice, we often know little about the kinetic parameters of an unfamiliar chemical reaction. Therefore, choosing appropriate initial parameters is really a problem that the D-optimum design should face. Naturally, one wants to seek other experimental methods that are non-sensitive to the location of the initial values of the parameters and can also obtain the parameters with satisfactory accuracy. Xu et al. (2000) considered the D-optimal design (DOD), orthogonal design (OD), and uniform designs (UD) for this chemical reaction and compared their performance. For nonlinear model, for example, the model from kinetics of a chemical reaction, two or three levels for each factor seem to be too few to characterize it, while the UD has an advantage that it can offer as many levels as you need for the factors with only a little increment of experimental runs. Heuristically, the UD is especially suitable for nonlinear model. Thus, it is expected to give better performance in the estimation of parameters of the kinetic model of a reversible reaction.

1.3.4 Nonparametric Regression Models

When the experimenter does not have any prior knowledge about the underlying model and wants to explore relationships between the response y and the factors $(x_1, \ldots, x_s)$, in this case, a nonparametric regression model

$$y(\boldsymbol{x}) = f(\boldsymbol{x}) + \varepsilon = f(x_1, \ldots, x_s) + \varepsilon, \tag{1.3.19}$$

can be employed, where function f is unknown, and the random error ε has $E(\epsilon) = 0$ and $\mathrm{Var}(\varepsilon) = \sigma^2$ with unknown σ^2. We want to estimate $y(\boldsymbol{x})$ at each $\boldsymbol{x}$. A natural thought is to spread experimental points uniformly on the experimental domain. This idea created terminologies: "*space-filling design*" that involves "*uniform design*" and "*Latin hypercube sampling*" (its definition can refer to Definition 2.1.1).
Let us see an example to show the above three kinds of designs for the same problem.

Example 1.3.1 (*Weibull growth model*) In a biological experiment, we wish to explore relationship between the growth time (x) and the response (y). Suppose that the underlying model

$$y = y(x) = 1 - e^{-2x^2}, \quad x \in [0, 2], \tag{1.3.20}$$

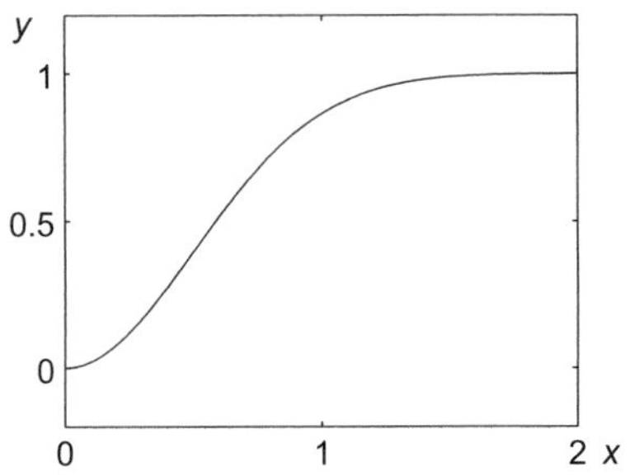

Fig. 1.5 Weibull growth curve model

is unknown. Figure 1.5 gives a plot of the growth curve. Note that random error will effect the response y. For this experiment, the underlying model becomes

$$y = y(x) = 1 - e^{-2x^2} + \varepsilon, \quad x \in [0, 2], \tag{1.3.21}$$

where random error $\varepsilon \sim N(0, \sigma^2)$ with unknown σ^2. Figure 1.6 shows plots of a factorial design, where four levels are chosen. The experiment repeats three times at each level. By this design, we can estimate the overall mean μ and the main effects x at these four levels as well as the variance σ^2. If the quadratic regression model below is chosen by the analyst of the experiment

$$y(x) = \beta_0 + \beta_1 x + \beta_2 x^2 + \varepsilon, \tag{1.3.22}$$

the corresponding D-optimal design is presented in Fig. 1.7, where the dash line is the fitting curve. We can see that the fitting is not well, as model (1.3.22) is far away from the true model. If we choose a cubic regression model and employ the related D-optimal design, the corresponding result will be much better. So the optimal design does not have the robustness against model changes. A uniform design with 12 runs is shown in Fig. 1.8, where the dash line is the fitting curve by a polynomial regression model. We can see that the fitting is well except in small intervals around 0 and 2. We emphasize again that modeling is a very important and difficult issue in experiments under model uncertainty.

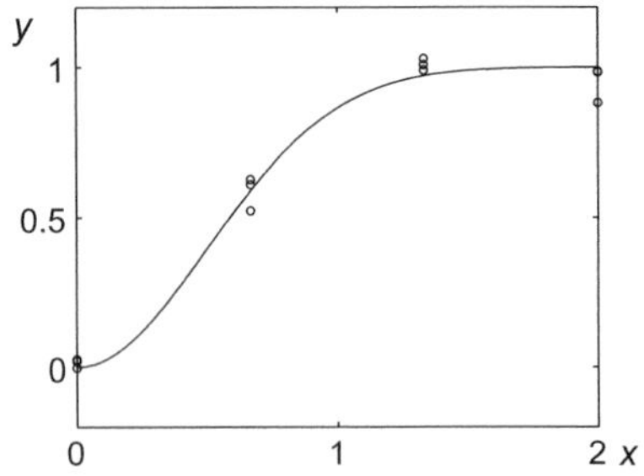

Fig. 1.6 Factorial design

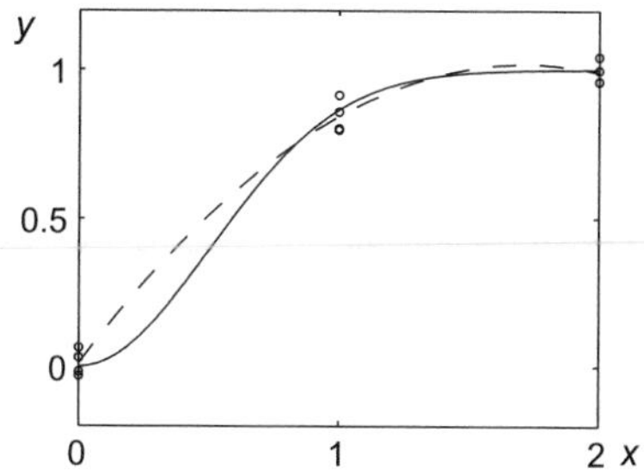

Fig. 1.7 Optimal design

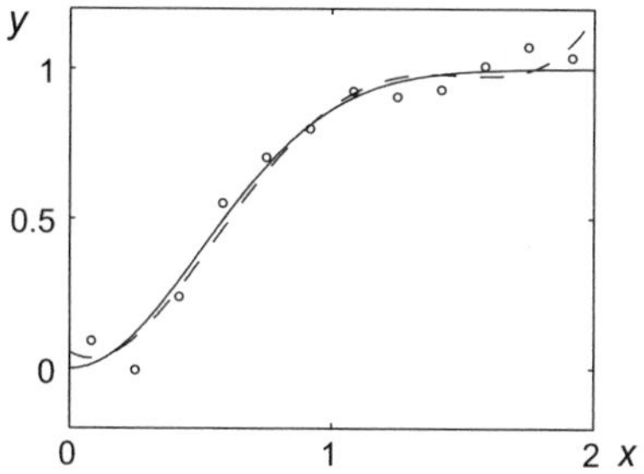

Fig. 1.8 Uniform design

1.3.5 Robustness of Regression Models

In practice, the underlying model is very often not completely known due to complexity of the problem. It is desirable that the design chosen in some way should be robust against small deviations from the assumed model. That is, a small change in the underlying model or in observations should cause only a small change in the performance of a statistical procedure. The common deviations may be from:

(a) The assumed model is not correct or not complete correct.
(b) The random error is not from the assumed distribution (e.g., a normal distribution).
(c) The variance of the random error is not constant over the experimental domain.
(d) There are wild observations (outliers).

A good experimental design should be robust against small deviations from (a) to (d). A design with a such property is called as a *robust design*. As robustness may have a quite different class of alternatives, a solution of robust designs depends on the specific class of alternatives. Huber (1975), Box and Draper (1959), Marcus and Sacks (1978), Li and Notz (1982), Wiens (1990, 1991, 1992), Yue and Hickernell (1999), and Xie and Fang (2000) gave a comprehensive studies on robustness of experimental designs and robust designs. In this book, we focus on the robustness against changes of the model; in particular, we consider the following case.

If the experimenter knows the underlying model is close to a linear model, a robust regression model

$$y(x) = \beta_1 f_1(x) + \cdots + \beta_m f_m(x) + h(x) + \varepsilon, \tag{1.3.23}$$

can be considered, where $h(x)$ denotes the departure of the model (1.3.10) from the true model. When function h belongs to some class of functions, we wish to find a design such that we can obtain the best estimators for $\beta_1, \ldots, \beta_m$ under a certain sense. For example, we define a loss function can consider the worst case and the average loss. It has been shown that the uniform design is a robust design against model changes.

Before giving introduction to theory of the uniform design, we will give a real case study for experiments with model uncertainty in Sect. 1.5.

1.4 Word-Length Pattern: Resolution and Minimum Aberration

A criterion for assessing designs on the design space $\mathcal{U}$ is a function of $\boldsymbol{U}$ and is denoted by $\phi(\boldsymbol{U})$. Each criterion has its own specification from statistical inference or other consideration. Sometimes a criterion can be a vector $(\phi_1(\boldsymbol{U}), \ldots, \phi_m(\boldsymbol{U}))$ with nonnegative components. Now, we meet a problem on how to ordering two nonnegative vectors $\boldsymbol{x} = (x_1, \ldots, x_m)$ and $\boldsymbol{y} = (y_1, \ldots, y_m)$ in $\mathbb{R}^m_+$. Here, we introduce two methods in the literature.

1.4.1 Ordering

(a) Dictionary ordering: Sequentially comparing x_1 and y_1, x_2 and y_2, ... by the following way: We write $\boldsymbol{x} \vdash \boldsymbol{y}$ if $x_1 < y_1$; otherwise if $x_1 = y_1$ and $x_2 < y_2$; otherwise if $x_i = y_i, i = 1, 2$ and $x_3 < y_3$;...; otherwise if $x_i = y_i, i = 1, \ldots, m-1$ and $x_m < y_m$. More compact statement is given by: We write $\boldsymbol{x} \vdash \boldsymbol{y}$ if there exists a k, $1 \leqslant k \leqslant m$ such that $x_i = y_i$ for $i < k$ and $x_k < y_k$; write $\boldsymbol{x} \models \boldsymbol{y}$ if $x_i \leqslant y_i$, $1 \leqslant i \leqslant m$. Note that "$\boldsymbol{x} \models \boldsymbol{y}$" implies "$\boldsymbol{x} = \boldsymbol{y}$" or "$\boldsymbol{x} \vdash \boldsymbol{y}$". Here, the relation "$\models$" and "$\vdash$" can be regarded as "$\leqslant$" and "$<$" in classical algebra, respectively.

(b) Majorization ordering: Let us now briefly review the *majorization theory* (see Marshall and Olkin 1979). For any positive constant c, define a set of nonnegative vectors

$$\mathcal{X}_c = \{\boldsymbol{x} = (x_1, \ldots, x_m) : \boldsymbol{x} \in \mathbb{R}^m_+, \sum_{i=1}^m x_i = c\}.$$

For a nonnegative vector $\boldsymbol{x} \in \mathcal{X}_c$, denote its increasing order statistics by $x_{(1)} \leqslant x_{(2)} \leqslant \cdots \leqslant x_{(m)}$. For any $\boldsymbol{x}, \boldsymbol{y} \in \mathcal{X}_c$, we say $\boldsymbol{x}$ is *majorized* by $\boldsymbol{y}$ and write $\boldsymbol{x} \preceq \boldsymbol{y}$ if

$$\sum_{r=1}^{k} x_{(r)} \geqslant \sum_{r=1}^{k} y_{(r)}, k = 1, 2, \ldots, m. \tag{1.4.1}$$

If there exists at least one strict inequality $\sum_{r=1}^{k} x_{(r)} > \sum_{r=1}^{k} y_{(r)}$ for some k, we write $\boldsymbol{x} \prec \boldsymbol{y}$ strictly. A real-valued function Ψ on $\mathbb{R}_{+}^{m}$ is called *Schur-convex* if $\Psi(\boldsymbol{x}) \leqslant \Psi(\boldsymbol{y})$ for every pair $\boldsymbol{x}, \boldsymbol{y} \in \mathbb{R}_{+}^{m}$ with $\boldsymbol{x} \preceq \boldsymbol{y}$. Necessarily, $\Psi(\boldsymbol{x})$ is symmetric in its arguments, i.e., invariant under permutating $x_1, \ldots, x_m$. We are mainly interested in the following separable *separable convex* class of *Schur-convex functions*

$$\Psi(\boldsymbol{x}) = \sum_{r=1}^{m} \psi(x_r), \quad \psi \text{ is convex on } R_{+} \tag{1.4.2}$$

as well as their monotonic mapping $g(\Psi(\boldsymbol{x}))$ for some g. The following is important for finding some lower bounds of a criterion.

Lemma 1.4.1 *The vector $\bar{\boldsymbol{x}} = (c/m, \ldots, c/m)$ belongs the set $\mathcal{X}_c$ and is the smallest under majorization ordering, i.e., $\bar{\boldsymbol{x}} \preceq \boldsymbol{x}$ for any $\boldsymbol{x} \in \mathcal{X}_c$. For any Schur-convex function Ψ, we have $\Psi(\bar{\boldsymbol{x}}) \leqslant \Psi(\boldsymbol{x})$ and $\Psi(\bar{\boldsymbol{x}})$ is a lower bound of the Ψ-criterion.*

For applying the above theory to experimental design, assume that $\boldsymbol{x}$ in $\mathcal{X}_c$ is an integer vector. If c/m is an integer, the lower bond obtained in Lemma 1.4.1 is tight, otherwise is not attachable. By Lemma 5.2.1 of Dey and Mukerjee (1999), we can obtain a lower bound for the latter case. Let θ and f be the integral part and fractional part of c/m, respectively, and let

$$\widetilde{\boldsymbol{x}} \equiv (\underbrace{\theta, \ldots, \theta}_{m(1-f)}, \underbrace{\theta+1, \ldots, \theta+1}_{mf})'.$$

Lemma 1.4.2 *Under the above notation, we have $\bar{\boldsymbol{x}} \preceq \widetilde{\boldsymbol{x}} \preceq \boldsymbol{x}$. Any separable convex function $\sum_{r=1}^{m} \psi(x_r)$ on the domain $\mathcal{X}_c$ has a tight lower bound*

$$m(1-f)\psi(\theta) + mf\psi(\theta+1). \tag{1.4.3}$$

When c/m is an integer, the above lower bound reduces into $m\psi(\theta)$.

Zhang et al. (2005) used the above theory to find some good lower bounds.

1.4.2 Defining Relation

There are many criteria for fractional factorial designs; for example, see Dey and Mukerjee (1999) and Wu and Hamada (2009). A regular fractional is determined by its defining relation. Let us see a typical example. Consider a two-level experiment involving four factors, denoted by A, B, C, D, by using the orthogonal design table

Table 1.6 Orthogonal design $L_8(2^7)$

No	I	1	2	3	4	5	6	7
1	1	1	1	1	1	1	1	1
2	1	1	1	1	−1	−1	−1	−1
3	1	1	−1	−1	1	1	−1	−1
4	1	1	−1	−1	−1	−1	1	1
5	1	−1	1	−1	1	−1	1	−1
6	1	−1	1	−1	−1	1	−1	1
7	1	−1	−1	1	1	−1	−1	1
8	1	−1	−1	1	−1	1	1	−1

$L_8(2^7)$ listed in the last seven columns of Table 1.6, where two levels are marked as -1 and 1 ($-$ and $+$ for short) and factors A, B, C, D are put on columns 1, 2, 4, and 7. Let us add one more column, I, with all elements one into Table $L_8(2^7)$ that is put into the first column of Table 1.6. This table has the following interesting facts:

- Except for column I, each column has an equal number of plus and minus signs.
- The sum of the dot product of any two columns is zero.
- Column I multiplying times any column leaves that column unchanged.
- The dot product of any two columns yields a column in the table.

For example, $12 = 3, 57 = 2, 34 = 7$; here, 12 means the dot product $1 \cdot 2$, and others are similar.

- The dot product of columns 1, 2, 4, and 7 is column I and denote this fact by $I = 1247$. This fact can also be expressed as $I = ABCD$ if we put factors A, B, C, D on columns 1, 2, 3, 4, respectively. Here, $ABCD$ is called the *generator* of this particular fraction and $I = ABCD$ is called the *defining relation*.
- To dot product A to both sides of $I = ABCD$ results in $A = BCD$. Similarly, we have

$$A = BCD, \ B = ACD, \ C = ABD, D = ABC \tag{1.4.4}$$

$$AB = CD, \ AC = BD, \ AD = BC \tag{1.4.5}$$

that means main effect A is aliased with interaction BCD, interaction AB is aliased with CD, etc. It can be seen that all the aliasing relations can be obtained by the defining relation $I = ABCD$. If all the interactions of order three can be ignored, the relations in (1.4.4) become

$$AB = CD, AC = BD, AD = BC.$$

In this case, all the main effects can be estimated, and all the interactions of order are confounded each other.

1.4.3 *Word-Length Pattern and Resolution*

Many designs are determined by more than one defining relations. Let us see an example.

Example 1.4.1 Consider a 2^{5-2} FFD in Table 1.7. A 2^{s-k} FFD is a fractional factorial design with $n = 2^{s-k}$ runs and k factors each having two levels. From the rule of dot product, we find the following relations

$$4 = 12, \quad 5 = 13. \tag{1.4.6}$$

These relations imply

$$I = 124 = 135 = 2345 \tag{1.4.7}$$

that are defining relations of the design. The relation $I = 2345$ can be obtained by relations $I = 124$ and $I = 135$. From these defining relations, all the aliasing relations are as follows

$$\begin{aligned} 1 &= 24 = 35 = 12345 \\ 2 &= 14 = 345 = 1235 \\ 3 &= 15 = 245 = 1234 \\ 4 &= 12 = 235 = 1345 \\ 5 &= 13 = 234 = 1245. \end{aligned}$$

Assume that a 2^{s-k} FFD, $\boldsymbol{U}$, is determined by k defining relations, or defining words, where a *word* consists of letters which stand for the factors denoted by $1, 2, \ldots, s$ or $A, B, \ldots$. The number of letters in a word is its *word-length*, and the group formed by the k defining words is called the *defining contrast subgroup*. The group consists of $2^k - 1$ words plus the identity element I.

Table 1.7 A 2^{5-2} FFD

No	1	2	3	4	5
1	1	1	1	1	1
2	1	1	−1	1	−1
3	1	−1	1	−1	1
4	1	−1	−1	−1	−1
5	−1	1	1	1	−1
6	−1	1	−1	1	1
7	−1	−1	1	−1	−1
8	−1	−1	−1	−1	1

Let $A_i(\boldsymbol{U})$ be the number of words with word-length i in the defining contrast subgroup of design $\boldsymbol{U}$. The vector

$$W(\boldsymbol{U}) = (A_1(\boldsymbol{U}), \ldots, A_s(\boldsymbol{U})) \tag{1.4.8}$$

is called *word-length pattern* of $\boldsymbol{U}$. The word-length pattern of a design $\boldsymbol{U}$ indicates its statistical inference ability. The *resolution* of $\boldsymbol{U}$ proposed by Box and Hunter (1961a, b) is defined to be the smallest t with positive $A_t(\boldsymbol{U})$. Resolution III and IV designs are more useful in practice:

Resolution III designs: No main effects are aliased with any other main effect, but main effects are aliased with two-factor interactions and two-factor interactions may be aliased with each other.

Resolution IV designs: No main effect is aliased with any other main effect or with any two-factor interaction, but two-factor interactions may be aliased with each other.

Example 1.4.2 A 2^{3-1} design with defining relation $I = 123$ is of resolution III with $W = (0, 0, 1)$; a 2^{4-1} design with defining relation $I = 1234$ is of resolution IV with $W = (0, 0, 0, 1)$; and design in Example 1.4.1 is of resolution III with $W = (0, 0, 2, 1, 0)$.

Example 1.4.3 Two 2^{5-1} orthogonal designs, denoted by $\boldsymbol{U}_1$ and $\boldsymbol{U}_2$, with respective defining relations

$$\boldsymbol{U}_1 : I = 4567 = 12346 = 12357$$
$$\boldsymbol{U}_2 : I = 1236 = 1457 = 234567$$

They have the same resolution IV, but they have different word-length patterns

$$W(\boldsymbol{U}_1) = (0, 0, 0, 1, 2, 0, 0), \quad W(\boldsymbol{U}_2) = (0, 0, 0, 2, 0, 1, 0).$$

Therefore, $\boldsymbol{U}_1$ has a less aberration than $\boldsymbol{U}_2$. In fact, there are three pairs of two-factor interactions being confounded, i.e., $45 = 67, 46 = 57, 47 = 56$ by using $\boldsymbol{U}_1$, but there are six pairs of two-factor interactions being confounded, i.e., $12 = 36$, $13 = 26$, $16 = 23$, $14 = 57$, $15 = 47$, $17 = 45$, by using $\boldsymbol{U}_2$. However, we need more mathematical criteria, one of which is the minimum aberration criterion.

1.4.4 Minimum Aberration Criterion and Its Extension

The *minimum aberration* criterion, proposed by Fries and Hunter (1980), has been popularly used for regular factorial designs. It is also defined based on the word-length pattern. We omit its definition here as its general version will be given below. The minimum aberration criterion assumes the *effect hierarchy principle*. It tends to minimize the contamination of non-negligible interactions sequentially from low to high dimensions. Many authors have studied on finding minimum aberration designs, for example, Cheng and Mukerjee (1998), Fang and Mukerjee (2000), Mukerjee and Wu (2001), and Cheng and Tang (2005).

Recently, the definition of generalized aberration for non-regular designs has become an hot research topic. Tang and Deng (1999) suggested a minimu G_2-aberration criterion for non-regular two-level designs; Xu and Wu (2001) and Ma and Fang (2001) independently proposed the *generalized word-length pattern* and related criteria. Other important works on this direction include Deng and Tang (2002), Cheng et al. (2002), Ye (2003), and Fang and Zhang (2004).

According to Xu and Wu (2001), consider the ANOVA model

$$\boldsymbol{y} = \alpha_0 \mathbf{1}_n + \boldsymbol{X}_{(1)}\boldsymbol{\alpha}_1 + \cdots + \boldsymbol{X}_{(s)}\boldsymbol{\alpha}_s + \boldsymbol{\epsilon},$$

where $\boldsymbol{y}$ is the vector of n observations, $\mathbf{1}_n$ is the n-vector of ones, α_0 is the intercept, $\boldsymbol{\alpha}_j$ is the vector of all j-factor interactions, $\boldsymbol{X}_{(j)}$ is the matrix of contrast coefficients for $\boldsymbol{\alpha}_j$ and $\boldsymbol{\epsilon}$ are the vectors of independent random errors. For a $U(n; q_1, \ldots, q_s)$ design, $\mathcal{P}$, let $\boldsymbol{X}_{(j)} = (x_{ik}^j)$ be the matrix consisting of all j-factor contrast coefficients, for $j = 0, \ldots, s$.

Definition 1.4.1 Define

$$A_j(\mathcal{P}) = n^{-2} \sum_k \left| \sum_{i=1}^n x_{ik}^j \right|^2. \tag{1.4.9}$$

The vector $W(\mathcal{P}) = (A_1(\mathcal{P}), \ldots, A_s(\mathcal{P}))$ is called the *generalized word-length pattern*

This definition is picked up from Xu and Wu (2001). Then, the *resolution* of $\mathcal{P}$ is the smallest j with positive $A_j(\mathcal{P})$ in $W(\mathcal{P})$. Let $\boldsymbol{U}_1$ and $\boldsymbol{U}_2$ be two designs and t be the smallest integer such that $A_t(\boldsymbol{U}_1) \neq A_t(\boldsymbol{U}_2)$. Then, $\boldsymbol{U}_1$ is said to have *less generalized aberration* than $\boldsymbol{U}_2$ if $A_t(\boldsymbol{U}_1) < A_t(\boldsymbol{U}_2)$. A design $\boldsymbol{U}$ is said to have *generalized minimum aberration* (GMA) if no other design in the design space has less generalized aberration than it.

Note that for a U-type design $\boldsymbol{U} \in \mathcal{U}(n; q^s)$, based on the coding theory, Ma and Fang (2001) proposed another definition of *generalized word-length pattern*, denoted by $W^g(\boldsymbol{U}) = \{A_1^g(\boldsymbol{U}), \ldots, A_s^g(\boldsymbol{U})\}$, and the corresponding *generalized minimum aberration* (GMA) criterion. Since for a U-type design $\boldsymbol{U} \in \mathcal{U}(n; q^s)$, in fact

$$A_j^g(\boldsymbol{U}) = A_j(\boldsymbol{U})/(q-1), \ \text{for } j = 1, \ldots, s,$$

we omit the details for this criterion here.

For the generalized word-length pattern, we have the following facts:

1. $A_j(\boldsymbol{U})$ reduces to the $A_j(\boldsymbol{U})$ in (1.4.8) for any regular fractional design $\boldsymbol{U}$.
2. $A_j(\boldsymbol{U}) \geqslant 0$ for any design $\boldsymbol{U}$ and $1 \leqslant j \leqslant s$.

3. $A_1(\boldsymbol{U}) = 0$ for any U-type design.
4. $A_j(\boldsymbol{U}) = 0$ for $j = 1, \ldots, t$ if $\boldsymbol{U}$ is an orthogonal array of strength t.

1.5 Implementation of Uniform Designs for Multifactor Experiments

In this section, a general procedure of implementation of uniform design for physical experiments under model uncertainty is given. A demonstration example for the use of the uniform design is represented. This experiment was carried out by Zeng (1994); later, Fang (2002) used this case study for the demonstration purpose. We reorganize the above material in this section.

The following steps give a general guideline for multifactor industrial/laboratory experiments:

Step 1. Choose factors and experimental domain, and determine suitable number of levels for each factor.
Step 2. Choose a suitable UD table to accommodate the number of factors and the corresponding levels. This can be easily done by visiting the UD Web or by a computer software.
Step 3. From the uniform design table, randomly determine the run order of experiments and conduct the experiments.
Step 4. Fit the data set by one or few suitable models. Regression analysis, neural networks, wavelets, multivariate splines, and empirical Kriging models are useful in modeling.
Step 5. Knowledge discovery from the built model. For example, we want to find the "best" combination of the factor values that maximizes/minimizes the response and verify the claim with further experiments.
Step 6. Further investigation: From the up-to-date information obtained in the data analysis, some additional experiments may be necessary.

Next, we describe these procedures step by step via a chemical engineering example.

Example 1.5.1 A chemical experiment is conducted in order to find the best setup to increase the yield.

Four factors, the amount of formaldehyde (x_1), the reaction temperature (x_2), the reaction time (x_3), and the amount of potassium carbolic acid (x_4), are under consideration. The response variable is designated as the yield (y). The experimental domain is chosen to be $\mathcal{X} = [1.0, 5.4] \times [5, 60] \times [1.0, 6.5] \times [15, 70]$ and each factor takes 12 levels in this domain. The factors and levels are listed in Table 1.8.

Choose a UD table of the form $U_n(12^4)$. Assign each factor for a column of the table such that the four factors occupy different columns. The 12 levels marked by $1, 2, \ldots, 12$ are transformed into the real levels for each factor. It results in a

Table 1.8 Factors and levels

Factor	Unit	Level
x_1, the amount of formaldehyde	mol	1.0, 1.4, 1.8, 2.2, 2.6, 3.0, 3.4, 3.8, 4.2, 4.6, 5.0, 5.4
x_2, the reaction temperature	hour	5, 10, 15, 20, 25, 30, 35, 40, 45, 50, 55, 60
x_3, the reaction time	hour	1.0, 1.5, 2.0, 2.5, 3.0, 3.5, 4.0, 4.5, 5.0, 5.5, 6.0, 6.5
x_4, the amount of potassium carbolic acid	ml	15, 20, 25, 30, 35, 40, 45, 50, 55, 60, 65, 70

Table 1.9 $U_{12}(12^4)$ and related design

No. of runs	x_1	x_2	x_3	x_4	y
5	1.0 (1)	50 (10)	2.5 (4)	45 (7)	0.0795
6	1.4 (2)	25 (5)	6.0 (11)	25 (3)	0.0118
10	1.8 (3)	5 (1)	4.0 (7)	55 (9)	0.0109
7	2.2 (4)	30 (6)	1.0 (1)	35 (5)	0.0991
11	2.6 (5)	55 (11)	5.5 (10)	65 (11)	0.1266
9	3.0 (6)	45 (9)	4.5 (8)	15 (1)	0.0717
8	3.4 (7)	20 (4)	3.0 (5)	70 (12)	0.1319
3	3.8 (8)	10 (2)	2.0 (3)	20 (2)	0.0900
2	4.2 (9)	35 (7)	6.5 (12)	50 (8)	0.1739
4	4.6 (10)	60 (12)	3.5 (6)	30 (4)	0.1176
1	5.0 (11)	40 (8)	1.5 (2)	60 (10)	0.1836*
12	5.4 (12)	15 (3)	5.0 (9)	40 (6)	0.1424

design listed in Table 1.9. For simplicity, we tentatively just use "the table" for Table 1.9. Randomize the order of these 12 level-combinations and list the order of runs in the first column of the table, implement 12 experiments according to the level-combinations in the table, and record the corresponding yield y in the very last column of the table, where the value marked by $*$ is the maximum.

Choosing a good metamodel will be very useful for various purpose, especially for finding a good level-combination of the four factors with the largest yield. The latter is the major goal of the data analysis in this experiment. The best result among the 12 responses is $y_1 = 18.36\%$ at $x_1 = 5.0$, $x_2 = 40$, $x_3 = 1.5$, and $x_4 = 60$. This can be served as a *benchmark*. Note that these 12 runs represent for $12^4 = 20736$ level-combinations or represent for any level-combination in the experimental domain. There is a high chance to find another level-combination that corresponds to a higher yield than $y_1 = 18.36\%$. For this task, if we can find a metamodel to approximate the true one, we may find a good level-combination corresponding a higher yield.

Due to lack of knowledge about the underlying model, a nonparametric regression model (1.3.19) is considered. There are many ways to search a good metamodel. The reader can refer to Fang et al. (2006) for a comprehensive review. For experiments with more factors, it is popular to consider a set of basis of functions, $\{B_0(\boldsymbol{x}), B_1(\boldsymbol{x}), \ldots\}$ and a *maximal model of interest*,

$$\hat{g}(\boldsymbol{x}) = \beta_1 B_1(\boldsymbol{x}) + \beta_2 B_2(\boldsymbol{x}) + \ldots + \varepsilon. \tag{1.5.1}$$

Then by techniques for variable selection one of submodel of (1.5.1) will be used as a metamodel. In this example, we consider only linear and quadratic regression models. At first, let us try the first-order regression model of the form

$$E(y) = \beta_0 + \beta_1 x_1 + \beta_2 x_2 + \beta_3 x_3 + \beta_4 x_4$$

as it is simple. Based on the data in Table 1.9, it turns out

$$\hat{y} = -0.0533 + 0.0281 x_1 + 0.0010 x_2 - 0.0035 x_3 + 0.0011 x_4. \tag{1.5.2}$$

Its ANOVA table is shown in Table 1.10. From the ANOVA table, we find that the model (1.5.2) involves an insignificant term "x_3" with p-value 0.4962. We have to remove this term from the model. By the backward elimination techniques in regression analysis (see Miller (2002), for example), the resulting model turns out to be

$$\hat{y} = 0.0107 + 0.0289 x_1$$

with $R^2 = 57.68\%$ and $s^2 = 0.0014$. This model is not consistent with the experience of the experimenter as there are three factors not being involved in the model. Therefore, a more flexible second-order quadratic regression of the form, as another *maximal model of interest*,

$$E(y) = \beta_0 + \sum_{i=1}^{4} \beta_i x_i + \sum_{i \leqslant j} \beta_{ij} x_i x_j \tag{1.5.3}$$

is considered. Here, the number of unknown parameters is greater than the number of runs and this model is inestimable. However, this model provides a base and some submodel may fit the purpose of the experiment well. The remaining study is going to find a submodel of (1.5.3) as a metamodel. With MAXR, a technique of selection of variables, we find a good subset model to be

$$\begin{aligned} \hat{y} = {} & 0.0446 + 0.0029 x_2 - 0.0260 x_3 + 0.0071 x_1 x_3 \\ & + 0.000036 x_2 x_4 - 0.000054 x_2^2 \end{aligned} \tag{1.5.4}$$

with $R^2 = 97.43\%$ and $s^2 = 0.0001$. The corresponding ANOVA table is shown in Table 1.11.

Statistical diagnostics are useful for checking metamodels. A regression model is usually under a certain assumption (normality, constant variance, etc). But sometimes the assumptions are in doubt. Therefore, we need statistical diagnostics to check the

Table 1.10 ANOVA table for model (1.5.2)

Analysis of Variance					
Source	DF	Sum of Squares	Mean Square	F Stat	Prob > F
Model	4	0.0274	0.0069	8.2973	0.0086
Error	7	0.0058	0.0008		
C Total	11	0.0332			

Type III Tests					
Source	DF	Sum of Squares	Mean Square	F Stat	Prob > F
X1	1	0.0180	0.0180	21.8021	0.0023
X2	1	0.0033	0.0033	3.9496	0.0872
X3	1	0.0004	0.0004	0.5150	0.4962
X4	1	0.0046	0.0046	5.6248	0.0495

Table 1.11 ANOVA table for model (1.5.4)

Analysis of Variance					
Source	DF	Sum of Squares	Mean Square	F Stat	Prob > F
Model	5	0.0323	0.0065	45.5461	0.0001
Error	6	0.0009	0.0001		
C Total	11	0.0332			

Type III Tests					
Source	DF	Sum of Squares	Mean Square	F Stat	Prob > F
X2	1	0.0014	0.0014	10.1949	0.0188
X3	1	0.0125	0.0125	88.0883	0.0001
X1X3	1	0.0193	0.0193	135.5636	0.0001
X2X2	1	0.0024	0.0024	16.8276	0.0063
X2X4	1	0.0062	0.0062	43.6923	0.0006

assumptions. For example, is the model correct? are there any outliers? is the variance constant? and is the error normally distributed? If some assumption fails, we have to make some modification. The reader can find some basic knowledge of the statistical diagnostics in Mayers (1990) or Cook (1986). By some statistical diagnostics, we conclude model (1.5.4) is acceptable. We omit the details.

In the literature, the centered quadratic regression model of the form

$$E(y) = \beta_0 + \sum_{i=1}^{4} \beta_i (x_i - \bar{x}_i) + \sum_{i \leqslant j} \beta_{ij} (x_i - \bar{x}_i)(x_j - \bar{x}_j), \qquad (1.5.5)$$

is also suggested for a maximal model of interest, where $\bar{x}_i$ is the sample mean of x_i. In this data set, $\bar{x}_1 = 3.2$, $\bar{x}_2 = 32.5$, $\bar{x}_3 = 3.75$, and $\bar{x}_4 = 42.5$. Once again, by using some model selection technique, the final model is

Table 1.12 ANOVA table for model (1.5.6)

Analysis of Variance					
Source	DF	Sum of Squares	Mean Square	F Stat	Prob > F
Model	5	0.0322	0.0064	39.5466	0.0002
Error	6	0.0010	0.0002		
C Total	11	0.0332			

Type III Tests					
Source	DF	Sum of Squares	Mean Square	F Stat	Prob > F
X1	1	0.0180	0.0180	110.7654	0.0001
X2	1	0.0031	0.0031	19.2816	0.0046
X4	1	0.0046	0.0046	28.5141	0.0018
X3X4	1	0.0023	0.0023	13.8349	0.0099
X2X2	1	0.0047	0.0047	29.0842	0.0017

$$\hat{y} = 0.1277 + 0.0281(x_1 - 3.2) + 0.000937(x_2 - 32.5) + 0.00114(x_4 - 42.5) \\ + 0.00058(x_3 - 3.75)(x_4 - 42.5) - 0.000082(x_2 - 32.5)^2 \quad (1.5.6)$$

with $R^2 = 97.05\%$ and $s^2 = 0.0002$. The corresponding ANOVA table is given in Table 1.12.

The metamodels (1.5.4) and (1.5.6) can be used to predict response at any point of the experimental domain. It also can be used for searching the "best" combination of the factor values. Maximize y with respect to x_i, $i = 1, \ldots, 4$ under models (1.5.4) or (1.5.6), respectively, over the domain $\mathcal{X}$ given in *Step 1*, that is to find x_i^*, $i = 1, \ldots, 4$ such that

$$\hat{y}(x_1^*, x_2^*, x_3^*, x_4^*) = \max_{\mathcal{X}} \hat{y}(x_1, x_2, x_3, x_4),$$

where $\hat{y}(x_1, x_2, x_3, x_4)$ is given by (1.5.4) or (1.5.6), respectively. By any optimization algorithm, it is easily found that under model (1.5.4), $x_1^* = 5.4$, $x_2^* = 50.2$, $x_3^* = 1$, $x_4^* = 70$ and the corresponding response $\hat{y}(5.4, 50.2, 1, 70) = 19.3\%$ is the maximum; and under model (1.5.6), $x_1^* = 5.4$, $x_2^* = 43.9$, $x_3^* = 6.5$, $x_4^* = 70$ and the corresponding response $\hat{y}(5.4, 43.9, 6.5, 70) = 26.5\%$ is the maximum. It looks that model (1.5.6) is better, but it needs some additional experiments to judge which metamodel is really better.

As two optimal points $\boldsymbol{x}_1^* = (5.4, 50.2, 1, 70)$ and $\boldsymbol{x}_2^* = (5.4, 43.9, 6.5, 70)$ do not appear in the plan (Table 1.9), some additional experiments are necessary. A simplest way is to implement m runs at these two optimal points $\boldsymbol{x}_1^*$ and $\boldsymbol{x}_2^*$ and to compare their mean yield. In this experiment, the experimenter implemented three runs at $\boldsymbol{x}_1^*$ and $\boldsymbol{x}_2^*$ and found that the mean of y is 20.1% at $\boldsymbol{x}_1^*$ and 26.3% at $\boldsymbol{x}_2^*$, respectively. Thus, we prefer (1.5.6), the centered quadratic regression model, as our final metamodel.

Note that both metamodels recommend $x_1 = 5.4$ and $x_4 = 70$. This fact implies that we should consider increasing upper bounds of the experimental levels for x_1 and x_4. The experimenter should consider a further investigation and arrange a consequent experiment.

1.6 Applications of the Uniform Design

A design that chooses experimental points uniformly scattered on the domain is called *uniform experimental design*, or *uniform design* for simplicity. The uniform design was proposed in 1980 by Fang and Wang (cf. Fang 1980 and Wang and Fang 1981) and has been widely used for thousands of physical and computer experiments with model uncertainty. In the literature, the uniform design can be utilized as

- A fractional factorial design with model unknown;
- A space-filling design for computer experiments;
- A robust design against the model specification;
- A design of experiments with mixtures;
- A supersaturated design.

Supersaturated designs are factorial designs in which the number of main effects and/or interactions is greater than the number of runs. Such designs are often called screening designs. Many authors appreciate the advantages of the uniform design for

- More choices of designs to the users;
- Designs have been tabulated;
- Both factorial and computer experiments can be applied to;
- Less information of the underlying model can be accepted.

There are various connections of the uniform design with other designs such as the fractional factorial design including the orthogonal array, supersaturated design, robust design, combinatorial design, and code theory. A number of uniform design tables can be found on the Web

http://www.math.hkbu.edu.hk/UniformDesign, or http://web.stat.nankai.edu.cn/cms-ud/.

Exercises

1.1

Compare the physical experiment and the computer experiment and list their difference.

1.2

What are metamodels in computer experiments? Give some requirements for metamodels.

1.3

In an experiment, there are many variables in general. Give the difference between variable and factor in experimental design. Give some examples for quantitative factors and quantitative factors. What is the difference between environmental variables and nuisance variables?

1.4

Consider model for the one-factor experiment and its statistical model (1.3.1)

$$y_{ij} = \mu_j + \varepsilon_{ij}, \quad j = 1, \ldots, q, i = 1, \ldots, n_j,$$

where ε_{ij} are i.i.d. distributed as $N(0, \sigma^2)$. Let $n = n_1 + \cdots + n_q$ be the number of runs. This model can be expressed as a linear model $\boldsymbol{y} = \boldsymbol{X}\boldsymbol{\beta} + \boldsymbol{\epsilon}$.

(1) Write down $\boldsymbol{y}$, $\boldsymbol{X}$, $\boldsymbol{\beta}$ in details.

(2) Find $\boldsymbol{X}'\boldsymbol{X}$ and $\boldsymbol{X}'\boldsymbol{y}$.

(3) Find the distributions of y_{ij} and $\boldsymbol{y}$.

1.5

Model in the previous exercise can be expressed as (1.3.2), i.e.,

$$y_{ij} = \mu + \alpha_j + \varepsilon_{ij}, \quad j = 1, \ldots, q, i = 1, \ldots, n_j.$$

(1) Express this model as a linear model $\boldsymbol{y} = \boldsymbol{X}\boldsymbol{\beta} + \boldsymbol{\epsilon}$, and give $\boldsymbol{y}$, $\boldsymbol{X}$, $\boldsymbol{\beta}$ in details.

(2) Let $SS_E = \sum_{i=1}^{q} \sum_{j=1}^{r} (y_{ij} - \bar{y}_i)^2$ be the error sum of squares, where $n_1 = \cdots = n_q = r$. Prove $E[SS_E] = q(r-1)\sigma^2$.

1.6

For a two-factor experiment, its ANOVA model is given by (1.3.3), i.e.,

$$\begin{aligned} y_{ijk} &= \mu + \alpha_i + \beta_j + (\alpha\beta)_{ij} + \varepsilon_{ijk}, \\ &\quad i = 1, \ldots, q_1; \; j = 1, \ldots, q_2; k = 1, \ldots, r, \end{aligned}$$

with constraints

$$\begin{aligned} &\alpha_1 + \cdots + \alpha_{q_1} = 0; \beta_1 + \cdots + \beta_{q_2} = 0; \\ &\sum_{i=1}^{q_1} (\alpha\beta)_{ij} = 0, \; j = 1, \ldots, q_2; \sum_{j=1}^{q_2} (\alpha\beta)_{ij} = 0, i = 1, \ldots, q_1. \end{aligned}$$

Prove

(1) When $q_1 = q_2 = 2$, there is one linearly independent interaction among $\{(\alpha\beta)_{11}, (\alpha\beta)_{12}, (\alpha\beta)_{21}, (\alpha\beta)_{22}\}$.

(2) There are $(q_1 - 1)(q_2 - 2)$ linearly independent interactions among $\{(\alpha\beta)_{ij}, i = 1, 2; j = 1, \ldots, q_i\}$.

1.7

Consider the following model of a three-factor experiment

$$y_{ijk} = \mu + \alpha_i + \beta_j + \gamma_k + (\beta\gamma)_{jk} + \varepsilon_{ijk}, \ i = 1, 2, 3, \ j = 1, 2, 3, \ k = 1, 2,$$

where A, B, C are the factors,
y_{ijk} is the response at $A = A_i$, $B = B_j$ and $C = C_k$,
μ is the overall mean,
α_i is the main effect of the factor A at the level A_i,
β_j is the main effect of the factor B at the level B_j,
γ_k is the main effect of the factor C at the level C_k,
$(\beta\gamma)_{jk}$ is the interaction of B and C at B_j and C_k,
ε_{ijk} is random error at the experiment with $A = A_i$, $B = B_j$ and $C = C_k$, and $\varepsilon_{ijk} \sim N(0, \sigma^2)$.

Answer the following questions:

(a) Give constrains on α_i, β_j, γ_k, and $(\beta\gamma)_{jk}$. How many independent parameters among α_i, β_j, γ_k, and $(\beta\gamma)_{jk}$?

(b) Express this model as of the form $\boldsymbol{y} = \boldsymbol{X}\boldsymbol{\beta} + \boldsymbol{\epsilon}$ where $\boldsymbol{\beta}$ is formed by independent parameters discussed in question a) and indicate $\boldsymbol{y}$, $\boldsymbol{X}$, $\boldsymbol{\beta}$, and $\boldsymbol{\epsilon}$.

(c) Give the degrees of freedom for the sum of squares: SS_A, SS_B, SS_C, $SS_{B\times C}$, SS_E, and SS_T.

(d) Give formulas for SS_A, SS_B, SS_C, $SS_{B\times C}$, SS_E, and SS_T.

1.8

The concept "orthogonality" has been appeared in different fields. Answer the following questions:

(1) Give definition for two line segments in R^d be orthogonal.

(2) Give definition for two planes be orthogonal.

(3) Give definition for two linear spaces be orthogonal.

(4) Let $\boldsymbol{X}$ be the matrix of the orthogonal design table $L_9(3^4)$. Denote by $\mathcal{L}_i, i = 1, 2, 3, 4$ the linear subspace generated by the ith column of $\boldsymbol{X}$; denote by $\mathcal{L}_{ij}$, $1 \leqslant i < j \leqslant 4$ the linear subspace generated by the ith and jth columns of $\boldsymbol{X}$. Prove that $\mathcal{L}_i$ and $\mathcal{L}_j$ are orthogonal if $i \neq j$ and $\mathcal{L}_{12}$ and $\mathcal{L}_{34}$ are orthogonal.

1.9

Give the word-length pattern and resolution for the following designs:

(a) A design 2^{6-2} with defining relations $I = ABCE = BCDF$;

(b) A design 2^{7-2} with defining relations $I = ABCDF = ABDEG$;

(c) A design 2^{7-3} with defining relations $I = ABCE = BCDF = ACDG$.

1.10

Answer the following questions:

1. There is a command "hadamard" to generate a Hadamard matrix of order n. Use this command to find a Hadamard matrix of order 8 by which we can obtain $L_8(2^7)$.
2. If $\boldsymbol{H}$ is a Hadamard matrix, then $-\boldsymbol{H}$ is a Hadamard matrix.
3. If $\boldsymbol{H}$ is a Hadamard matrix, let

$$\boldsymbol{V} = \begin{bmatrix} \boldsymbol{H} & -\boldsymbol{H} \\ \boldsymbol{H} & \boldsymbol{H} \end{bmatrix}.$$

   Prove that $\boldsymbol{V}$ is a Hadamard matrix.

1.11

Calculate the Hamming distances between any two different runs designed by $L_9(3^4)$ below. Give your finding and conjecture.

No	1	2	3	4
1	1	1	1	1
2	1	2	2	2
3	1	3	3	3
4	2	1	2	3
5	2	2	3	1
6	2	3	1	2
7	3	1	3	2
8	3	2	1	3
9	3	3	2	1

1.12

In economics, the so-called Lorenz curve is a graphical representation of the distribution of income. Give a review on the Lorenz curve and relationship between the Lorenz curve and the majorization theory.

References

Atkinson, A.C., Donev, A.N.: Optimum Experimental Designs. Oxford Science Publications, Oxford (1992)

Atkinson, A.C., Bogacka, B., Bagacki, M.B.: D- and T-optimum designs for the kinetics of a reversible chemical reaction. Chemometr. Intell. Lab. **43**, 185–198 (1998)

Box, G.E.P., Draper, N.R.: A basis for the selection of a response surface design. J. Amer. Statist. Assoc. **54**, 622–654 (1959)

Box, G.E.P., Hunter, J.S.: The 2^{k-p} fractional factorial designs i. Technometrics **3**, 311–351 (1961a)

Box, G.E.P., Hunter, J.S.: The 2^{k-p} fractional factorial designs ii. Technometrics **3**, 449–458 (1961b)
Cheng, C.S., Mukerjee, R.: Regular fractional factorial designs with minimum aberration and maximum estimation capacity. Ann. Statist. **26**, 2289–2300 (1998)
Cheng, C.S., Tang, B.: A general theory of minimum aberration and its applications. Ann. Statist. **33**, 944–958 (2005)
Cheng, C.S., Deng, L.W., Tang, B.: Generalized minimum aberration and design efficiency for non-regular fractional factorial designs. Statist. Sinica **12**, 991–1000 (2002)
Cook, R.D.: Assessment of local influence (with discussion). J. R. Stat. Soc. Ser. B **48**, 133–169 (1986)
Dean, A., Morris, M., Stufken, J., Bingham, D.: Handbook of Design and Analysis of Experiments. Chapman and Hall/CRC, Boca Raton (2015)
Deng, L.Y., Tang, B.: Design selection and classification for hadamard matrices using generalized minimum aberration criteria. Technometrics **44**, 173–184 (2002)
Dey, A., Mukerjee, R.: Fractional Factorial Plans. Wiley, New York (1999)
Elsayed, E.A.: Reliability Engineering. Addison Wesley, Massachusetts (1996)
Fang, K.T.: The uniform design: application of number-theoretic methods in experimental design. Acta Math. Appl. Sinica **3**, 363–372 (1980)
Fang, K.T.: Experimental designs for computer experiments and for industrial experiments with model unknown. J. Korean Statist. Soc. **31**, 277–299 (2002)
Fang, K.T., Chan, L.Y.: Uniform design and its industrial applications. In: Pham, H. (ed.) Springer Handbook of Engineering Statistics, pp. 229–247. Springer, New York (2006)
Fang, K.T., Mukerjee, R.: A connection between uniformity and aberration in regular fractions of two-level factorials. Biometrika **87**, 1993–198 (2000)
Fang, K.T., Wang, Y.: Number-Theoretic Methods in Statistics. Chapman and Hall, London (1994)
Fang, K.T., Zhang, A.: Minimum aberration majorization in non-isomorphic saturated designs. J. Statist. Plann. Inference **126**, 337–346 (2004)
Fang, K.T., Li, R., Sudjianto, A.: Design and Modeling for Computer Experiments. Chapman and Hall/CRC, New York (2006)
Fries, A., Hunter, W.G.: Minimum aberration 2^{k-p} designs. Technometrics **22**, 601–608 (1980)
Ghosh, S., Rao, C.R.: Handbook of Statistics, 13. North-Holland, New York (1996)
Hedayat, A.S., Sloane, N.J.A., Stufken, J.: Orthogonal Arrays: Theory and Applications. Springer, New York (1999)
Ho, W.M.: Case studies in computer experiments, applications of uniform design and modern modeling techniques, Master Thesis, Hong Kong Baptist University (2001)
Huber, J.: Robustness and designs. In: Srivastave, J.N. (ed.) A Survey of Statistical Design and Linear Models, pp. 287–303. North-Holland, Amsterdam (1975)
Kiefer, J.: Optimaml experimental designs (with discussion). J. Roy. Statist. Soc. Ser. B **21**, 272–319 (1959)
Li, K.C., Notz, W.: Robust designs for nearly linear regression. J. Statist. Plann. Inference **6**, 135–151 (1982)
Ma, C.X., Fang, K.T.: A note on generalized aberration in factorial designs. Metrika **53**, 85–93 (2001)
Marcus, M.B., Sacks, J.: Robust designs for regression problems. Statistical Decision Theory and Related Topics, vol. 2, pp. 245–268. Academic Press, New York (1978)
Marshall, A.W., Olkin, I.: Inequalities: Theory of Majorization and Its Applications. Academic Press, New York (1979)
Mayers, R.H.: Classical and Modern Regression with Applications. Duxbury Classic Series, Amsterdam (1990)
Miller, A.: Subset Selection in Regression, 2nd edn. Chapman & Hall/CRC, London (2002)
Mukerjee, R., Wu, C.F.J.: Minimum aberration designs for mixed factorials in terms of complementary sets. Statist. Sinica **11**, 225–239 (2001)
Pukelsheim, F.: Optimum Design of Experiments. Wiley, New York (1993)

Santner, T.J., Williams, B.J., Notz, W.I.: The Design and Analysis of Computer Experiments. Springer, New York (2003)
Tang, B., Deng, L.Y.: Minimum G_2-aberration for nonregular fractional designs. Ann. Statist. **27**, 1914–1926 (1999)
Wang, Y., Fang, K.T.: A note on uniform distribution and experimental design. Chinese Sci. Bull. **26**, 485–489 (1981)
Wiens, D.P.: Robust minimax designs for multiple linear regression. Linear Algebra Appl. **127**, 327–340 (1990)
Wiens, D.P.: Designs for approximately linear regression: two optimality properties of uniform designs. Statist. Probab. Lett. **12**, 217–221 (1991)
Wiens, D.P.: Minimax designs for approximately linear regression. J. Statist. Plann. Inference **31**, 353–371 (1992)
Wu, C.F.J., Hamada, M.: Experiments: Planning, Analysis, and Optimization, 2nd edn. Wiley, New York (2009)
Xie, M.Y., Fang, K.T.: Admissibility and minimaxity of the uniform design in nonparametric regression model. J. Statist. Plann. Inference **83**, 101–111 (2000)
Xu, H.Q., Wu, C.F.J.: Generalized minimum aberration for asymmetrical fractional factorial designs. Ann. Statist. **29**, 1066–1077 (2001)
Xu, Q.S., Liang, Y.Z., Fang, K.T.: The effects of different experimental designs on parameter estimation in the kinetics of a reversible chemical reaction. Chemometr. Intell. lab. **52**, 155–166 (2000)
Ye, K.Q.: Indicator function and its application in two-level factorial design. Ann. Statist. **31**(984–994) (2003)
Yue, R.X., Hickernell, F.J.: Robust designs for fitting liner models with misspecification. Statist. Sinica **9** (1999)
Zeng, Z.J.: The uniform design and its applications. Liaoning People's Publishing House, Shenyang, China (in Chinese) (1994)
Zhang, A., Fang, K.T., Li, R., Sudjianto, A.: Majorization framework for balanced lattice designs. Ann. Statist. **33**, 2837–2853 (2005)

Chapter 2
Uniformity Criteria

Motivated by the overall mean model and the famous Koksma–Hlawka inequality, the main idea of uniform experimental design is to scatter the experimental design points uniformly over the experimental domain. In this regard, uniformity measures constitute the main concept of the uniform experimental design. Discrepancy is a measure which is defined as the deviation between the empirical and the theoretical uniform distribution. Therefore, discrepancy is a measure of uniformity which provides a way of construction of uniform designs. However, there are several discrepancies under different considerations. This chapter introduces the definitions and derives lower bounds for different discrepancies, which can be used to construct uniform designs.

Section 2.1 introduces the overall mean model and the Koksma–Hlawka inequality. Sections 2.1–2.5 give the definitions and properties of different discrepancies including star discrepancy, centered L_2-discrepancy, wrap-around L_2-discrepancy, mixture discrepancy, discrete discrepancy, and Lee discrepancy. Most of them can be defined by the tool of reproducing kernel Hilbert space. In Sect. 2.6, the lower bounds of different discrepancies are given, which can be used as a benchmark for searching uniform designs.

2.1 Overall Mean Model

Assume

$$y = f(\boldsymbol{x}) \tag{2.1.1}$$

be the true model of a system on a domain $\mathcal{X}$, where $\boldsymbol{x} = (x_1, \ldots, x_s)$ are variables/factors and y is response. Very often, we can assume the domain to be a hypercube $[a_1, b_1] \times \cdots \times [a_s, b_s]$. Without loss of any generality, we can assume that the hypercube is the unit hypercube $C^s = [0, 1]^s = [0, 1] \times \cdots \times [0, 1]$. Let

K.-T. Fang et al., *Theory and Application of Uniform Experimental Designs*, Lecture Notes in Statistics 221,
https://doi.org/10.1007/978-981-13-2041-5_2

$\mathcal{P} = \{\boldsymbol{x}_1, \ldots, \boldsymbol{x}_n\}$ be a set of n design points on C^s. One important issue is to estimate the true model $f(\boldsymbol{x})$ based on $\mathcal{P}$ by some metamodel

$$\hat{y} = g(\boldsymbol{x}). \tag{2.1.2}$$

There are many alternative metamodels in general, how to find a criterion to assess the metamodel becomes an important problem. One may consider the overall mean model, which aims at finding the best estimator of the overall mean of y, that is

$$E(y) = \int_{C^s} f(\boldsymbol{x}) d\boldsymbol{x}. \tag{2.1.3}$$

This means that the most preliminary aim of the design $\mathcal{P}$ is to obtain the best estimator of the overall mean of y in a certain sense. To estimate $E(y)$, a natural idea is to use the sample mean of $\mathcal{P}$

$$\bar{y}(\mathcal{P}) = \frac{1}{n} \sum_{i=1}^{n} y_i, \tag{2.1.4}$$

where $y_i = f(\boldsymbol{x}_i), i = 1, \ldots, n$. Given the number of design points n, one may search a set $\mathcal{P}$ with n points on $\mathcal{X}$ to minimize the difference

$$\text{diff-mean} = |\bar{y}(\mathcal{P}) - E(y)|. \tag{2.1.5}$$

There are two types of methods to choose $\mathcal{P}$: *stochastic approach* and *deterministic approach*. The main idea of stochastic approach is to find a design $\mathcal{P}$ such that the sample mean $\bar{y}(\mathcal{P})$ is an unbiased or asymptotically unbiased estimator of $E(y)$ and has the smallest possible estimation variance. On the other hand, the deterministic approach aims at finding a sampling scenario so that the difference in (2.1.5) can be as small as possible.

The simplest stochastic approach is the Monte Carlo method, where the n design points in $\mathcal{P}$ are independent samples from $U(C^s)$, the uniform distribution on C^s, and the corresponding sample mean $\bar{y}(\mathcal{P})$ is an unbiased estimator of $E(y)$ with the estimation variance $\text{Var}(f(\boldsymbol{x}))/n$, where random variable $\boldsymbol{x}$ follows the uniform distribution on C^s. From the central limited theorem, the difference

$$|\bar{y}(\mathcal{P}) - E(y)| \leqslant 1.96\sqrt{\text{Var}(f(\boldsymbol{x}))/n},$$

with 95% confidence. However, the estimation variance $\text{Var}(f(\boldsymbol{x}))/n$ is too large for many cases and should be reduced. *Latin hypercube sampling* (LHS), proposed by McKay et al. (1979), has been widely used to reduce the estimation variance. Its main idea is to randomly choose $\boldsymbol{x}_1, \ldots, \boldsymbol{x}_n$ such that they are dependent and have the same marginal distribution. The construction method of LHS is to divide the domain C^s of each x_k into n strata with equal marginal probability $1/n$ and sample once from each stratum. The more detailed definition is given as follows:

Definition 2.1.1 Let $\mathcal{X} = [0, 1]^s$ be the experimental domain and $\mathcal{U}$ be the set of a grid of n^s equally spaced points on $\mathcal{X}$. In fact, the set of $\mathcal{U}$ is all the U-type designs, $U(n, n^s)$ each column being a permutation of $1/(2n), 3/(2n), \dots, (2n-1)/(2n)$. A Latin hypercube design (LHD) or a midpoint Latin hypercube sampling (MLHS) is a random sample on $\mathcal{U}$, or equivalent, it is a U-type design $U(n, n^s)$ each column being a random permutation of $1/(2n), 3/(2n), \dots, (2n-1)/(2n)$.

Let $\mathcal{P} = \{\boldsymbol{x}_1, \dots, \boldsymbol{x}_n\}$ be a Latin hypercube design. For each $\boldsymbol{x}_j$, there is a sub-cube, denoted by $C_{\boldsymbol{x}_j}$, with side-length $1/n$ and center $\boldsymbol{x}_j$. Let $\boldsymbol{y}_j$ be a random sample on $C_{\boldsymbol{x}_j}$, and the set $\{\boldsymbol{y}_1, \dots, \boldsymbol{y}_n\}$ is called a Latin hypercube sampling (LHS).

McKay et al. (1979) showed that the covariance between any two points is negative if $f(\boldsymbol{x})$ is monotonic in each variable x_k. Denote the sample mean of the responses of the LHS be $\bar{y}_{\text{LHS}}$. Stein (1987) and Owen (1992) found an expression for the variance of $\bar{y}_{\text{LHS}}$ and showed that

$$\text{Var}(\bar{y}_{\text{LHS}}) = \frac{\text{Var}(f(\boldsymbol{x}))}{n} - \frac{c}{n} + o\left(\frac{1}{n}\right),$$

where c is a positive constant. This indicates that the estimation variance is smaller than that of Monte Carlo method. However, it was also shown that LHS does not reach the smallest possible variance for the sample mean. Many modifications of LHS have been proposed to improve performance of LHS, such as orthogonal array-based Latin hypercube design proposed by Tang (1993), and a comprehensive discussion can be found in Koehler and Owen (1996).

The deterministic approach is another widely used way to minimize the difference in (2.1.5). The famous *Koksma–Hlawka inequality* in quasi-Monte Carlo method shows that

$$|\bar{y}(\mathcal{P}) - E(y)| \leqslant V(f)D^*(\mathcal{P}), \tag{2.1.6}$$

where $V(f)$ is the total variation of the function f in the sense of Hardy and Krause (see Niederreiter 1992 and Hua and Wang 1981) and $D^*(\mathcal{P})$ is the star discrepancy of $\mathcal{P}$, which does not depend on f. Section 2.2 will give its definition and detailed discussion. The upper bound in Koksma–Hlawka inequality (2.1.6) is tight in some cases. Given the function f and the experimental domain, $V(f)$ remains invariant. If $V(f)$ is bounded in the experimental domain, then one may choose $\mathcal{P}$ with n design points on C^s such that its star discrepancy $D^*(\mathcal{P})$ is as small as possible and that we can minimize the upper bound of the difference in (2.1.5).

The star discrepancy, proposed by Weyl (1916), is a popularly used measure of uniformity in quasi-Monte Carlo methods. The lower the star discrepancy, the better will be the uniformity of the set of points under consideration. Then, from the Koksma–Hlawka inequality, one may find a set of n points uniformly scattered on C^s, which is called a uniform design (UD), proposed by Fang (1980) and Fang and Wang (1981). The uniform design is robust against model specification. For example, if two functions $f_1(\boldsymbol{x})$ and $f_2(\boldsymbol{x})$ have the same total variation, then the uniform design indicates that the difference in (2.1.5) for the two functions has the same upper bound.

Note that both LHS and UD are motivated by the over mean model. The best design for the over mean model may be not enough to estimate the true model $f(\boldsymbol{x})$ over C^s. Fortunately, both LHS and UD are space-filling designs, which have a good performance not only for estimation of the overall mean but also for finding a good approximate model.

2.2 Star Discrepancy

The star discrepancy, proposed by Weyl (1916), is a widely used uniformity measure in the uniform design and quasi-Monte Carlo method. It is defined to measure the difference between the uniform probability distribution on C^s and the empirical distribution function of a set of n points in C^s.

2.2.1 *Definition*

For any $\boldsymbol{x} = (x_1, \ldots, x_s) \in C^s$, let $F(\boldsymbol{x})$ be the distribution function of uniform distribution on C^s, i.e.,

$$F(\boldsymbol{x}) = \begin{cases} x_1 \ldots x_s, & \text{if } 0 \leqslant x_i \leqslant 1, i = 1, \ldots, s \\ 0, & \text{otherwise.} \end{cases} \tag{2.2.1}$$

Let $F_{\mathcal{P}}(\boldsymbol{x})$ be the empirical distribution function corresponding to the design $\mathcal{P} = \{\boldsymbol{x}_1, \ldots, \boldsymbol{x}_n\}, \boldsymbol{x}_i \in C^s$, i.e.,

$$F_{\mathcal{P}}(\boldsymbol{x}) = \frac{1}{n} \sum_{i=1}^{n} 1_{[\boldsymbol{x}_i, \boldsymbol{\infty})}(\boldsymbol{x}), \tag{2.2.2}$$

where $\boldsymbol{x}_i = (x_{i1}, \ldots, x_{is})$, $\boldsymbol{\infty} = (\infty, \ldots, \infty)$, $1_{[\boldsymbol{x}_i, \boldsymbol{\infty})}(\boldsymbol{x})$ is the indicator function, whose value equals 1 if $\boldsymbol{x} \in [\boldsymbol{x}_i, \boldsymbol{\infty})$, otherwise 0, and $[\boldsymbol{x}_i, \boldsymbol{\infty}) = [x_{i1}, \infty) \times \cdots \times [x_{is}, \infty)$.

Definition 2.2.1 The *star L_p-discrepancy* of a design $\mathcal{P}$ is defined as the L_p-norm of the difference between $F_{\mathcal{P}}(\boldsymbol{x})$ and $F(\boldsymbol{x})$, i.e.,

$$D_p^*(\mathcal{P}) = \| F_{\mathcal{P}}(\boldsymbol{x}) - F(\boldsymbol{x}) \|_p = \left\{ \int_{C^s} |F_{\mathcal{P}}(\boldsymbol{x}) - F(\boldsymbol{x})|^p \, \mathrm{d}\boldsymbol{x} \right\}^{1/p}. \tag{2.2.3}$$

From geometric point view, the empirical distribution can be represented by $F_{\mathcal{P}}(\boldsymbol{x}) = N(\mathcal{P} \cap [\boldsymbol{0}, \boldsymbol{x}))/n$, where $N(\mathcal{P} \cap [\boldsymbol{0}, \boldsymbol{x}))$ is the number of points of $\mathcal{P}$ falling in the rectangle $[\boldsymbol{0}, \boldsymbol{x}) = [0, x_1) \times \cdots \times [0, x_s)$. The uniform probability distribution $F(\boldsymbol{x})$ is the volume of the rectangle $[\boldsymbol{0}, \boldsymbol{x})$, denoted as $\mathrm{vol}([\boldsymbol{0}, \boldsymbol{x}))$. Then, the star L_p-discrepancy in (2.2.3) can be expressed as

$$D_p^*(\mathcal{P}) = \left\{ \int_{C^s} \left| \frac{N(\mathcal{P} \cap [\boldsymbol{0}, \boldsymbol{x}))}{n} - \mathrm{vol}([\boldsymbol{0}, \boldsymbol{x})) \right|^p \mathrm{d}\boldsymbol{x} \right\}^{1/p}. \tag{2.2.4}$$

The most useful value of p is $p = \infty$ and $p = 2$.

(1) When $p = \infty$, the star L_∞-discrepancy, simplified by star discrepancy, can be expressed as

$$D^*(\mathcal{P}) = \sup_{\boldsymbol{x} \in C^s} \left| \frac{N(\mathcal{P} \cap [\boldsymbol{0}, \boldsymbol{x}))}{n} - \mathrm{vol}([\boldsymbol{0}, \boldsymbol{x})) \right|, \tag{2.2.5}$$

which is known as the *Kolmogorov–Smirnov statistic* in goodness-of-fit testing. The star discrepancy plays an important role in quasi-Monte Carlo methods as well as in the field of statistics, but it cannot be computed in polynomial time (Winker and Fang 1997).

(2) When $p = 2$, Warnock (1972) showed that the star L_2-discrepancy has a simple formula

$$\begin{aligned} [D_2^*(\mathcal{P})]^2 = {} & \left(\frac{1}{3}\right)^2 - \frac{2}{n} \sum_{i=1}^n \prod_{j=1}^s \frac{(1 - x_{ij}^2)}{2} \\ & + \frac{1}{n^2} \sum_{i,l=1}^n \prod_{j=1}^s [1 - \max(x_{ij}, x_{lj})], \end{aligned} \tag{2.2.6}$$

and its computational complexity is $O(n^2 s)$. The star L_2-discrepancy is known as the *Cramér–Von Mises goodness-of-fit statistic* (D'Agostino and Stephens 1986). However, when $p \neq 2$, the star L_p-discrepancy does not have explicit expression.

Define the *local discrepancy function* by

$$\mathrm{disc}^*(\boldsymbol{x}) = \frac{N(\mathcal{P} \cap [\boldsymbol{0}, \boldsymbol{x}))}{n} - \mathrm{vol}([\boldsymbol{0}, \boldsymbol{x})), \tag{2.2.7}$$

which is the difference between the uniform probability distribution function and empirical distribution function at the point $\boldsymbol{x}$. Then, the star L_p-discrepancy is

$$D_p^*(\mathcal{P}) = \left\| \mathrm{disc}^* \right\|_p = \begin{cases} \left\{ \int_{C^s} |\mathrm{disc}^*(\boldsymbol{x})|^p \, \mathrm{d}\boldsymbol{x} \right\}^{1/p}, & 1 \leqslant p < \infty, \\ \sup_{\boldsymbol{x} \in C^s} |\mathrm{disc}^*(\boldsymbol{x})|, & p = \infty, \end{cases} \tag{2.2.8}$$

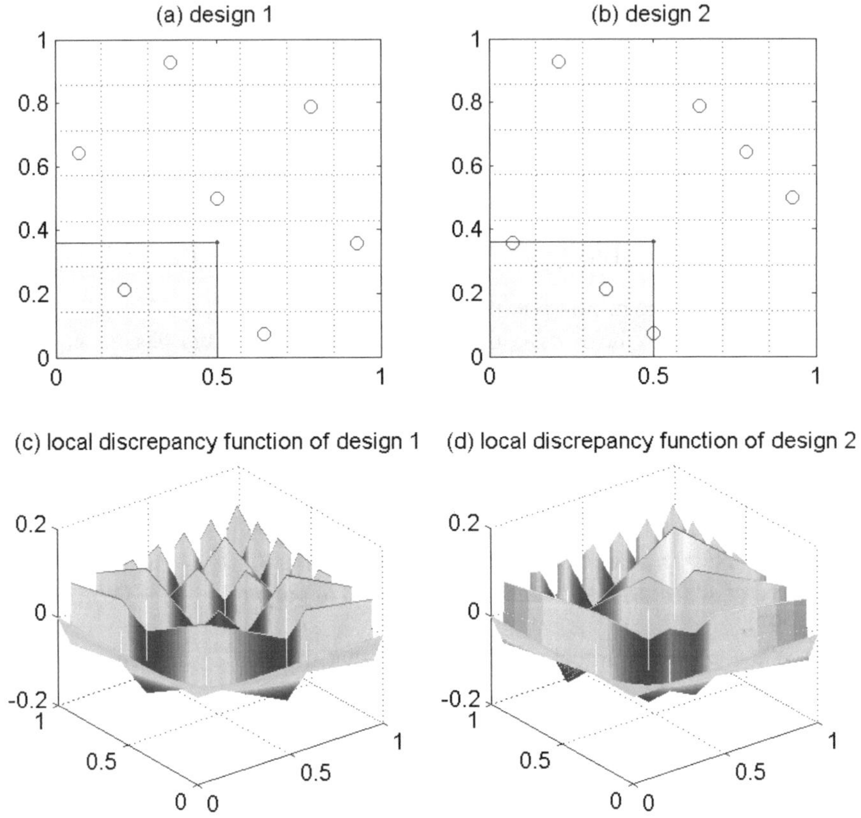

Fig. 2.1 Two designs and their corresponding local discrepancy functions

which can be considered as the average local discrepancy function under L_p-norm on C^s if $p < \infty$ and as the supremum of the absolute of local discrepancy function if $p = \infty$. Figure 2.1 shows two designs and their corresponding local discrepancy functions on C^s. For example, when $\boldsymbol{x} = (0.5, 0.36)$, $\text{disc}^*(\boldsymbol{x}) = 1/7 - 0.5 \times 0.36 = -0.0371$ for design 1 and $\text{disc}^*(\boldsymbol{x}) = 3/7 - 0.5 \times 0.36 = 0.2486$ for design 2. It seems that the first design is more uniform than the second one, as well as the fluctuation of local discrepancy function.

2.2.2 Properties

For an n-run design $\mathcal{P} = \{\boldsymbol{x}_1, \ldots, \boldsymbol{x}_n\}$ on C^s, define $\boldsymbol{X} = [\boldsymbol{x}_1, \ldots, \boldsymbol{x}_n]^T$ be a $n \times s$ design matrix, where $\boldsymbol{x}_i = (x_{i1}, \ldots, x_{is})^T$, $i = 1, \ldots, n$. For any type of discrepancy D, let $D(\mathcal{P}) = D(\boldsymbol{X})$. A reasonable measure of uniformity should satisfy:

C_1 **Invariant under permuting factors and/or runs**. It is obvious that the permutation of factors and/or runs does not change the property of the design. Accordingly, the uniformity measure should not be changed when the factors and/or runs are permuted. Let $\boldsymbol{P} : n \times n$ and $\boldsymbol{Q} : s \times s$ be two permutation matrices, i.e., their elements are zeroes or ones, with exactly one 1 in each row and column. A discrepancy D is called *permutation invariant* if D satisfies

$$D(\boldsymbol{P}\boldsymbol{X}\boldsymbol{Q}) = D(\boldsymbol{X}) \quad \text{for any design matrix } \boldsymbol{X}.$$

Here, multiplication by $\boldsymbol{P}$ permutes the runs, and multiplication by $\boldsymbol{Q}$ permutes the factors.

C_2 **Invariant under the coordinates rotation or reflection**. Intuitively, the uniformity of a design should remain unchanged, when a design is rotated around one coordinate or reflected. Let $\boldsymbol{Q}_\theta$ be an $s \times s$ rotation matrix in $x_i - x_j$ plane. For example, the clockwise rotation θ degree matrix in $x_1 - x_2$ plane is

$$\boldsymbol{Q}_\theta = \begin{pmatrix} \cos\theta & -\sin\theta & & & \\ \sin\theta & \cos\theta & & & \\ & & 1 & & \\ & & & \ddots & \\ & & & & 1 \end{pmatrix}. \tag{2.2.9}$$

Then, rotating the design $\boldsymbol{X}$ in $x_i - x_j$ plane can be expressed as

$$\boldsymbol{X}_\theta = (\boldsymbol{X} - 0.5)\boldsymbol{Q}_\theta + 0.5, \tag{2.2.10}$$

where for a matrix $\boldsymbol{A} = (a_{ij})$, the matrix $\boldsymbol{A} \pm 0.5 = (a_{ij} \pm 0.5)$. A *rotation-invariant discrepancy* is one that satisfies

$$D(\boldsymbol{X}_\theta) = D(\boldsymbol{X}), \text{ for } \theta = \pi/2, \pi, 3\pi/2, \text{ and any design } \boldsymbol{X}.$$

Moreover, define the reflection of the design $\mathcal{P}$ through plane $x_j = 1/2$ passing through the center of the domain as

$$\mathcal{P}_{\text{ref},j} = \{(x_{i1}, \ldots, x_{i,j-1}, 1 - x_{ij}, x_{i,j+1}, \ldots, x_{is})^T : i = 1, \ldots, n\}.$$

The *reflection-invariant discrepancy* requests

$$D(\mathcal{P}_{\text{ref},j}) = D(\mathcal{P}), \quad \forall j = 1, \ldots, s, \text{ for any design } \mathcal{P}.$$

It is easily known that both rotating $x_i - x_j$ plane with clockwise $\pi/2$, π or $3\pi/2$ angle and reflecting the experimental domain $[0, 1]^s$ through the plane $x_j = 1/2$ give the same experimental domain. Rotation-invariant discrepancy and reflection-invariant discrepancy require that these rotation and reflection do not change the value of uniformity criterion.

C_3 **Measure the projection uniformity**. According to the *hierarchical ordering principle* in Sect. 1.3.2, lower-order effects are more likely to be important than higher-order effects, and effects of the same order are equally likely to be important. We need to consider uniformity of the design projected on each coordinator for the main effect estimation, uniformity of the design on $x_i - x_j$ plane for estimation of interaction between factor i and factor j, and so on. A good uniformity criterion should measure not only uniformity of $\mathcal{P}$ on C^s but also projection uniformity of $\mathcal{P}$ on C^u, where u is a non-empty subset of $\{1, \ldots, s\}$.

C_4 **Have some geometric meaning**. The uniform design has its own geometric meaning, and thus uniformity criterion should also process some geometric meaning.

C_5 **Easy to compute**. Given a uniformity criterion, it is not an easy job to search a uniform design under the uniformity criterion. Usually, some optimization algorithms are employed to find a uniform design. In the searching procedure, the criterion value of each design must be computed a lot of time, and it is better that the uniformity criterion can be computed in polynomial time of the number of runs n.

C_6 **Satisfy Koksma–Hlawka inequality**. The inequality (2.1.6) provides a major support of the uniform design, and thus each uniformity criterion should satisfy this requirement.

Now, consider the properties of the star discrepancy in (2.2.5) and the star L_2-discrepancy in (2.2.6). It is well known that both the discrepancies satisfy C_1, C_4, and C_6. With reference to the other criteria, we have the following comments.

For C_2. Neither the star discrepancy nor the star L_2-discrepancy satisfies C_2. This is because when one rotates or reflects the design $\mathcal{P}$, the values of the two discrepancies may be changed. The reason for the change is that the origin $\mathbf{0}$ plays a special role in defining the star L_p-discrepancy by anchoring the box $[\mathbf{0}, \boldsymbol{x})$. For illustration, one example is shown for the star discrepancy. Consider to rotate and reflect a two-dimensional design in $\boldsymbol{C}^s$ with the clockwise rotation matrix $\boldsymbol{Q}_\theta$ in (2.2.9)

$$\boldsymbol{Q}_\theta = \begin{pmatrix} \cos\theta & -\sin\theta \\ \sin\theta & \cos\theta \end{pmatrix}, \quad \theta = \pi/2, \pi, 3\pi/2,$$

and reflection axial $x_1 = 1/2$ and $x_2 = 1/2$. Figure 2.2 shows that rotating and reflecting a design can change the star discrepancy, i.e., the star discrepancy is neither a rotation invariant nor a reflection invariant. Note that the star discrepancy of the rotated design in Fig. 2.2b equals that of the reflected design in Fig. 2.2e. Moreover, the star discrepancies of the designs in Fig. 2.2d, f are also equal, since these designs can be obtained

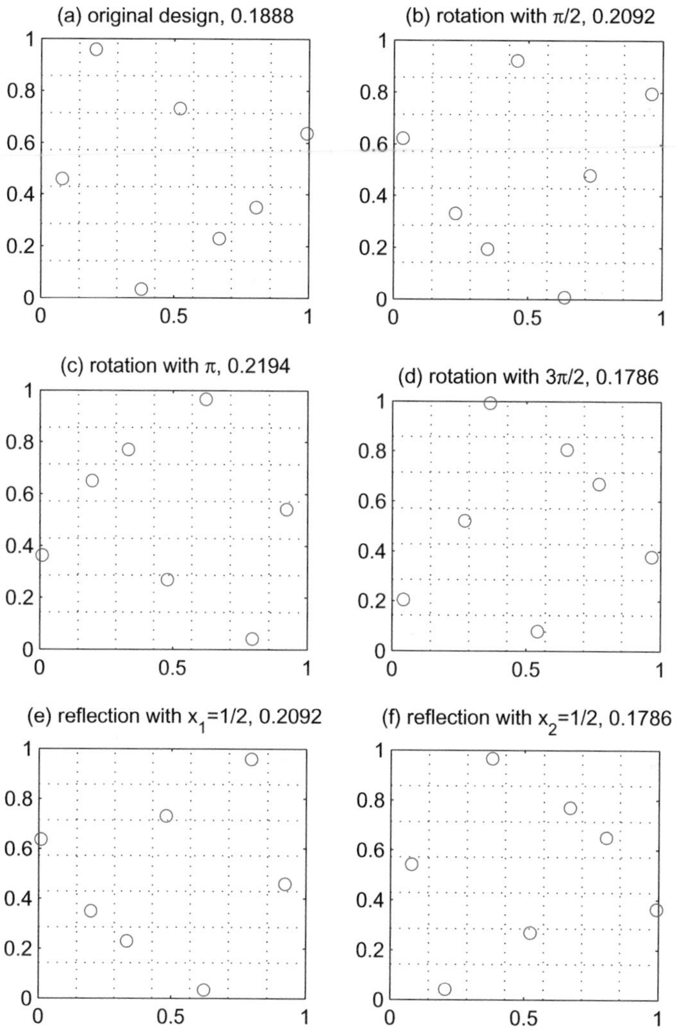

Fig. 2.2 Star discrepancy of a design and the designs after clockwise rotation and reflection

from reflecting one design through the line $y = x$ to other. The star L_2-discrepancy has a similar property.

For C_3. The star L_p-discrepancy does not measure uniformity well for projection designs with less factors except $p = \infty$, i.e., the star discrepancy satisfies C_3, while the star L_2-discrepancy does not. From the definition (2.2.8), the local discrepancy function $\text{disc}^*(\boldsymbol{x})$ of the star L_p-discrepancy ignores the low-dimensional subregion when $p < \infty$.

Let $u \subseteq \{1, \ldots, s\}$ be a set indexing factors of interest and $[0, 1]^u$ be the corresponding experimental domain. Define the local discrepancy function restricted to only the factors indexed by u to be

$$\mathrm{disc}_u^*(\boldsymbol{x}) \triangleq \mathrm{disc}^*(\boldsymbol{x}_u, \mathbf{1}), \tag{2.2.11}$$

where $\boldsymbol{x}_u$ denotes the projection of $\boldsymbol{x}$ onto $[0, 1]^u$ and $\mathrm{disc}^*(\boldsymbol{x})$ is defined in (2.2.7). For example, for $\boldsymbol{x} = (x_1, \ldots, x_s)$, $(\boldsymbol{x}_u, \mathbf{1}) = (1, x_2, 1)$ when $s = 3$ and $u = \{2\}$; and $(\boldsymbol{x}_u, \mathbf{1}) = (x_1, 1, x_3, 1)$ when $s = 4$ and $u = \{1, 3\}$. Note that $\mathrm{disc}_\emptyset^*(\boldsymbol{x}) = 0$. The star L_p-discrepancy of a design restricted to the factors indexed by u is written as $D_{p,u}^*(\mathcal{P}) = \left\|\mathrm{disc}_u^*\right\|_p$. The star discrepancy satisfies

$$D^*(\mathcal{P}) = \sup_u D_{\infty,u}^*(\mathcal{P}).$$

Then, the projection uniformity is already considered by star discrepancy.

For C_5. The star L_2-discrepancy has an explicit expression, while the star discrepancy does not have, and thus only the star L_2-discrepancy satisfies C_5. For computing the value of star discrepancy, one at least needs to compute the limiting values of $|\mathrm{disc}^*(\boldsymbol{x})|$ as $\boldsymbol{x}$ approaches points in the set $\{0, 1, x_{11}, \ldots, x_{n1}\} \times \cdots \times \{0, 1, x_{1s}, \ldots, x_{ns}\}$ from all possible directions, since the local discrepancy is a piecewise multi-linear function of $\boldsymbol{x}$, as depicted in Fig. 2.1. This requires $\mathrm{O}(n^s)$ operations (Winker and Fang 1997), which is prohibitive even for a moderate number of factors. However, it has been observed that if such n^s design points for $\boldsymbol{x}$ are not enough, then one needs more points to compute the approximate value of star discrepancy. Efficient calculation of the star discrepancy has been discussed for one factor (Niederreiter 1973), two factors (Clerk 1986), small numbers of factors with $s \leqslant 10$, $n \leqslant 100$ (Bundschuh and Zhu 1993), and large n (Winker and Fang 1997).

In summary, the star discrepancy satisfies C_1, C_3, C_4, and C_6, while the star L_2-discrepancy satisfies C_1, C_4, C_5, and C_6. Moreover, for star L_p-discrepancy with $p \neq 2$, there is no explicit expression, and it also cannot measure the projection uniformity, i.e., it only satisfies C_1, C_4 and C_6. Next section will give some generalized discrepancies to overcome the shortcoming of star L_p-discrepancy.

2.3 Generalized L_2-Discrepancy

To overcome the shortcomings of star L_p-discrepancy mentioned in the last section, some generalized L_2-discrepancy proposed by Hickernell (1998a, b) and Zhou et al. (2013) is given in this section.

Since the star L_2-discrepancy has an explicit expression, the generalized discrepancy prefers to use the L_2-norm. However, the star L_2-discrepancy fails to measure the projection uniformity, and a generalized discrepancy should measure the uniformity for any projection intervals (one dimension), rectangles (two or higher dimensions). Moreover, the origin $\mathbf{0}$ plays a key role in computing star L_p-discrepancy and results in some unreasonable phenomenon, and then a generalized discrepancy may eliminate the effect of the origin $\mathbf{0}$ such that every corner point of the unit cube plays the same role as the origin $\mathbf{0}$.

2.3.1 Definition

Let $\mathcal{P}$ be a design with n runs and s-factors on C^s. Denote $\{1 : s\} = \{1, \dots, s\}$, and let $u \subset \{1 : s\}$ be the set used for indexing the factors of interest, $[0, 1]^u$ be the unit cube in the coordinates indexed by u, $\boldsymbol{x}_u = (x_j)_{j \in u}$ be the projection of $\boldsymbol{x}$ onto $[0, 1]^u$, and $\mathcal{P}_u$ be the projection of the design $\mathcal{P}$ onto $[0, 1]^u$. Denote $R_u(\boldsymbol{x}_u) \subseteq [0, 1]^u$ to be a pre-defined region for all $\boldsymbol{x}_u \in [0, 1]^u$. The region $R_u(\boldsymbol{x}_u)$ will be given at the end of this subsection. Define *local projection discrepancy* for the factors indexed by u as

$$\mathrm{disc}_u^R(\boldsymbol{x}_u) = \mathrm{Vol}(R_u(\boldsymbol{x}_u)) - \frac{|\mathcal{P}_u \cap R_u(\boldsymbol{x}_u)|}{n}, \tag{2.3.1}$$

which may be considered as a function of $\boldsymbol{x}_u$. The L_2-norm of disc_u^R is defined as

$$\left\|\mathrm{disc}_u^R\right\|_2 = \left\{\int_{[0,1]^u} \left|\mathrm{disc}_u^R(\boldsymbol{x}_u)\right|^2 \mathrm{d}\boldsymbol{x}_u\right\}^{1/2} = \left\{\int_{[0,1]^u} \left|\mathrm{disc}_u^R(\boldsymbol{x}_u)\right|^2 \mathrm{d}\boldsymbol{x}\right\}^{1/2}.$$

A generalized L_2-discrepancy is defined in terms of all of these local projection discrepancies as follows:

$$\begin{aligned} D_2^R(\mathcal{P}) &= \left\|\left(\left\|\mathrm{disc}_u^R\right\|_2\right)_{u \subseteq \{1:s\}}\right\|_2 \\ &= \left\{\sum_{u \subseteq \{1:s\}} \int_{[0,1]^u} \left|\mathrm{disc}_u^R(\boldsymbol{x}_u)\right|^2 \mathrm{d}\boldsymbol{x}_u\right\}^{1/2} \\ &= \left\{\int_{[0,1]^s} \sum_{u \subseteq \{1:s\}} \left|\mathrm{disc}_u^R(\boldsymbol{x}_u)\right|^2 \mathrm{d}\boldsymbol{x}\right\}^{1/2}. \end{aligned} \tag{2.3.2}$$

Surely, the generalized L_2-discrepancy in (2.3.2) considers the projection uniformity and satisfies the criterion C_3. When the number of factors increases from s to $\tilde{s}$, the discrepancy may increase since it contains additional contributions from $\left\|\mathrm{disc}_u^R\right\|_2$ for all u having a non-empty intersection with $\{(s + 1) : \tilde{s}\}$.

Moreover, to satisfy C_4, the definition of $R_u(\boldsymbol{x}_u)$ should have some geometric interpretation. Different definitions of $R_u(\boldsymbol{x}_u)$ in (2.3.1) give different generalized discrepancies. Hickernell (1998a, b) proposed a family of generalized L_2-discrepancies, and among them the *centered L_2-discrepancy* (CD) and *wrap-around L_2-discrepancy* (WD) have been widely used in theoretical study and practical applications. Zhou et al. (2013) pointed out that CD and WD also have some shortcomings and proposed a new discrepancy, known as *mixture discrepancy* (MD). In the following subsections, a detailed introduction of various definitions of $R_u(\boldsymbol{x}_u)$ and the expressions of CD, WD, and MD are given.

2.3.2 Centered L_2-Discrepancy

Denote the 2^s vertices of the unit cube $[0, 1]^s$ by the set $\{0, 1\}^s$. For every point $\boldsymbol{x} \in [0, 1]^s$, let $\boldsymbol{a}_{\boldsymbol{x}} \in \{0, 1\}^s$ denote the vertex closest to $\boldsymbol{x}$, i.e., $\boldsymbol{a}_{\boldsymbol{x}} = (a_{x_1}, \ldots, a_{x_s})$ is defined by $a_{x_j} = 0$ for $0 \leqslant x_j \leqslant 1/2$ and $a_{x_j} = 1$ for $1/2 < x_j \leqslant 1$. For any projection $u \subset \{1 : s\}$, let $a_{\boldsymbol{x}_u}$ be the vertex in the unit cube $[0, 1]^u$ which is closest to $\boldsymbol{x}_u$. Define $R_u^C(\boldsymbol{x}_u)$ be the hyperrectangle between the points $\boldsymbol{a}_{\boldsymbol{x}_u}$ and $\boldsymbol{x}_u$.

For any two points $\boldsymbol{x}, \boldsymbol{y} \in [0, 1]^s$, let $J(\boldsymbol{x}, \boldsymbol{y})$ denote the hypercube containing the points between $\boldsymbol{x}$ and $\boldsymbol{y}$, i.e.,

$$J(\boldsymbol{x}, \boldsymbol{y}) = \{(t_1, \ldots, t_s) : \min(x_j, y_j) \leqslant t_j \leqslant \max(x_j, y_j)\ \forall j = 1, \ldots, s\}.$$

If all the elements of $\boldsymbol{x}$ are less than or equal to the corresponding elements in $\boldsymbol{y}$, then $J(\boldsymbol{x}, \boldsymbol{y}) = [\boldsymbol{x}, \boldsymbol{y}]$. For centered L_2-discrepancy, define $R_u^C(\boldsymbol{x}_u) = J(\boldsymbol{a}_{\boldsymbol{x}_u}, \boldsymbol{x}_u)$. For example, when $u = \{1, 2\}$, the unit cube $[0, 1]^2$ is split into four square cells. Corner points of each cell involve one corner point of the original unit cube, the central point $(1/2, 1/2)$, and others. Then, $\boldsymbol{a}_{\boldsymbol{x}_u}$ and $\boldsymbol{x}_u$ are the two diagonal vertexes of the corresponding rectangle region $R_u^C(\boldsymbol{x}_u)$. A plot of the local projection discrepancy of CD for $u = \{1, 2\}$ is given in Fig. 2.3.

The local projection discrepancy in (2.3.1) becomes

$$\begin{aligned} \mathrm{disc}_u^C(\boldsymbol{x}_u) &= \mathrm{Vol}(R_u^C(\boldsymbol{x}_u)) - \frac{\left|\mathcal{P}_u \cap R_u^C(\boldsymbol{x}_u)\right|}{n} \\ &= \mathrm{Vol}(J(\boldsymbol{a}_{\boldsymbol{x}_u}, \boldsymbol{x}_u)) - \frac{\left|\mathcal{P}_u \cap J(\boldsymbol{a}_{\boldsymbol{x}_u}, \boldsymbol{x}_u)\right|}{n}. \end{aligned} \tag{2.3.3}$$

From (2.3.2), the centered L_2-discrepancy becomes

$$\mathrm{CD}(\mathcal{P}) = \left\| \left(\left\| \mathrm{disc}_u^C \right\|_2 \right)_{u \subseteq \{1:s\}} \right\|_2.$$

Hickernell (1998b) showed that the centered L_2-discrepancy possesses a convenient formula for computation:

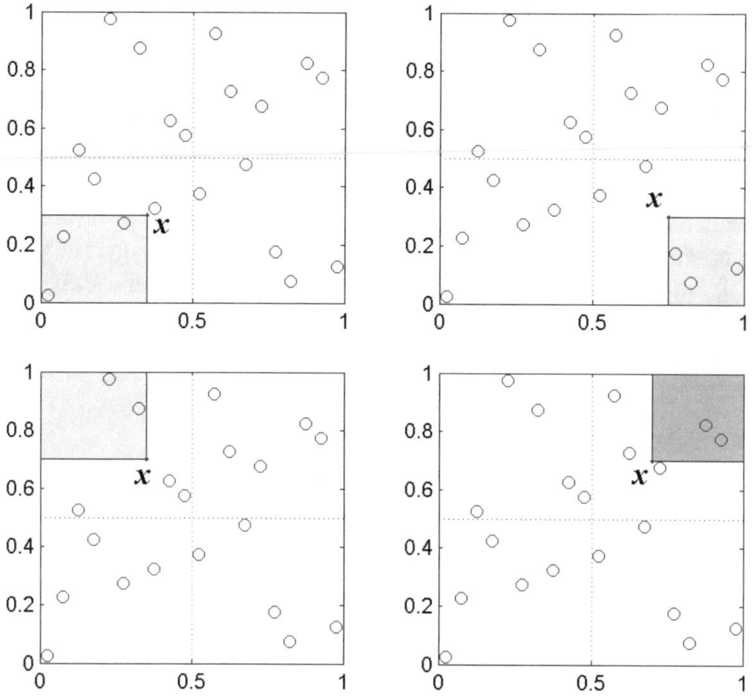

Fig. 2.3 $R^C_{\{1,2\}}(\boldsymbol{x})$ (shaded) used to define the centered discrepancy

$$\mathrm{CD}(\mathcal{P}) = \left\{ \left(\frac{13}{12}\right)^s - \frac{2}{n}\sum_{i=1}^{n}\prod_{j=1}^{s}\left(1+\frac{1}{2}|x_{ij}-0.5|-\frac{1}{2}|x_{ij}-0.5|^2\right) \right.$$
$$\left. +\frac{1}{n^2}\sum_{i,k=1}^{n}\prod_{j=1}^{s}\left(1+\frac{1}{2}|x_{ij}-0.5|+\frac{1}{2}|x_{kj}-0.5|-\frac{1}{2}|x_{ij}-x_{kj}|\right)\right\}^{1/2}. \tag{2.3.4}$$

Obviously, CD satisfies the criteria C_1, C_3, C_4, and C_5. Since the region $R_u^C(\boldsymbol{x}_u)$ treats the function of 2^s corner points as same as the origin $\mathbf{0}$, the reflection and rotation of the design $\mathcal{P}$ do not change the CD-value, i.e., CD is a rotation- and reflection-invariant discrepancy which implies CD satisfies C_2. Hickernell (1998b) showed that CD also satisfies Koksma–Hlawka inequality (C_6), where the total variation of the function should be modified.

2.3.3 Wrap-around L_2-Discrepancy

Both the star L_p-discrepancy and the centered L_2-discrepancy require one or more corner points of the unit cube in definition of $R_u(\boldsymbol{x}_u)$. A natural extension for the pre-defined region in the definition of generalized discrepancy (2.3.1) is to fall inside of the unit cube and not to involve any corner point of the unit cube, which leads to the so-called unanchored discrepancy in the literature. Then, the pre-defined region can be chosen as a rectangle determined by two points $\boldsymbol{x}_1$ and $\boldsymbol{x}_2$ in $[0, 1]^u$. Moreover, for satisfying the property C_2, we can wrap the unit cube for each coordinate.

Consider the unit cube like a torus, i.e., $x_j = 0$ and $x_j = 1$ are treated as the same point. Let the region $R_u^W(\boldsymbol{y}_u, \boldsymbol{x}_u)$ and the local discrepancy function be

$$R_j^W(y_j, x_j) = \begin{cases} [y_j, x_j], & y_j \leqslant x_j, \\ [0, x_j] \cup [y_j, 1], & x_j < y_j, \end{cases}$$

$$R_u^W(\boldsymbol{y}_u, \boldsymbol{x}_u) = \bigotimes_{j \in u} R_j^W(y_j, x_j),$$

$$\mathrm{disc}_u^W(\boldsymbol{y}_u, \boldsymbol{x}_u) = \mathrm{Vol}(R_u^W(\boldsymbol{y}_u, \boldsymbol{x}_u)) - \frac{|\mathcal{P} \cap R_u^W(\boldsymbol{y}_u, \boldsymbol{x}_u)|}{n}, \quad (2.3.5)$$

where $\bigotimes$ denotes for the Kronecker product. Figure 2.4 shows a plot of some regions $R_u^W(\boldsymbol{y}_u, \boldsymbol{x}_u)$, where the two shaded rectangles are treated as one region.

Since the region has a wrap-around property, the corresponding discrepancy is called as wrap-around L_2-discrepancy. From (2.3.2), the WD is defined as

$$\mathrm{WD}(\mathcal{P}) = \left(\sum_{u \subseteq \{1:s\}} \int_{[0,1]^{2u}} \left|\mathrm{disc}_u^W(\boldsymbol{y}_u, \boldsymbol{x}_u)\right|^2 \mathrm{d}\boldsymbol{x}_u \mathrm{d}\boldsymbol{y}_u\right)^{1/2}.$$

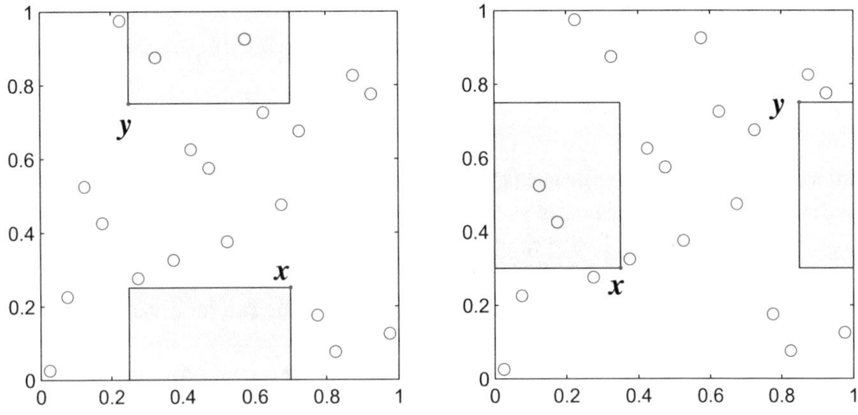

Fig. 2.4 $R_{\{1,2\}}^W(\boldsymbol{y}, \boldsymbol{x})$ (shaded) in wrap-around L_2-discrepancy

The expression of the squared WD-value can be written as (Hickernell 1998a)

$$\mathrm{WD}^2(\mathcal{P}) = -\left(\frac{4}{3}\right)^s + \frac{1}{n^2}\sum_{i,k=1}^{n}\prod_{j=1}^{s}\left[\frac{3}{2} - |x_{ij} - x_{kj}| + |x_{ij} - x_{kj}|^2\right]. \quad (2.3.6)$$

It is easy to check that WD satisfies the criteria C_1, C_3, C_4, and C_5. Since the wrap-around property of WD eliminates the effect of origin $\mathbf{0}$, the criterion C_2 is also satisfied, i.e., WD is also a rotation- and reflection-invariant discrepancy. Hickernell (1998a) showed that WD satisfies the criterion C_6, where the total variation in Koksma–Hlawka inequality needs some modification.

2.3.4 Some Discussion on CD and WD

We have shown that CD and WD satisfy criteria $C_1, \ldots, C_6$. As a consequence, CD and WD are widely used for the construction of uniform designs. Most existing uniform designs are obtained under CD or WD.

However, the uniformity criterion CD may derive some result which violates the intuitive view.

Example 2.3.1 Consider two designs $U(24, 3^3)$ below

$$\boldsymbol{X}_1 = \begin{bmatrix} 1\,1\,1\,1\,1\,1\,1\,1\,2\,2\,2\,2\,2\,2\,2\,2\,3\,3\,3\,3\,3\,3\,3\,3 \\ 1\,1\,2\,2\,2\,3\,3\,3\,1\,1\,1\,2\,2\,2\,3\,3\,1\,1\,1\,2\,2\,3\,3\,3 \\ 1\,3\,1\,2\,3\,1\,2\,3\,1\,2\,3\,1\,2\,3\,1\,2\,1\,2\,3\,2\,3\,1\,2\,3 \end{bmatrix}^T$$

and

$$\boldsymbol{X}_2 = \begin{bmatrix} 1\,1\,1\,1\,1\,1\,1\,1\,2\,2\,2\,2\,2\,2\,2\,2\,3\,3\,3\,3\,3\,3\,3\,3 \\ 1\,1\,1\,2\,2\,3\,3\,3\,1\,1\,2\,2\,2\,2\,3\,3\,1\,1\,1\,2\,2\,3\,3\,3 \\ 1\,2\,3\,1\,3\,1\,2\,3\,1\,3\,2\,2\,2\,2\,1\,3\,1\,2\,3\,1\,3\,1\,2\,3 \end{bmatrix}^T,$$

where $\boldsymbol{A}^T$ denotes the transpose of the matrix $\boldsymbol{A}$. Consider the three-level three-factor full design with 27 runs. We also consider all its 24-run subdesigns without repeat point. There are total $\binom{27}{24} = 2925$ such subdesigns. Here, $\boldsymbol{X}_1$ is the design with lowest CD or WD among these subdesigns and is the U-type design without repeat point. $\boldsymbol{X}_2$ is from the list as the uniform design under CD on the Web site http://www.math.hkbu.edu.hk/UniformDesign/. The design $\boldsymbol{X}_2$ is more uniform than $\boldsymbol{X}_1$ under CD as

$$\mathrm{CD}^2(\boldsymbol{X}_1) = 0.032779, \quad \mathrm{CD}^2(\boldsymbol{X}_2) = 0.032586.$$

But $\mathrm{WD}^2(\boldsymbol{X}_1) = 0.100852$, $\mathrm{WD}^2(\boldsymbol{X}_2) = 0.101732$, i.e., $\boldsymbol{X}_1$ is more uniform than $\boldsymbol{X}_2$ under WD. This example shows that the uniform design may be different under

Table 2.1 Mean and standard deviation of r for the best design $\mathcal{P}_1'$

s	10	20	30	40	50	60	70	80	90	100
$\mathcal{P}_1'$: $\bar{r}$	0.974	0.894	0.775	0.660	0.569	0.495	0.440	0.390	0.352	0.320
(std)	(0.005)	(0.005)	(0.005)	(0.003)	(0.003)	(0.005)	(0.002)	(0.000)	(0.004)	(0.000)

different discrepancies. Moreover, the center point (2 2 2) repeats 4 times in $\boldsymbol{X}_2$. In computer experiments, the repeated point does not carry any additional information. It is reasonable to require that there is no repeat point in computer experiments. From this example, Zhou et al. (2013) pointed out that CD does not care much about points located in the center region according to the definition of $R_u^C(\boldsymbol{x}_u)$. They were concerned with the behavior of CD in high-dimensional cases and pointed out that CD is not suitable as a discrepancy in such scenarios.

Consider an experiment to test the dimensionality effect on CD. Let $\mathcal{P}_0' = \{\boldsymbol{x}_1, \ldots, \boldsymbol{x}_n\}$ be a set of n points in the unit hypercube C^s generated by the midpoint Latin hypercube sampling (see Fang et al. 2006a). Let

$$\mathcal{Y}(r) = \{\boldsymbol{y}_j = r(\boldsymbol{x}_j - \boldsymbol{m}) + \boldsymbol{m},\ j = 1, \ldots, n\}, \tag{2.3.7}$$

where $\boldsymbol{m} = \frac{1}{2}\mathbf{1}_s$ is the center point of C^s and $0 < r < 1$. When the ratio r decreases from 1 to 0, point $\boldsymbol{y}_j$ from $\boldsymbol{x}_j$ converges to $\boldsymbol{m}$ along the direction between $\boldsymbol{x}_j$ and $\boldsymbol{m}$. Minimize CD-value of $\mathcal{Y}(r)$ with respect to r on $0 < r \leqslant 1$ and let $\mathcal{P}_1'$ be the $\mathcal{Y}(r)$ with the minimum CD-value. The average mean of r, $\bar{r}$ say, and the average standard deviation of the ratio r of $\mathcal{P}_1'$, from 100 midpoint Latin hypercube samples, are listed in Table 2.1.

Table 2.1 shows that when the number of dimensions increases from 10 to 100, the mean of the ratio, $\bar{r}$ for $\mathcal{P}_1'$, significantly decreases and the standard deviation of r for $\mathcal{P}_1'$ also decreases to zero. For a more intuitive view, for $s = 10, 40, 70, 100$, a point set $\mathcal{P}_0'$ is generated by midpoint Latin hypercube sampling (see Fang et al. 2006a), and the first two-dimensional projection of $\mathcal{P}_0'$ and the corresponding $\mathcal{P}_1'$ are shown in Fig. 2.5, which shows that the ratio r of $\mathcal{P}_1'$ under CD decreases with the increase of the number of factors s. There are $\binom{s}{2}$ such projection plots, but they give the same conclusion. Moreover, for a sample of $\mathcal{P}_0'$ when $s = 100$, the CD-values of $\mathcal{P}_0'$ and $\mathcal{P}_1'$ are, respectively, 56411746.8 and 2962.0, which means $\mathcal{P}_1'$ is more uniform than $\mathcal{P}_0'$. Therefore, with the increase in the number of factors, more design points are set near the center point $\boldsymbol{m}$ and are more uniform under CD. We can conclude that CD prefers points close to the center $\boldsymbol{m}$. In other words, as a criterion of uniformity, CD has a significant dimensionality effect. Therefore, CD is not suitable for assessing the uniformity of a set of points, especially when the number of dimensions is large.

Fang and Ma (2001b) pointed out that a location shift of a one-dimensional design does not change its WD. Furthermore, from the expression of WD in (2.3.6), it is easily seen that any level shift in (2.3.8) does not change the WD-value. This flexibility may result in some unreasonable results. Figure 2.6 gives four designs that

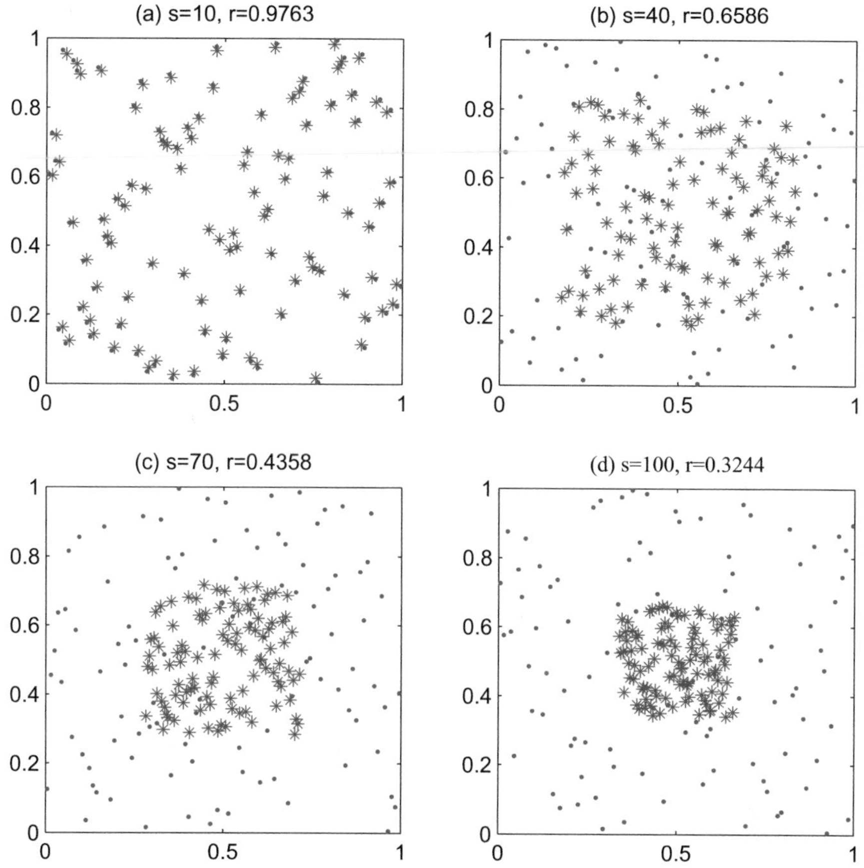

Fig. 2.5 Scale ratio r for a different number of factors under CD. Here '.' and '$*$', respectively, denote the first two-dimensional projections of $\mathcal{P}_0'$ and $\mathcal{P}_1'$

have the same WD-value, but D_1 is more uniform by intuition. This indicates that WD is not sensitive about a level shift of one or more factors. Therefore, we need to add more criteria for measures of uniformity below.

C_7 **Sensitivity on a shift for one or more dimensions**. The uniformity criterion D is sensitive to small point shift on the design, i.e., some shift may change the D-value.

Consider all the points of $\mathcal{P}$ to be shifted as

$$\mathcal{P}' = \{(x_{i1} + a_1, \ldots, x_{is} + a_s)^T \text{ (mod1)}, \ i = 1, \ldots, n\}, \qquad (2.3.8)$$

for some $a_j \in [0, 1), \ j = 1, 2, \ldots, s$, where the operator (mod 1) means that each element of the ith point modulo 1, which guarantees the design points of

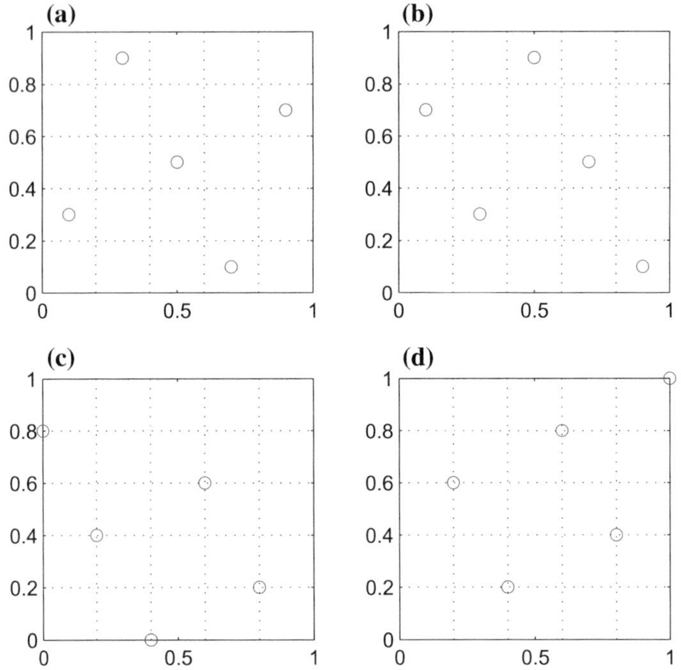

Fig. 2.6 Four designs constructed by level shift have the same WD-value

$\mathcal{P}'$ fall in $[0, 1]^u$. Note that this criterion does not require that the discrepancy D must be changed for any $a_j \in [0, 1),\ \ j = 1, 2, \ldots, s$.

C_8 **Less curse of dimensionality**. It is known that many problems have quite different behaviors between low- and high-dimensional cases, i.e., the same problem from a lower-dimensional case to the corresponding higher-dimensional case may have significant variant of behavior. This phenomenon is called *curse of dimensionality*. Many uniformity criteria are strongly related to the volume, and then the behavior of their criteria in high-dimensional cases should be studied. It is required that each point on the unit cube plays the same role.

Let us look at the star L_p-discrepancy and check whether they satisfy C_7 and C_8.

For C_7. When the design points are shifted for one or more dimensions in (2.3.8), the value $N(\mathcal{P} \cap [\mathbf{0}, \boldsymbol{x}])$ in the definition (2.2.4) or (2.2.5) may be changed for given $\boldsymbol{x}$, as well as the value of star discrepancy or star L_2-discrepancy. Then, the two discrepancies are sensitive on a shift for one or more dimensions.

For C_8. The star L_p-discrepancy has some unreasonable phenomenon in high-dimensional case. For example, the uniform design with one run under the star discrepancy takes the form $\mathcal{P} = \{(z, \ldots, z)\}$, where z satisfies $z^s + z - 1 = 0$. Then, $z = 1/2$ when $s = 1$; $z = (\sqrt{5} - 1)/2$ when $s = 2$; and z tends to 1 when the number of factors increases. However, in intuitive view, the uniform design with one run should be located at the center of C^s. The reason for such phenomenon is that the origin $\mathbf{0}$ plays the special role in L_p-discrepancy.

For CD, it is shown that CD does not satisfy C_8 according to Fig. 2.5 and Table 2.1. But, the CD satisfies C_7. For example, consider the $\boldsymbol{X}_1$ in Example 2.3.1. Let $\boldsymbol{U} = (\boldsymbol{X}_1 - 0.5)/3$ be the induced matrix. Let $(a_1, a_2, a_3) = (0.1, 0, 0)$ in (2.3.8), i.e., we only shift the first dimension. Then, the squared CD-value of the resulted design is 0.0443, which is different with that of $\boldsymbol{U}$.

For WD, we showed that it does not satisfy C_7 since the WD-value does not change for any $a_1, \ldots, a_s$ in (2.3.8). For C_8, consider the same experiment as that in Table 2.1. It shows that the best ratio r for WD is much close to 1, and then WD may have less curse of dimensionality in this sense.

2.3.5 *Mixture Discrepancy*

Although WD and CD overcome most of the shortcomings of the star L_p-discrepancy, CD does not satisfy the criterion C_8 and WD does not satisfy the criterion C_7. It is thus required to find some new uniformity criterion which can satisfy all the eight requests C_1–C_8. Since WD and CD have many goodnesses, one may want to keep their goodness and avoid their shortcomings. A natural idea is to develop a criterion through mixing WD and CD measure in some way.

As we know, the definition of the region $R_u(\boldsymbol{x})$ determines the property of the corresponding discrepancy, and then it is reasonable to get a better $R_u(\boldsymbol{x})$ from some modifications of that of CD and WD. The discrepancy proposed by Zhou et al. (2013), called as *mixture discrepancy*, satisfies this requirement. The definition of $R_j^W(x_i, y_i)$ for WD only considers whether x_i is larger than y_i or not, and does not care about the distance between x_i and y_i. However, it is more reasonable to define $R(\boldsymbol{x}_u)$ as a larger region, i.e., the difference between the empirical and uniform distributions on a larger region is considered. Thus, one may modify the definition of $R_u^W(\boldsymbol{x}_u, \boldsymbol{y}_u)$ in WD as follows,

$$R_1^M(x_i, y_i) = \begin{cases} [\min(x_i, y_i), \max(x_i, y_i)], & |x_i - y_i| \geqslant \frac{1}{2}, \\ [0, \min(x_i, y_i)] \cup [\max(x_i, y_i), 1], & |x_i - y_i| < \frac{1}{2}, \end{cases}$$
$$R_1^M(\boldsymbol{x}_u, \boldsymbol{y}_u) = \otimes_{i \in u} R_1^M(x_i, y_i). \tag{2.3.9}$$

Moreover, the region $R_u^C(\boldsymbol{x}_u)$ for CD only considers a small part of [0,1] and ignores the center region. Then, one may modify $R_u^C(\boldsymbol{x}_u)$ as follows

$$R_2^M(x_i) = \begin{cases} [x_i, 1], \ x_i \leqslant \frac{1}{2}, \\ [0, x_i], \ x_i > \frac{1}{2}, \end{cases} \quad R_2^M(\boldsymbol{x}_u) = \otimes_{i \in u} R_2^M(x_i), \tag{2.3.10}$$

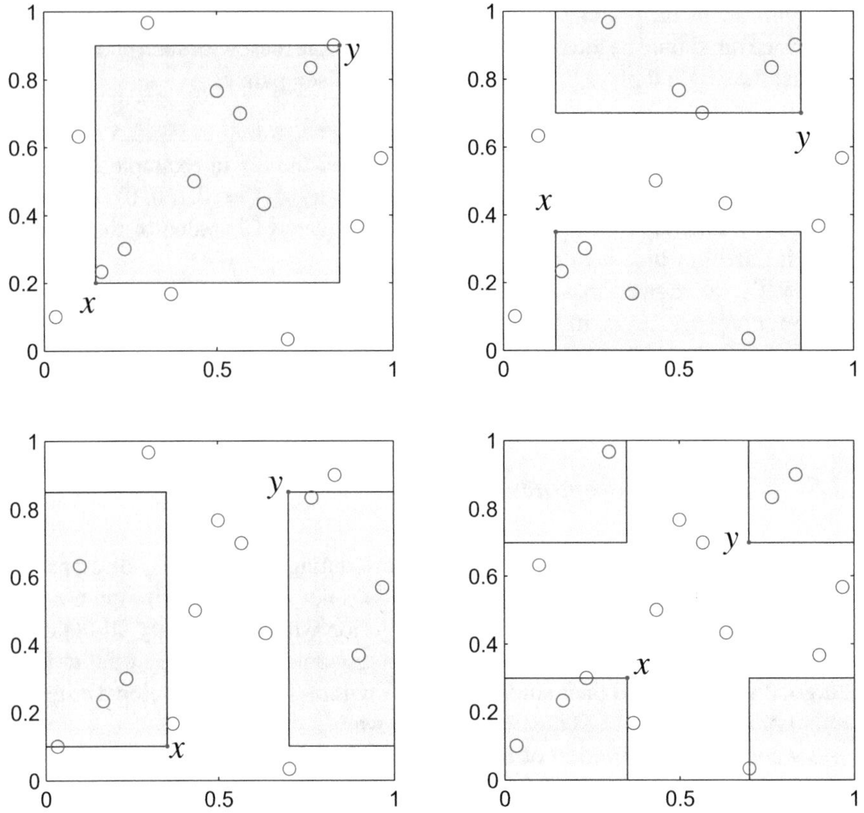

Fig. 2.7 Illustration of $R_1^M(\boldsymbol{x}, \boldsymbol{y})$ in two-dimensional case

which includes the center point of every dimension. Figures 2.7 and 2.8 give some illustrations of $R_1^M(\boldsymbol{x}, \boldsymbol{y})$ and $R_2^M(\boldsymbol{x})$ in two-dimensional case, where the area of $R_1^M(\boldsymbol{x}, \boldsymbol{y})$ is larger than 1/4, while the area of $R^W(\boldsymbol{x}, \boldsymbol{y})$ in wrap-around L_2-discrepancy is often less than 1/4. The points in C^2 may have the same probability to be counted in $R_1^M(\boldsymbol{x}, \boldsymbol{y})$, and the center region is always counted in $R_2^M(\boldsymbol{x})$.

It can be shown that the discrepancy with respect to $R_1^M(\boldsymbol{x}_u, \boldsymbol{y}_u)$ also has the special characteristic as that of WD, i.e., when one shifts the points in $\mathcal{P}_0$ with one direction $(a_1, \ldots, a_s)$, the value of its corresponding discrepancy remains unchanged. Similarly, $R_2^M(\boldsymbol{x})$ may pay more attention to the center region. Therefore, it is reasonable to combine the two aspects together, which may retain the good properties and overcome the unreasonable phenomena. The simplest way is to define the $R^M(x_j, y_j)$ as follows:

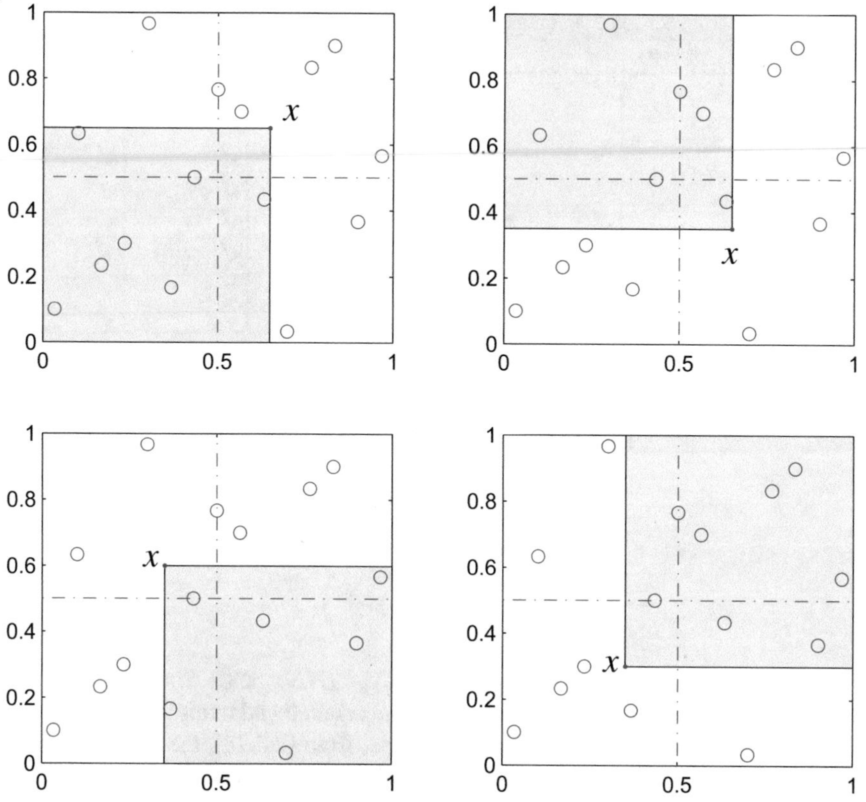

Fig. 2.8 Illustration of $R_2^M(\boldsymbol{x})$ in two-dimensional case

$$
\begin{aligned}
R^M(x_i, y_j) &= \frac{1}{2} R_1^M(x_i, y_j) + \frac{1}{2} R_2^M(x_i), \qquad (2.3.11)\\
R^M(\boldsymbol{x}_u, \boldsymbol{y}_u) &= \otimes_{i \in u} R^M(x_j, y_j),\\
\mathrm{disc}_u^M(\boldsymbol{x}_u, \boldsymbol{y}_u) &= \frac{1}{2}\left(\mathrm{Vol}(R_1^M(\boldsymbol{x}_u, \boldsymbol{y}_u)) - \frac{\left|\mathcal{P} \cap R_1^M(\boldsymbol{x}_u, \boldsymbol{y}_u)\right|}{n}\right)\\
&\quad + \frac{1}{2}\left(\mathrm{Vol}(R_2^M(\boldsymbol{x}_u)) - \frac{\left|\mathcal{P} \cap R_2^M(\boldsymbol{x}_u)\right|}{n}\right), \qquad (2.3.12)
\end{aligned}
$$

For a set of points $\mathcal{P} = \{\boldsymbol{x}_1, \ldots, \boldsymbol{x}_n\}$ on C^s, Zhou et al. (2013) derived the expression of MD as follows.

Table 2.2 Properties for some commonly used discrepancies

Criteria	$L_p (p \neq 2)$	L_2	L_∞	CD	WD	MD
C_1	√	√	√	√	√	√
C_2	–	–	–	√	√	√
C_3	–	–	√	√	√	√
C_4	√	√	√	√	√	√
C_5	–	√	–	√	√	√
C_6	√	√	√	√	√	√
C_7	√	√	√	√	–	√
C_8	–	–	–	–	√	√

$$\begin{aligned} \mathrm{MD}^2(\mathcal{P}) = &\left(\frac{19}{12}\right)^s - \frac{2}{n}\sum_{i=1}^{n}\prod_{j=1}^{s}\left(\frac{5}{3} - \frac{1}{4}|x_{ij} - \frac{1}{2}| - \frac{1}{4}|x_{ij} - \frac{1}{2}|^2\right) \\ &+ \frac{1}{n^2}\sum_{i=1}^{n}\sum_{k=1}^{n}\prod_{j=1}^{s}\left(\frac{15}{8} - \frac{1}{4}|x_{ij} - \frac{1}{2}| - \frac{1}{4}|x_{kj} - \frac{1}{2}|\right. \\ &\left. - \frac{3}{4}|x_{ij} - x_{kj}| + \frac{1}{2}|x_{ij} - x_{kj}|^2\right). \end{aligned} \tag{2.3.13}$$

It can be easily shown that MD satisfies C_1, C_3, C_4, C_5, and C_7. Since the definition of $R^M(\boldsymbol{x}_u, \boldsymbol{y}_u)$ eliminates the special role of the origin $\mathbf{0}$ and treats the 2^s vertices to be the same role, MD satisfies C_2. For example, from (2.3.13), the MD-value does not change for the reflection transformation $x_i' = 1 - x_i$. According to Zhou et al. (2013), the MD can also be defined from the kernel function (see Sect. 2.4 for the details). Hence, it also satisfies Koksma–Hlawka inequality (2.1.6), as well as the criterion C_6; see Hickernell (1998a). For C_8, considering the same experiment as that in Table 2.1, it can also be shown that the best ratio r for MD is much close to 1, and then MD does not have curse of dimensionality in this sense. Therefore, MD satisfies all of the eight criteria $C_1 - C_8$ and may be a better uniformity criterion than star L_p-discrepancy, CD and WD. In summary, the properties of the commonly used discrepancies star L_p-discrepancy, CD, WD, and MD are listed in Table 2.2.

2.4 Reproducing Kernel for Discrepancies

The analytical expressions for the computation of the generalized L_2-discrepancies for CD, WD, and MD are provided in the last preceding section without any proof. In this section, we introduce the tool of *reproducing kernel Hilbert space* and give the procedure to derive the analytical expression for the computation of these discrepancies.

Let $\mathcal{X}$ be an experimental domain on R^s and $\mathcal{K}(\boldsymbol{z}, \boldsymbol{t})$ be a real-valued function defined on $\mathcal{X}^2 = \mathcal{X} \times \mathcal{X}$ satisfying two properties: (i) symmetric,

$$\mathcal{K}(zt) = \mathcal{K}(t, z), \ \forall zt \in \mathcal{X},$$

and (ii) nonnegative definite,

$$\sum_{i,k=1}^{n} c_i \mathcal{K}(z_i, z_k) c_k \geqslant 0, \ \forall z_i \in \mathcal{X},$$

and c_i are real numbers. Such function $\mathcal{K}(\cdot, \cdot)$ is called a *kernel function*. Let $\mathcal{W}$ be the space of real-valued functions on $\mathcal{X}$ with the kernel function $\mathcal{K}$ as follows:

$$\mathcal{W} = \left\{ F(\boldsymbol{x}) : \int_{\mathcal{X}^2} \mathcal{K}(zt) \mathrm{d}F(z) dF(t) < \infty \right\}. \tag{2.4.1}$$

Define the inner product of two arbitrary functions $F, G \in \mathcal{W}$ as

$$\langle F, G \rangle_{\mathcal{W}} = \int_{\mathcal{X}^2} \mathcal{K}(\boldsymbol{x}, \boldsymbol{y}) \mathrm{d}F(\boldsymbol{x}) \mathrm{d}G(\boldsymbol{y}). \tag{2.4.2}$$

The norm $\|\cdot\|_{\mathcal{W}}$ is induced from $\|F\|_{\mathcal{W}} = [\langle F, F \rangle_{\mathcal{W}}]^{1/2}$ for any function $F \in \mathcal{W}$. Then, $\mathcal{W}$ is a Hilbert space of real-valued functions on $\mathcal{X}$ and $\langle F, G \rangle_{\mathcal{W}}$ is also finite according to the Cauchy–Schwarz inequality. If the kernel function $\mathcal{K}$ satisfies the following property

$$\begin{cases} \mathcal{K}(\cdot, \boldsymbol{x}) \in \mathcal{W}, \ \forall \, \boldsymbol{x} \in \mathcal{X}, \\ F(x) = \langle \mathcal{K}(\cdot, \boldsymbol{x}), F \rangle_{\mathcal{W}}, \ \forall \, F \in \mathcal{W}, \boldsymbol{x} \in \mathcal{X}, \end{cases} \tag{2.4.3}$$

then this kernel is called a *reproducing kernel* for $\mathcal{W}$ and $\mathcal{W}$ is called a *reproducing kernel Hilbert space*. The Moore–Aronszajn theorem (Aronszajn 1950) showed that for any given kernel function $\mathcal{K}$, there is a Hilbert space of functions on $\mathcal{X}$ such that $\mathcal{K}$ is a reproducing kernel.

Using the tool of reproducing kernel Hilbert space, one can define the L_2-discrepancy for measuring uniformity of a design. Let $\mathcal{X}$ be an experimental domain, F be the uniform distribution function on $\mathcal{X}$, and $F_{\mathcal{P}}$ be the empirical distribution function of a design $\mathcal{P} = \{\boldsymbol{x}_1, \ldots, \boldsymbol{x}_n\}$ on $\mathcal{X}$.

Definition 2.4.1 A *discrepancy* for measuring uniformity of $\mathcal{P}$ is defined as

$$D(\mathcal{P}, \mathcal{K}) = \|F - F_{\mathcal{P}}\|_{\mathcal{W}}, \tag{2.4.4}$$

where $\mathcal{W}$ is defined in (2.4.1).

The discrepancy in (2.4.4) measures the difference between F and $F_{\mathcal{P}}$, which is determined by the kernel function $\mathcal{K}$. A kernel function $\mathcal{K}$ derives a type of discrepancy. Expanding out the integrals in (2.4.4) yields an expression resembling the computational formula of the discrepancy:

$$
\begin{aligned}
D^2(\mathcal{P}, \mathcal{K}) &= \langle F - F_{\mathcal{P}}, F - F_{\mathcal{P}} \rangle_{\mathcal{W}} \\
&= \int_{\mathcal{X}^2} \mathcal{K}(\boldsymbol{t}, \boldsymbol{z}) \mathrm{d}(F - F_{\mathcal{P}})(\boldsymbol{t}) \mathrm{d}(F - F_{\mathcal{P}})(\boldsymbol{z}) \\
&= \int_{\mathcal{X}^2} \mathcal{K}(\boldsymbol{t}, \boldsymbol{z})\, \mathrm{d}F(\boldsymbol{t}) \mathrm{d}F(\boldsymbol{z}) - \frac{2}{n} \sum_{i=1}^{n} \int_{\mathcal{X}} \mathcal{K}(\boldsymbol{t}, \boldsymbol{x}_i)\, \mathrm{d}F(\boldsymbol{t}) \\
&\quad + \frac{1}{n^2} \sum_{i=1}^{n} \sum_{k=1}^{n} \mathcal{K}(\boldsymbol{x}_i, \boldsymbol{x}_k). \qquad (2.4.5)
\end{aligned}
$$

When the levels of each of the s-factors can be set independently, the experimental domain $\mathcal{X}$ can be chosen as the unit hypercube $[0, 1]^u$. In this case, $dF(\boldsymbol{t}) = d\boldsymbol{t}$, and the discrepancy with respect to $\mathcal{K}$ is as follows.

$$
\begin{aligned}
D^2(\mathcal{P}, \mathcal{K}) = & \int_{[0,1]^{2s}} \mathcal{K}(\boldsymbol{t}, \boldsymbol{z})\, \mathrm{d}\boldsymbol{t} \mathrm{d}\boldsymbol{z} - \frac{2}{n} \sum_{i=1}^{n} \int_{[0,1]^s} \mathcal{K}(\boldsymbol{t}, \boldsymbol{x}_i)\, \mathrm{d}\boldsymbol{t} \\
& + \frac{1}{n^2} \sum_{i=1}^{n} \sum_{k=1}^{n} \mathcal{K}(\boldsymbol{x}_i, \boldsymbol{x}_k). \qquad (2.4.6)
\end{aligned}
$$

Given a pre-defined region $R_u(\boldsymbol{x})$, it can determine a reproducing kernel $\mathcal{K}$, as well as the corresponding L_2-discrepancy.

Theorem 2.4.1 *The generalized L_2-discrepancy in* (2.3.2) *can be defined in the form* (2.4.4)*, and its kernel function $\mathcal{K}$ can be expressed by*

$$
\mathcal{K}^R(\boldsymbol{t}, \boldsymbol{z}) = \sum_{u \subseteq \{1:s\}} \mathcal{K}_u^R(\boldsymbol{t}_u, \boldsymbol{z}_u), \qquad (2.4.7)
$$

where

$$
\mathcal{K}_u^R(\boldsymbol{t}_u, \boldsymbol{z}_u) = \int_{[0,1]^u} 1_{R_u(\boldsymbol{x})}(\boldsymbol{t}_u) 1_{R_u(\boldsymbol{x})}(\boldsymbol{z}_u)\, \mathrm{d}\boldsymbol{x}. \qquad (2.4.8)
$$

Proof From (2.3.1), the local projection discrepancy can be rewritten in terms of the integral of the index function of the set $R_u(\boldsymbol{x})$:

$$
\begin{aligned}
\mathrm{disc}_u^R(\boldsymbol{x}_u) &= \mathrm{Vol}(R_u(\boldsymbol{x})) - \frac{|\mathcal{P}_u \cap R_u(\boldsymbol{x})|}{n} \\
&= \int_{[0,1]^u} 1_{R_u(\boldsymbol{x})}(\boldsymbol{t}_u)\, \mathrm{d}F(\boldsymbol{t}) - \int_{[0,1]^u} 1_{R_u(\boldsymbol{x})}(\boldsymbol{t}_u)\, \mathrm{d}F_{\mathcal{P}}(\boldsymbol{t}) \\
&= \int_{[0,1]^u} 1_{R_u(\boldsymbol{x})}(\boldsymbol{t}_u)\, \mathrm{d}(F - F_{\mathcal{P}})(\boldsymbol{t}),
\end{aligned}
$$

where F and $F_{\mathcal{P}}$ are the uniform distribution function and empirical distribution function of the design $\mathcal{P}$, respectively, and $\boldsymbol{t} = (t_1, \ldots, t_s)$ and $\boldsymbol{t}_u = (t_j)_{j \in u}$ are the projection of $\boldsymbol{t}$ onto $[0,1]^u$. This implies that the squared L_2-norm of the local projection discrepancy is

$$
\begin{aligned}
\left\|\mathrm{disc}_u^R\right\|_2^2 &= \int_{[0,1]^u} \left|\mathrm{disc}_u^R(\boldsymbol{x}_u)\right|^2 \mathrm{d}\boldsymbol{x} \\
&= \int_{[0,1]^u}\int_{[0,1]^{2s}} 1_{R_u(\boldsymbol{x})}(\boldsymbol{t}_u) 1_{R_u(\boldsymbol{x})}(\boldsymbol{z}_u) \\
&\qquad \times \mathrm{d}(F-F_{\mathcal{P}})(\boldsymbol{t})\mathrm{d}(F-F_{\mathcal{P}})(\boldsymbol{z})\,\mathrm{d}\boldsymbol{x} \\
&= \int_{[0,1]^{2s}}\int_{[0,1]^u} 1_{R_u(\boldsymbol{x})}(\boldsymbol{t}_u) 1_{R_u(\boldsymbol{x})}(\boldsymbol{z}_u) \\
&\qquad \times \mathrm{d}\boldsymbol{x}\,\mathrm{d}(F-F_{\mathcal{P}})(\boldsymbol{t})\mathrm{d}(F-F_{\mathcal{P}})(\boldsymbol{z}) \\
&= \int_{[0,1]^{2s}} \mathcal{K}_u(\boldsymbol{t}_u, \boldsymbol{z}_u)\,\mathrm{d}(F-F_{\mathcal{P}})(\boldsymbol{t})\mathrm{d}(F-F_{\mathcal{P}})(\boldsymbol{z}),
\end{aligned}
$$

where $\mathcal{K}_u^R(\boldsymbol{t}_u, \boldsymbol{z}_u) = \int_{[0,1]^u} 1_{R_u(\boldsymbol{x})}(\boldsymbol{t}_u) 1_{R_u(\boldsymbol{x})}(\boldsymbol{z}_u)\,\mathrm{d}\boldsymbol{x}$, as shown in (2.4.8). Then, the generalized L_2-discrepancy in (2.3.2) can be rewritten as follows.

$$
\begin{aligned}
D_2^R(\mathcal{P}) &= \left\|\left(\left\|\mathrm{disc}_u^R\right\|_2\right)_{u\subseteq\{1:s\}}\right\|_2 \\
&= \left\{\sum_{u\subseteq\{1:s\}} \int_{[0,1]^{2s}} \mathcal{K}_u^R(\boldsymbol{t}_u, \boldsymbol{z}_u)\,\mathrm{d}(F-F_{\mathcal{P}})(\boldsymbol{t})\mathrm{d}(F-F_{\mathcal{P}})(\boldsymbol{z})\right\}^{1/2} \\
&= \left\{\int_{[0,1]^{2s}} \sum_{u\subseteq\{1:s\}} \mathcal{K}_u^R(\boldsymbol{t}_u, \boldsymbol{z}_u)\,\mathrm{d}(F-F_{\mathcal{P}})(\boldsymbol{t})\mathrm{d}(F-F_{\mathcal{P}})(\boldsymbol{z})\right\}^{1/2} \\
&= \left\{\int_{[0,1]^{2s}} \mathcal{K}^R(\boldsymbol{t}, \boldsymbol{z})\,\mathrm{d}(F-F_{\mathcal{P}})(\boldsymbol{t})\mathrm{d}(F-F_{\mathcal{P}})(\boldsymbol{z})\right\}^{1/2},
\end{aligned}
$$

where $\mathcal{K}^R(\boldsymbol{t}, \boldsymbol{z}) = \sum_{u\subseteq\{1:s\}} \mathcal{K}_u^R(\boldsymbol{t}_u, \boldsymbol{z}_u)$, as shown in (2.4.7). Then, $R_u(\boldsymbol{x})$ determines the function $\mathcal{K}_u^R(\boldsymbol{t}_u, \boldsymbol{z}_u)$ in (2.4.8), which determines the expression of generalized L_2-discrepancy. The proof is finished.

Given a pre-defined region $R_u(\boldsymbol{x})$, it may need more requirement of kernel function $\mathcal{K}^R(\boldsymbol{t}, \boldsymbol{z})$ for calculating the analytic expression of the corresponding discrepancy, especially for WD, CD, and MD. Usually, the kernel function is assumed as a *separated kernel*, i.e.,

$$
\mathcal{K}(\boldsymbol{t}, \boldsymbol{z}) = \prod_{j=1}^{s} \mathcal{K}_j(t_j, z_j), \text{ for any } \boldsymbol{t}, \boldsymbol{z} \in \mathcal{X}. \tag{2.4.9}
$$

The separated kernel can be obtained when the s-factor experimental domain is a Cartesian product of marginal domains, i.e., $\mathcal{X} = \mathcal{X}_1 \times \cdots \times \mathcal{X}_s$ (refer Sect. 1.2).

Actually, the function $\mathcal{K}^R(\boldsymbol{t}, \boldsymbol{z})$ in (2.4.7) is a separated kernel function, since the region $R_u(\boldsymbol{x}_u)$ is a Cartesian product of marginal domains, and the kernel function $\mathcal{K}_u^R$ also has a product form:

$$\mathcal{K}_u^R(\boldsymbol{t}_u, \boldsymbol{z}_u) = \prod_{j \in u} \tilde{\mathcal{K}}_1^R(t_j, z_j),$$

where

$$\tilde{\mathcal{K}}_1^R(t_j, z_j) = \int_0^1 1_{R_j(x)}(t_j) 1_{R_j(x)}(z_j)\,\mathrm{d}x, \tag{2.4.10a}$$

if the one-factor region $R_1(x)$ is only determined by one point, and

$$\tilde{\mathcal{K}}_1^R(t_j, z_j) = \int_0^1 \int_0^1 1_{R_j(x,y)}(t_j) 1_{R_j(x,y)}(z_j)\,\mathrm{d}x\mathrm{d}y, \tag{2.4.10b}$$

if the one-factor region $R_1(x, y)$ is determined by two points. Then, by the binomial theorem the kernel function $\mathcal{K}^R$ defined in (2.4.7) is also of product form:

$$\mathcal{K}^R(\boldsymbol{t}, \boldsymbol{z}) = \prod_{j=1}^{s} [1 + \tilde{\mathcal{K}}_1^R(t_j, z_j)]. \tag{2.4.11}$$

Therefore, given a region $R_u(\boldsymbol{x}_u)$ or $R_u(\boldsymbol{x}_u, \boldsymbol{y}_u)$, one can obtain $\tilde{\mathcal{K}}_1^R(t_j, z_j)$ from (2.4.10) and substitute the product form of (2.4.11) into (2.4.6) to yield the analytic formula of the corresponding discrepancy. For example, the kernel functions corresponding to the measures CD, WD, and MD are given as follows.

(a) CD. According to the region $R_u^C(\boldsymbol{x}_u)$ defined in last section, the $\tilde{\mathcal{K}}_1^C(t_j, z_j)$ and $\mathcal{K}^C(\boldsymbol{t}, \boldsymbol{z})$ are as follows.

$$\begin{aligned}
\tilde{\mathcal{K}}_1^C(t_j, z_j) &= \int_0^1 1_{R_j^C(x)}(t_j) 1_{R_j^C(x)}(z_j)\,\mathrm{d}x \\
&= \begin{cases} 1/2 - \max(t_j, z_j), & t_j, z_j \leqslant 1/2, \\ \min(t_j, z_j) - 1/2, & 1/2 < t_j, z_j, \\ 0, & \text{otherwise}, \end{cases} \\
&= \frac{1}{2}\left|t_j - \frac{1}{2}\right| + \frac{1}{2}\left|z_j - \frac{1}{2}\right| - \frac{1}{2}\left|t_j - z_j\right|, \\
\mathcal{K}^C(\boldsymbol{t}, \boldsymbol{z}) &= \prod_{j=1}^{s}\left[1 + \frac{1}{2}\left|t_j - \frac{1}{2}\right| + \frac{1}{2}\left|z_j - \frac{1}{2}\right| - \frac{1}{2}\left|t_j - z_j\right|\right].
\end{aligned} \tag{2.4.12}$$

(b) WD. The $\tilde{\mathcal{K}}_1^W(t_j, z_j)$ and $\mathcal{K}^W(\boldsymbol{t}, \boldsymbol{z})$ are as follows.

$$\begin{aligned}
\tilde{\mathcal{K}}_1^W(t_j, z_j) &= \int_0^1 \int_0^1 1_{R_j^W(x,y)}(t_j) 1_{R_j^W(x,y)}(z_j)\, \mathrm{d}x\mathrm{d}y \\
&= \frac{1}{2} - \left|t_j - z_j\right| + \left|t_j - z_j\right|^2, \\
\mathcal{K}^W(\boldsymbol{t}, \boldsymbol{z}) &= \prod_{j=1}^{s} \left[\frac{3}{2} - \left|t_j - z_j\right| + \left|t_j - z_j\right|^2\right]. \qquad (2.4.13)
\end{aligned}$$

(c) MD. According to (2.3.11), let

$$\begin{aligned}
k_j^1(z_j, t_j) &= \int_0^1 \int_0^1 1_{R_1^M(x_j,y_j)}(z_j) 1_{R_1^M(x_j,y_j)}(t_j) dx_j dy_j \\
&= \frac{3}{4} - |z_j - t_j| + |z_j - t_j|^2, \\
k_j^2(z_j, t_j) &= \int_0^1 1_{R_2^M(x_j)}(z_j) 1_{R_2^M(x_j)}(t_j) dx_j \\
&= 1 - \frac{1}{2}|z_j - \frac{1}{2}| - \frac{1}{2}|t_j - \frac{1}{2}| - \frac{1}{2}|z_j - t_j|,
\end{aligned}$$

then the $\tilde{\mathcal{K}}_1^M(t, z)$ and $\mathcal{K}^M(\boldsymbol{t}, \boldsymbol{z})$ are as follows.

$$\begin{aligned}
\tilde{\mathcal{K}}_1^M(t, z) &= \frac{1}{2} k_j^1(z_j, t_j) + \frac{1}{2} k_j^2(z_j, t_j) \\
&= \frac{7}{8} - \frac{1}{4}|x_{ij} - \frac{1}{2}| - \frac{1}{4}|x_{kj} - \frac{1}{2}| - \frac{3}{4}|x_{ij} - x_{kj}| + \frac{1}{2}|x_{ij} - x_{kj}|^2,
\end{aligned}$$

$$\begin{aligned}
\mathcal{K}^M(\boldsymbol{t}, \boldsymbol{z}) = \prod_{j=1}^{s} \Big[& \frac{15}{8} - \frac{1}{4}|x_{ij} - \frac{1}{2}| - \frac{1}{4}|x_{kj} - \frac{1}{2}| \\
& - \frac{3}{4}|x_{ij} - x_{kj}| + \frac{1}{2}|x_{ij} - x_{kj}|^2 \Big].
\end{aligned}$$

Then, the corresponding analytic formulas of CD, WD, and MD are shown in (2.3.4), (2.3.6), and (2.3.13), respectively.

From the kernel functions of CD, WD, and MD, the corresponding expressions can be obtained by (2.4.6). Hickernell (1998a) gave other type of discrepancies such as full star discrepancy, symmetrical discrepancy, and unanchored discrepancy. Their discrepancies can also be defined by kernel functions.

Note that the CD, WD, and MD are defined on a unit hypercube. In practical application, one may consider the discrepancy on other experimental domain $\mathcal{X}$ such as simplex and hypersphere. Let $R(\boldsymbol{x})$ be a pre-decided region on $\mathcal{X}$ and $\mathcal{P}$ be a design on $\mathcal{X}$. Suppose the volume of $\mathcal{X}$ is finite, i.e., $\mathrm{Vol}(\mathcal{X}) < \infty$. The local discrepancy function is

$$\begin{aligned}
\mathrm{disc}^R(\boldsymbol{x}) &= \frac{\mathrm{Vol}(R(\boldsymbol{x}))}{\mathrm{Vol}(\mathcal{X})} - \frac{|\mathcal{P} \cap R(\boldsymbol{x})|}{n} \\
&= \int_{\mathcal{X}} 1_{R(\boldsymbol{x})}(\boldsymbol{t})\,\mathrm{d}F(\boldsymbol{t}) - \int_{\mathcal{X}} 1_{R(\boldsymbol{x})}(\boldsymbol{t})\,\mathrm{d}F_{\mathcal{P}}(\boldsymbol{t}) \\
&= \int_{\mathcal{X}} 1_{R(\boldsymbol{x})}(\boldsymbol{t})\,\mathrm{d}(F - F_{\mathcal{P}})(\boldsymbol{t}),
\end{aligned} \tag{2.4.14}$$

where F is the uniform distribution function on $\mathcal{X}$ and $F_{\mathcal{P}}$ is the empirical distribution function of the design $\mathcal{P}$. This implies that the squared L_2-norm of the local projection discrepancy is

$$\begin{aligned}
\left\|\mathrm{disc}^R\right\|_2^2 &= \int_{\mathcal{X}} \left|\mathrm{disc}^R(\boldsymbol{x})\right|^2 \mathrm{d}\boldsymbol{x} \\
&= \int_{\mathcal{X}} \int_{\mathcal{X}^2} 1_{R(\boldsymbol{x})}(\boldsymbol{t}) 1_{R(\boldsymbol{x})}(\boldsymbol{z}) \mathrm{d}(F - F_{\mathcal{P}})(\boldsymbol{t}) \mathrm{d}(F - F_{\mathcal{P}})(\boldsymbol{z})\,\mathrm{d}\boldsymbol{x} \\
&= \int_{\mathcal{X}^2} \int_{\mathcal{X}} 1_{R(\boldsymbol{x})}(\boldsymbol{t}) 1_{R(\boldsymbol{x})}(\boldsymbol{z}) \mathrm{d}\boldsymbol{x}\ \mathrm{d}(F - F_{\mathcal{P}})(\boldsymbol{t}) \mathrm{d}(F - F_{\mathcal{P}})(\boldsymbol{z}) \\
&= \int_{\mathcal{X}^2} \mathcal{K}(\boldsymbol{t}, \boldsymbol{z})\,\mathrm{d}(F - F_{\mathcal{P}})(\boldsymbol{t}) \mathrm{d}(F - F_{\mathcal{P}})(\boldsymbol{z}),
\end{aligned} \tag{2.4.15}$$

where the function $\mathcal{K} : \mathcal{X}^2 \to \mathbb{R}$ is defined by

$$\mathcal{K}(\boldsymbol{t}, \boldsymbol{z}) = \int_{\mathcal{X}} 1_{R(\boldsymbol{x})}(\boldsymbol{t}) 1_{R(\boldsymbol{x})}(\boldsymbol{z})\,\mathrm{d}\boldsymbol{x}. \tag{2.4.16}$$

Then, the generalized L_2-discrepancy in general region $\mathcal{X}$ can be shown as that in (2.4.5). Usually, the general region $\mathcal{X}$ cannot be separated and the kernel $\mathcal{K}$ also cannot be separated.

2.5 Discrepancies for Finite Numbers of Levels

In previous sections, the experimental domain $\mathcal{X}$ is assumed to be continuous, such as the unit cube $[0, 1]^s$. However, in many physical or practical situations, it prefers to have an experimental domain with a finite number of levels. For example, categorical factors have only a few levels and quantitative factors may also choose a few levels because of convenience of experiments. Especially, in fractional factorial designs, the number of levels of each factor j is finite, labeled $\tau_1, \ldots, \tau_{q_j}$. Then, the experimental domain for an s-factor fractional factorial design may be written as a Cartesian product

$$\mathcal{X} = \mathcal{X}_1 \times \cdots \times \mathcal{X}_s, \qquad \text{where } \mathcal{X}_j = \{\tau_1, \ldots, \tau_{q_j}\}. \tag{2.5.1}$$

It is called an asymmetrical design if at least two factors have different numbers of levels and symmetrical design if $q_1 = \cdots = q_s = q$. Then, it is requested to give some discrepancies for experimental domain with finite candidates directly. Hickernell and Liu (2002), Liu and Hickernell (2002) and Fang et al. (2003a) considered a discrepancy, called *discrete discrepancy* or *categorical discrepancy*, and Zhou et al. (2008) proposed *Lee discrepancy* for finite numbers of levels. The discrete discrepancy is better for two-level designs, and the Lee discrepancy can be used for multi-level designs.

2.5.1 Discrete Discrepancy

For a fractional factorial design $\mathcal{P} = \{\boldsymbol{x}_1, \ldots, \boldsymbol{x}_n\}$ with n runs and s-factors, define the function $\delta_{ik}(\mathcal{P})$ as the *coincidence number* of the ith and kth runs:

$$\delta_{ik}(\mathcal{P}) = \delta(\boldsymbol{x}_i, \boldsymbol{x}_k) = \sum_{j=1}^{s} \delta_{x_{ij}\, x_{kj}}, \tag{2.5.2}$$

where δ_{tz} denotes the Kronecker delta function, i.e., $\delta_{tz} = 1$ if $t = z$ and $\delta_{tz} = 0$ otherwise. Note that $s - \delta_{ik}(\mathcal{P})$ is the *Hamming distance* between the ith and kth runs, which is defined as the number of positions where these two points differ.

According to (2.4.6), one can define a kernel function to obtain the corresponding discrepancy. The kernel function $\mathcal{K}$ of *discrete discrepancy* (DD) is defined as

$$\mathcal{K}(\boldsymbol{t}, \boldsymbol{z}) = \prod_{j=1}^{s} \mathcal{K}_j(t_j, z_j), \tag{2.5.3}$$

where

$$\mathcal{K}_j(t, z) = a^{\delta_{tz}} b^{1-\delta_{tz}} = \begin{cases} a & \text{if } t = z, \\ b & \text{if } t \neq z, \end{cases} \quad -a/(q-1) < b < a, \tag{2.5.4}$$

and the condition $-a/(q-1) < b < a$ ensures that $\mathcal{K}_j$ is positive definite. Therefore, $\mathcal{K}(\boldsymbol{t}, \boldsymbol{z}) = a^{\delta(\boldsymbol{t},\boldsymbol{z})} b^{s-\delta(\boldsymbol{t},\boldsymbol{z})}$. From (2.4.6), it is easy to obtain the expression of DD as

$$\mathrm{DD}^2(\mathcal{P}) = -\prod_{j=1}^{s} \left[\frac{a + (q_j - 1)b}{q_j} \right] + \frac{1}{n^2} \sum_{i,k=1}^{n} a^{\delta_{ik}(\mathcal{P})} b^{s-\delta_{ik}(\mathcal{P})}. \tag{2.5.5}$$

Then, we have

$$\mathrm{DD}^2(\mathcal{P}) = -\prod_{j=1}^{s} \left[\frac{a + (q_j - 1)b}{q_j} \right] + \frac{a^s}{n} + \frac{2}{n^2} \sum_{1 \leqslant i < k \leqslant n} a^{\delta_{ik}(\mathcal{P})} b^{s-\delta_{ik}(\mathcal{P})}. \tag{2.5.6}$$

From (2.5.6), it can be seen that one would like to make the coincidence numbers of different runs, $\delta_{ik}(\mathcal{P})$, as small as possible since $a > b$. This means that a uniform design under DD spreads the runs out so that they are as dissimilar as possible.

For symmetrical designs, $q_1 = \cdots = q_s = q$, the expression of DD in (2.5.6) reduces to

$$\text{DD}^2(\mathcal{P}) = -\left[\frac{a+(q-1)b}{q}\right]^s + \frac{a^s}{n} + \frac{2b^s}{n^2} \sum_{1\leqslant i<k\leqslant n} \left(\frac{a}{b}\right)^{\delta_{ik}(\mathcal{P})}. \tag{2.5.7}$$

For two-level design $\mathcal{P}$, which appears in most of the fractional factorial design literature, this discrepancy becomes

$$\text{DD}^2(\mathcal{P}) = -\left(\frac{a+b}{2}\right)^s + \frac{a^s}{n} + \frac{2b^s}{n^2} \sum_{1\leqslant i<k\leqslant n} \left(\frac{a}{b}\right)^{\delta_{ik}(\mathcal{P})}. \tag{2.5.8}$$

From (2.5.8), it is shown that the discrete discrepancy of the two-level design $\mathcal{P}$ can be expressed as function of their Hamming distances. Moreover, one can prove that the CD, WD, and MD defined in Sect. 2.3 can also be expressed as function of their Hamming distances, which is left as an exercise.

The DD makes no assumptions about the order of the levels, $\tau_1, \tau_2, \ldots, \tau_j$, say. This means that the discrepancy in (2.5.5) remains invariant under any arbitrary permutation of the levels, which is not held for the discrepancies defined on C^s such as CD, WD, and MD. Thus, the discrete discrepancy may be termed a categorical discrepancy because it makes sense for categorical factors.

Since DD is based on Hamming distances, it may not be effective in constructing uniform designs with multi-levels. For example, consider the following U-type design $U(10; 5^9)$

$$\begin{pmatrix}
1 & 5 & 5 & 3 & 4 & 3 & 3 & 1 & 1 \\
5 & 1 & 4 & 3 & 1 & 2 & 5 & 2 & 2 \\
5 & 3 & 2 & 1 & 4 & 4 & 2 & 4 & 4 \\
3 & 5 & 1 & 2 & 5 & 2 & 2 & 3 & 3 \\
3 & 2 & 5 & 4 & 3 & 1 & 1 & 2 & 4 \\
4 & 3 & 4 & 5 & 2 & 5 & 1 & 3 & 1 \\
4 & 4 & 3 & 1 & 5 & 1 & 4 & 1 & 2 \\
2 & 2 & 3 & 5 & 1 & 4 & 3 & 5 & 3 \\
2 & 4 & 1 & 4 & 2 & 3 & 5 & 4 & 5 \\
1 & 1 & 2 & 2 & 3 & 5 & 4 & 5 & 5
\end{pmatrix},$$

which is a uniform design under the DD (see Fang et al. 2002). The WD-value of this design is 1.5442. Now, suppose we exchange the levels 1 and 5 of the first factor. The WD-value of this new design is 1.5417 which is smaller than the original design. This illustrative example recommends that the discrete discrepancy is not suitable for constructing uniform designs with multi-level quantitative factors.

2.5.2 *Lee Discrepancy*

As we pointed out that the Hamming distance does not take into account the absolute distance between the levels of a factor, the same is true for the Lee distance, which has been widely used in coding theory (see Roth 2006, for example). For $x, y \in Z_q = \{1, \ldots, q\}$, the Lee distance between x and y is defined by

$$Lee(x, y) = \min\{|x - y|, q - |x - y|\}.$$

For example, if $q = 4$, then $Lee(1, 2) = Lee(2, 3) = Lee(3, 4) = Lee(4, 1) = 1$ and $Lee(1, 3) = Lee(2, 4) = 2$. It is clear that Lee distance involves a wrap-around function.

Then, from Lee distance, we define a kernel function as

$$\mathcal{K}_j(t, z) = 1 - \min\{|t - z|, 1 - |t - z|\}, \tag{2.5.9a}$$

$$\mathcal{K}(\boldsymbol{t}, \boldsymbol{z}) = \prod_{j=1}^{s} \mathcal{K}_j(t_j, z_j), \text{ for any } \boldsymbol{t}, \boldsymbol{z} \in \mathcal{X}. \tag{2.5.9b}$$

The discrepancy with this kernel function is called as *Lee discrepancy* (LD), which can be regarded as an extension of the discrete discrepancy. The expression of Lee discrepancy for symmetrical and asymmetrical designs is derived as follows.

First, consider symmetrical designs. Let $\mathcal{U}(n; q^s)$ be the set of all the U-type symmetrical designs with n runs, s-factors, and each column equally often takes values from a set of q integers, labeled as $1, \ldots, q$. For any design $\boldsymbol{U} = (u_{ij}) \in \mathcal{U}(n; q^s)$, its experimental domain is $\{1, \ldots, q\}^s$, which comprises all possible level-combinations of the s-factors. By mapping:

$$x_{ij} = \frac{2u_{ij} - 1}{2q}, i = 1, \ldots, n;\ j = 1, \ldots, s, \tag{2.5.10}$$

the n runs of $\boldsymbol{U}$ are transformed into n points in $[0, 1]^s$, and denote the corresponding design as $\boldsymbol{X} = (x_{ij})$. Then, the experimental domain becomes $\mathcal{X} = \{\frac{1}{2q}, \ldots, \frac{2q-1}{2q}\}^s$. For convenience, define

$$\alpha_{ij}^k = \min\{|x_{ik} - x_{jk}|, 1 - |x_{ik} - x_{jk}|\}, \tag{2.5.11}$$

then the squared LD for odd q and even q are given by

$$\mathrm{LD}^2_{odd}(\boldsymbol{U}) = \frac{1}{n} - \left(\frac{3}{4} + \frac{1}{4q^2}\right)^s + \frac{2}{n^2} \sum_{i=1}^{n-1} \sum_{j=i+1}^{n} \prod_{k=1}^{s} (1 - \alpha_{ij}^k). \tag{2.5.12a}$$

and

$$\mathrm{LD}^2_{even}(\boldsymbol{U}) = \frac{1}{n} - \left(\frac{3}{4}\right)^s + \frac{2}{n^2}\sum_{i=1}^{n-1}\sum_{j=i+1}^{n}\prod_{k=1}^{s}(1-\alpha_{ij}^k). \tag{2.5.12b}$$

respectively. Note that the Lee discrepancy of the design $\boldsymbol{U}$ is calculated through $\boldsymbol{X}$, the induced matrix of $\boldsymbol{U}$, and $\mathrm{MD}(\boldsymbol{U}) = \mathrm{MD}(\boldsymbol{X})$.

Next, consider the expression of LD for U-type asymmetrical designs $U(n; q_1 \times \cdots \times q_s)$ (see Definition 1.3.6). If some q_i's are equal, we denote the asymmetrical design by $U(n; q_1^{s_1}, \ldots, q_t^{s_t})$, where $s = \sum_{i=1}^t s_i$. Denote $\mathcal{U}(n; q_1 \times \cdots \times q_s)$ and $\mathcal{U}(n; q_1^{s_1} \times \cdots \times q_t^{s_t})$ respectively be all of the designs $U(n; q_1 \times \cdots \times q_s)$ and $U(n; q_1^{s_1} \times \cdots \times q_t^{s_t})$. Without loss of any generality, we assume that the levels $q_1, \ldots, q_r$ are odd and $q_{r+1}, \ldots, q_t$ are even, where $0 \leqslant r \leqslant t$, $r = 0$ means all of the levels are even, and $r = t$ means the levels are all odd. For each factor j, by mapping:

$$f : l \to (2l-1)/(2q_i), l = 1, \ldots, q_i, \tag{2.5.13}$$

the n runs of $\boldsymbol{U}$ are transformed into n points in $[0, 1]^s$, and denote the corresponding design as $\boldsymbol{X} = (x_{ij})$. Then, the squared Lee discrepancy of $\boldsymbol{U} \in \mathcal{U}(n; q_1^{s_1}, \ldots, q_t^{s_t})$ can be expressed by

$$\mathrm{LD}^2(\boldsymbol{U}) = \frac{1}{n} - \left(\frac{3}{4}\right)^{\sum_{i=r+1}^t s_i} \prod_{i=1}^{r}\left(\frac{3}{4} + \frac{1}{4q_i^2}\right)^{s_i} + \frac{2}{n^2}\sum_{i=1}^{n-1}\sum_{j=i+1}^{n}\prod_{k=1}^{s}(1-\alpha_{ij}^k), \tag{2.5.14}$$

where α_{ij}^k is defined in (2.5.11).

For distinguishing the difference between multi-levels, Lee distance is better than Hamming distance, and then Lee discrepancy is better than discrete discrepancy for multi-level designs. In next section, we provide lower bounds of different discrepancy measures.

2.6 Lower Bounds of Discrepancies

In the last section, we explained in detail that the discrepancy plays an important role in measuring the uniformity of designs. In this regard, we introduced several useful discrepancies. The issue of lower bounds for discrepancies has been much considered. Many authors have invested much effort in finding some lower bounds for different discrepancies. It is well known that the lower bounds for discrepancy can be used as a benchmark not only in searching for uniform designs but also in helping to validate that some good designs are in fact uniform. A design whose discrepancy value achieves a strict lower bound is a uniform design with respect to this discrepancy. The word 'strict' means the lower bound can be reached in some cases. In this section, we will provide some important results in this direction. For convenience, we first introduce the results about two-level designs and then those about high-level and mixed-level designs. We focus on the popular discrepancies, such as CD, WD, MD, DD, and LD.

For ease in reference, we now briefly describe notations and preliminaries. Let $D(n; q_1 \times \cdots \times q_s)$ be a $n \times s$ matrix with entries $\{1, \ldots, q_j\}$ at the jth columns. When some q_j are equal, we denote it by $D(n; q_1^{r_1} \times \cdots \times q_m^{r_m})$ with $r_1 + \cdots + r_m = s$ and by $D(n; q^s)$ when all q_j's are equal to q. Let $\mathcal{D}(n; q_1 \times \cdots \times q_s)$ be the set of all designs $D(n; q_1 \times \cdots \times q_s)$. Similarly, we have notations $\mathcal{D}(q_1^{r_1} \times \cdots \times q_m^{r_m})$ and $\mathcal{D}(n; q^s)$. In some cases, the lower bounds are obtained under the U-type constraint, i.e., the design set may be, respectively, changed from $\mathcal{D}(n; q_1 \times \cdots \times q_s)$, $\mathcal{D}(q_1^{r_1} \times \cdots \times q_m^{r_m})$, and $\mathcal{D}(n; q^s)$ into $\mathcal{U}(n; q_1 \times \cdots \times q_s)$, $\mathcal{U}(q_1^{r_1} \times \cdots \times q_m^{r_m})$ and $\mathcal{U}(n; q^s)$, which are defined in Definition 1.3.6.

A typical treatment combination of a design $\mathcal{P} \in \mathcal{D}(n; q^s)$ is represented as $\boldsymbol{x} = (x_1, x_2, \ldots, x_s)$, where $x_j \in \{0, \ldots, q-1\}, 1 \leqslant j \leqslant s$. Let $\mathcal{V}^*$ be the set of all N $(= q^s)$ treatment combinations written in the lexicographic order. For any $\boldsymbol{x} \in \mathcal{V}^*$ and $\mathcal{P} \in \mathcal{D}(n; q^s)$, let $\boldsymbol{y}_{\mathcal{P}}(x)$ be the number of times the treatment combination x occurs in $\mathcal{P}$ and $\boldsymbol{y}_{\mathcal{P}}$ be the $N \times 1$ vector with elements $\boldsymbol{y}_{\mathcal{P}}(x)$ arranged in the lexicographic order.

Let $\boldsymbol{I}_q$ and $\mathbf{1}_q$ respectively be the $q \times q$ identity matrix and the $q \times 1$ vector with all elements unity. Define

$$\boldsymbol{L}(0) = \mathbf{1}_q^T, \ \boldsymbol{L}(1) = \boldsymbol{I}_q, \ \boldsymbol{J}_q = \mathbf{1}_q \mathbf{1}_q^T. \tag{2.6.1}$$

The t-fold Kronecker products of $\mathbf{1}_q$, $\boldsymbol{I}_q$, and $\boldsymbol{J}_q$ will, henceforth, be denoted by $\mathbf{1}_q^{(t)}$, $\boldsymbol{I}_q^{(t)}$, and $\boldsymbol{J}_q^{(t)}$, respectively. Let Ω be the set of all binary q tuples. For any $\boldsymbol{u} = (u_1, u_2, \ldots, u_s) \in \Omega$, define the matrix

$$\boldsymbol{G}(\boldsymbol{u}) = \bigotimes_{j=1}^{s} \boldsymbol{L}(u_j). \tag{2.6.2}$$

It is to be noted that $\boldsymbol{G}(\boldsymbol{x})$ is of order $q^{\sum_j u_j} \times N$. Here, the symbol $\bigotimes$ represents the Kronecker product.

For every m columns of $\mathcal{P} \in \mathcal{D}(n; q^s)$, $(\mathcal{P}_{l_1}, \mathcal{P}_{l_2}, \ldots, \mathcal{P}_{l_m})$, define

$$B_{l_1 \ldots l_m}(\mathcal{P}) = \sum_{\alpha_1, \ldots, \alpha_m} \left(n_{\alpha_1 \ldots \alpha_m}^{(l_1 \ldots l_m)} - n/q^m\right)^2,$$

where $n_{\alpha_1 \ldots \alpha_m}^{(l_1 \ldots l_m)}$ is the number of runs in which $(\mathcal{P}_{l_1}, \mathcal{P}_{l_2}, \ldots, \mathcal{P}_{l_m})$ takes the level combination $(\alpha_1, \ldots, \alpha_m)$, and the summation is taken over all q^m level-combinations. If $B_{l_1 \ldots l_m}(d) = 0$, the subdesign formed by the columns $(d_{l_1}, d_{l_2}, \ldots, d_{l_m})$ is an orthogonal array of strength m. Furthermore, define

$$B_m(\mathcal{P}) = \sum_{1 \leqslant l_1 < \cdots < l_m \leqslant s} B_{l_1 \ldots l_m}(\mathcal{P}) \Big/ \binom{s}{m} \tag{2.6.3}$$

for $1 \leqslant m \leqslant s$. It is evident that $B_m(\mathcal{P}) = 0$ if and only if $\mathcal{P} \in \mathcal{D}(n; q^s)$ is an orthogonal array of strength m. Consequently, $B_m(\mathcal{P})$ measures the closeness to orthogonality of strength m of $\mathcal{P}$.

2.6.1 Lower Bounds of the Centered L_2-Discrepancy

The work of Fang and Mukerjee (2000) on regular fractions with two levels was a first attempt toward providing a lower bound for CD. Later, many works have been studied along this direction. For regular two-level U-type designs, Fang and Mukerjee (2000) gave the following lower bound.

Theorem 2.6.1 *Let $\mathcal{P}$ be any fraction, involving $n = 2^{s-k}$ runs, of a 2^s factorial. Then, $CD^2(\mathcal{P}) \geqslant LB_{CD}^{(1)}$, where*

$$LB_{CD}^{(1)} = \left(\frac{13}{12}\right)^s - 2\left(\frac{35}{32}\right)^s + \sum_{r=0}^{s-k} \binom{s}{r} \frac{1}{8^r} + \frac{1}{n} \sum_{r=s-k+1}^{s} \frac{1}{4^r}.$$

Proof Define $\boldsymbol{B}_0 = \frac{1}{4}\boldsymbol{I}_2 + \boldsymbol{J}_2$, and $\boldsymbol{B}_s$ is the s-fold Kronecker products of $\boldsymbol{B}_0$. It is easy to check that $\boldsymbol{B}_s = \sum_{u \in \Omega} \frac{1}{4^{\sum u_j}} \boldsymbol{G}(\boldsymbol{u})^T \boldsymbol{G}(\boldsymbol{u})$. Therefore,

$$\boldsymbol{y}_{\mathcal{P}}^T \boldsymbol{B}_s \boldsymbol{y}_{\mathcal{P}} = \sum_{u \in \Omega} \frac{1}{4^{\sum u_j}} \boldsymbol{y}_{\mathcal{P}}^T \boldsymbol{G}(\boldsymbol{u})^T \boldsymbol{G}(\boldsymbol{u}) \boldsymbol{y}_{\mathcal{P}}. \tag{2.6.4}$$

For every $u \in \Omega$, the vector $\boldsymbol{y}_{\mathcal{P}}^T$ is of the order $2^{\sum u_j} \times 1$; furthermore, the elements of $\boldsymbol{y}_{\mathcal{P}}^T$ are integers with sum n^{s-k}. Therefore, $\boldsymbol{y}_{\mathcal{P}}^T \boldsymbol{G}(\boldsymbol{u})^T \boldsymbol{G}(\boldsymbol{u}) \boldsymbol{y}_{\mathcal{P}} \geqslant n^2/2^{\sum u_j}$ if $\sum u_j \leqslant s-k$, and $\boldsymbol{y}_{\mathcal{P}}^T G(\boldsymbol{u})^T \boldsymbol{G}(\boldsymbol{u}) \boldsymbol{y}_{\mathcal{P}} \geqslant n$ if $\sum u_j > s-k$. By (2.6.4), we have

$$\boldsymbol{y}_{\mathcal{P}}^T \boldsymbol{B}_s \boldsymbol{y}_{\mathcal{P}} \geqslant n^2 \left[\sum_{r=0}^{s-k} \binom{s}{r} \frac{1}{8^r} + \frac{1}{n} \sum_{r=s-k+1}^{s} \frac{1}{4^r} \right]. \tag{2.6.5}$$

Note that

$$\mathrm{CD}^2(\mathcal{P}) = \left(\frac{13}{12}\right)^s - 2\left(\frac{35}{32}\right)^s + \frac{1}{n^2} \boldsymbol{y}_{\mathcal{P}}^T \boldsymbol{B}_s \boldsymbol{y}_{\mathcal{P}}. \tag{2.6.6}$$

The proof completes from (2.6.5) and (2.6.6).

Observe that the lower bound is attained in Theorem 2.6.1 if and only if the 2^{s-k} runs in $\mathcal{P}$ form a two-symbol orthogonal array of strength $s-k$. Since such an orthogonal array is often nonexistent, the lower bound is also often non-attainable.

Fang et al. (2002) extended the result in Theorem 2.6.1 to non-regular two-level fractional factorials and gave a lower bound of CD, which can be applied to both regular and non-regular fractions.

Theorem 2.6.2 *Let $\mathcal{P} \in \mathcal{D}(n; 2^s)$, then $CD^2(\mathcal{P}) \geqslant LB_{CD}^{(2)}$, where*

$$LB_{CD}^{(2)} = \left(\frac{13}{12}\right)^s - 2\left(\frac{35}{32}\right)^s + \frac{1}{n^2}\sum_{r=0}^{s}\binom{s}{r}\frac{1}{4^r}[nf_r + z_r(f_r+1)], \tag{2.6.7}$$

f_r is the largest integer contained in $n/2^r$ and $z_r = n - 2^r f_r$. In particular, if $n = 2^{s-k}$ $(1 \leqslant k \leqslant s-1)$, then $CD^2(\mathcal{P}) \geqslant LB_{CD}^{(1)}$.

Proof Let Ω_r be the set of Ω consisting of those binary s-tuples which have exactly r elements of u unity. From (2.6.4), we have

$$\boldsymbol{y}_{\mathcal{P}}^T \boldsymbol{B}_s \boldsymbol{y}_{\mathcal{P}} = \sum_{r=0}^{s}\frac{1}{4^i}\left\{\sum_{u\in\Omega_r}\boldsymbol{y}_{\mathcal{P}}^T \boldsymbol{G}(\boldsymbol{u})^T\boldsymbol{G}(\boldsymbol{u})\boldsymbol{y}_{\mathcal{P}}\right\}. \tag{2.6.8}$$

Note that for every $\boldsymbol{u} \in \Omega_r$, the elements of the order $2^r \times 1$ vector $\boldsymbol{y}_{\mathcal{P}}^T$ are nonnegative integers with sum n. Hence,

$$\boldsymbol{y}_{\mathcal{P}}^T G(\boldsymbol{u})^T G(\boldsymbol{u})\boldsymbol{y}_{\mathcal{P}} \geqslant f_r^2(2^r - z_r) + (f_r+1)^2 z_r = nf_r + z_r(f_r+1),$$

and $LB_{CD}^{(2)}$ follows from (2.6.6) and (2.6.8).

Moreover, if $n = 2^{s-k}$, then $f_r = n/2^r$, $z_r = 0$ for $r \leqslant s-k$ and $f_r = 0$, $z_r = n$ for $r > s-k$. From (2.6.7), we know that $\mathrm{CD}^2(\mathcal{P}) \geqslant LB_{CD}^{(1)}$ holds. The proof is complete.

In order to improve the result in Theorem 2.6.2, Fang et al. (2003b) found the CD-value of $\mathcal{P}$ can be reexpressed from two aspects: One is related to the distribution of all the level-combinations among columns of $\mathcal{P}$, and the other is based on Hamming distances between any two runs of $\mathcal{P}$. These new representations allow to obtain two kinds of lower bounds for U-type designs. We state them as the following theorem.

Theorem 2.6.3 *Let $\mathcal{P} \in \mathcal{U}(n; 2^s)$, then $CD^2(\mathcal{P}) \geqslant max\{LB_{CD}^{(c)}, LB_{CD}^{(r)}\}$, where*

$$LB_{CD}^{(c)} = \left(\frac{13}{12}\right)^s - 2\left(\frac{35}{32}\right)^s + \left(\frac{9}{8}\right)^s + \frac{1}{n^2}\sum_{r=1}^{s}\frac{1}{4^r}\binom{s}{r}s_{n,r,2}\left(1 - \frac{s_{n,r,2}}{2^r}\right) \tag{2.6.9}$$

and

$$LB_{CD}^{(r)} = \left(\frac{13}{12}\right)^s - 2\left(\frac{35}{32}\right)^s + \frac{1}{n}\left(\frac{5}{4}\right)^s + \frac{n-1}{n}\left(\frac{5}{4}\right)^\lambda, \tag{2.6.10}$$

where $s_{n,r,2}$ is the remainder at division of n by 2^r (n mod 2^r) and $\lambda = s(n-2)/[2(n-1)]$.

Proof Following Fang et al. (2003b), we know

$$\mathrm{CD}^2(\mathcal{P}) = \left(\frac{13}{12}\right)^s - 2\left(\frac{35}{32}\right)^s + \left(\frac{9}{8}\right)^s + \frac{1}{n^2}\sum_{r=1}^{s}\frac{1}{4^r}\binom{s}{r}B_r(\mathcal{P}) \tag{2.6.11}$$

$$= \left(\frac{13}{12}\right)^s - 2\left(\frac{35}{32}\right)^s + \frac{1}{n^2}\sum_{i=1}^{s}\sum_{j=1}^{s}\left(\frac{5}{4}\right)^{\delta_{ij}(\mathcal{P})}. \tag{2.6.12}$$

Let $n = t2^m + s_{n,r,2}$, where t is a nonnegative integer. For every r columns $(\mathcal{P}_{l_1}, \mathcal{P}_{l_2}, \ldots, \mathcal{P}_{l_r})$ of $\mathcal{P} \in \mathcal{D}(n; q^s)$, according to the definition of $B_{l_1\ldots l_r}(\mathcal{P})$, we can see that $B_{l_1\ldots l_r}(\mathcal{P})$ is minimized if there are $s_{n,r,2}$ level-combinations that occur $t+1$ times, while the other $2^r - s_{n,r,2}$ level-combinations that occur t times in the n rows. The minimum is

$$s_{n,r,2}\left(1 - \frac{s_{n,r,2}}{2^r}\right)^2 + (2^r - s_{n,r,2})\left(-\frac{s_{n,r,2}}{2^r}\right)^2 = s_{n,r,2}\left(1 - \frac{s_{n,r,2}}{2^r}\right).$$

Thus, $B_r(\mathcal{P}) \geqslant s_{n,r,2}\left(1 - \frac{s_{n,r,2}}{2^r}\right)$. By (2.6.11), we obtain the lower bound $LB_{CD}^{(c)}$ in (2.6.9).

To prove the lower bound $LB_{CD}^{(r)}$, we define a random variable Y which is uniformly distributed on the set $\{\delta_{ij}(\mathcal{P}), 1 \leqslant i \neq j \leqslant n\}$. Define a convex function $f(y) = \left(\frac{5}{4}\right)^y$, and then we have $E(Y) = \lambda$. Thus, by Jenssen's inequality, we obtain

$$E[f(Y)] = \frac{1}{n(n-1)}\sum_{1\leqslant i\neq j\leqslant n}\left(\frac{5}{4}\right)^{\delta_{ij}(\mathcal{P})} \geqslant f[E(Y)] = \left(\frac{5}{4}\right)^{\lambda}.$$

Noting that $\delta_{ii}(\mathcal{P}) = s$, we obtain the lower bound $LB_{CD}^{(r)}$ in (2.6.10) from (2.6.12). The proof is complete.

Note that in order to attend the lower bound in (2.6.10), λ must be an integer and all Hamming distances between any two distinct runs of $\mathcal{P}$ are equal. However, many two-level designs do not meet these requirements. Hence, Chatterjee et al. (2012a) gave the following improved lower bound for U-type designs.

Theorem 2.6.4 *Let $\mathcal{P} \in \mathcal{U}(n; 2^s)$, then $CD^2(\mathcal{P}) \geqslant LB_{CD}^{(3)}$, where*

$$LB_{CD}^{(3)} = \left(\frac{13}{12}\right)^s - 2\left(\frac{35}{32}\right)^s + \frac{1}{n}\left(\frac{5}{4}\right)^s + \frac{1}{n^2}\left(\frac{5}{4}\right)^w\left(p + \frac{5}{4}q\right), \tag{2.6.13}$$

$p + q = n(n-1)$, $pw + q(w+1) = n(n-2)s/2$, w is the largest integer contained in $s(n-2)/[2(n-1)]$.

Lemma 2.6.1 (Chatterjee et al. 2012a) *For any U-type design $\mathcal{P} \in \mathcal{U}(n; 2^s)$ and any integer l, we have*

$$\sum_{i=1}^{n} \sum_{j(\neq i)=1}^{n} [\delta_{ij}(\mathcal{P})]^l \geqslant pw^l + q(w+1)^l. \quad (2.6.14)$$

Proof of Theorem 2.6.4 From (2.6.12), we get

$$\begin{aligned}
\mathrm{CD}^2(\mathcal{P}) &= \left(\frac{13}{12}\right)^s - 2\left(\frac{35}{32}\right)^s + \frac{1}{n}\left(\frac{5}{4}\right)^s + \frac{1}{n^2}\sum_{i=1}^{n}\sum_{j(\neq i)=1}^{n} e^{\delta_{ij}(\mathcal{P})} \\
&= \left(\frac{13}{12}\right)^s - 2\left(\frac{35}{32}\right)^s + \frac{1}{n}\left(\frac{5}{4}\right)^m + \frac{1}{n^2}\sum_{i=1}^{n}\sum_{j(\neq i)=1}^{n}\sum_{l=0}^{\infty} \frac{\alpha^l[\delta_{ij}(\mathcal{P})]^l}{l!},
\end{aligned}$$

where $\alpha = ln(5/4)$. Now, from Lemma 2.6.1, we get

$$\begin{aligned}
\mathrm{CD}^2(\mathcal{P}) &\geqslant \left(\frac{13}{12}\right)^s - 2\left(\frac{35}{32}\right)^s + \frac{1}{n}\left(\frac{4}{5}\right)^s + \frac{p}{n^2}\sum_{l=0}^{\infty}\frac{\alpha^l w^l}{l!} + \frac{q}{n^2}\sum_{l=0}^{\infty}\frac{\alpha^l (w+1)^l}{l!} \\
&= \left(\frac{13}{12}\right)^s - 2\left(\frac{35}{32}\right)^s + \frac{1}{n}\left(\frac{5}{4}\right)^s + \frac{p}{n^2}e^{\alpha w} + \frac{q}{n^2}e^{\alpha(w+1)} \\
&= \left(\frac{13}{12}\right)^s - 2\left(\frac{35}{32}\right)^s + \frac{1}{n}\left(\frac{5}{4}\right)^s + \frac{p}{n^2}\left(\frac{5}{4}\right)^w + \frac{q}{n^2}\left(\frac{5}{4}\right)^{w+1}.
\end{aligned}$$

The proof is finished.

Figure 2.9 displays the difference among $LB_{CD}^{(c)}$, $LB_{CD}^{(r)}$, and $LB_{CD}^{(3)}$. It is noted that none of them can dominate others for all combination of the number of runs n and of factors s. Generally, $LB_{CD}^{(3)}$ is better than others for small n and large s, which means $LB_{CD}^{(3)}$ is more suitable for evaluating the uniformity of saturated or supersaturated designs. For large n and small s, $LB_{CD}^{(c)}$ is better than others.

For high-level design, there is some difficulty to obtain the lower bound of CD. This is because we need to check the roots of some function and make sure that the root is unique. Fang et al. (2006b) found some lower bounds of CD for $\mathcal{P} \in \mathcal{D}(n; 3^s)$ and $\mathcal{P} \in \mathcal{D}(n; 4^s)$, respectively. Elsawah and Qin (2014) improved the result in Fang et al. (2006b) and gave a sharper lower bound of CD for $\mathcal{P} \in \mathcal{D}(n; 4^s)$. The reader can find the details in their papers.

For mixed-level designs, Chatterjee et al. (2005) initiated a research and reported some new lower bounds for CD for a set of asymmetric fractional factorials $\mathcal{D}(n; p \times 2^s)$, where $p \geqslant 3$.

2.6.2 Lower Bounds of the Wrap-around L_2-Discrepancy

For two-level designs, Fang et al. (2002) initiated a research and obtained the following lower bound of WD, which can be applied to regular and non-regular fractions.

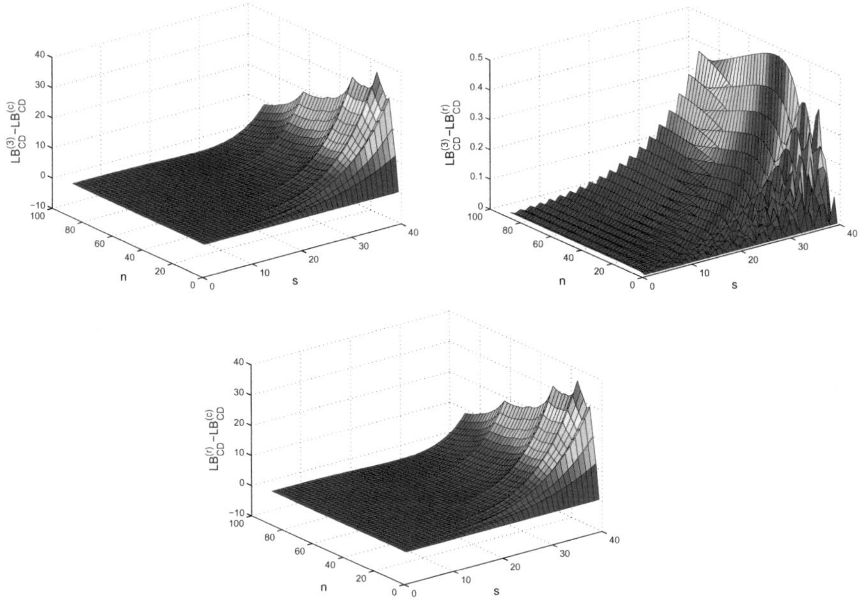

Fig. 2.9 Comparison among $LB_{CD}^{(c)}$, $LB_{CD}^{(r)}$, and $LB_{CD}^{(3)}$

Theorem 2.6.5 *Let $\mathcal{P} \in \mathcal{D}(n; 2^s)$, then $WD^2(\mathcal{P}) \geqslant LB_{WD}^{(21)}$, where*

$$LB_{WD}^{(21)} = -\left(\frac{4}{3}\right)^s + \frac{1}{n^2}\left(\frac{5}{4}\right)^s \sum_{r=0}^{s} \binom{s}{r} \frac{1}{5^r}[nf_r + z_r(f_r+1)], \tag{2.6.15}$$

f_r and z_r are defined in Theorem 2.6.2. In particular, if $n = 2^{s-k}$ $(1 \leqslant k \leqslant s-1)$, then $WD^2(\mathcal{P}) \geqslant LB_{WD}^{(22)}$, where

$$LB_{WD}^{(22)} = -\left(\frac{4}{3}\right)^s + \left(\frac{5}{4}\right)^s \sum_{r=0}^{s-k} \binom{s}{r} \frac{1}{10^r} + \frac{1}{n}\left(\frac{5}{4}\right)^s \sum_{r=s-k+1}^{s} \binom{s}{r} \frac{1}{5^r}.$$

Proof Note that

$$\mathrm{WD}^2(\mathcal{P}) = -\left(\frac{4}{3}\right)^s + \frac{1}{n^2}\boldsymbol{y}_{\mathcal{P}}^T \boldsymbol{D}_s \boldsymbol{y}_{\mathcal{P}}, \tag{2.6.16}$$

where $\boldsymbol{D}_s$ is the s-fold Kronecker products of $\frac{1}{4}\boldsymbol{I}_2 + \frac{5}{4}\boldsymbol{J}_2$. Being along lines of the proof of Theorem 2.6.2, we can obtain the lower bound $LB_{WD}^{(21)}$.

The lower bounds in Theorem 2.6.5 are often non-attainable and, therefore, conservative. In order to improve the lower bounds in Theorem 2.6.5, Fang et al. (2003b)

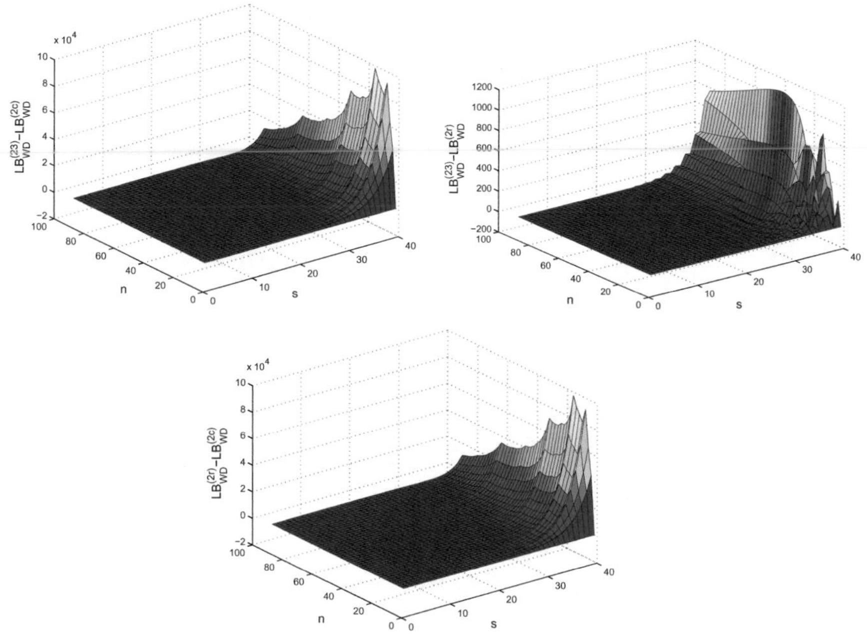

Fig. 2.10 Comparison among $LB_{WD}^{(2c)}$, $LB_{WD}^{(2r)}$, and $LB_{WD}^{(23)}$

and Chatterjee et al. (2012a) gave new lower bounds of WD for U-type designs, respectively. We summarize them as the following theorem.

Theorem 2.6.6 *Let* $\mathcal{P} \in \mathcal{U}(n; 2^s)$, *then*

$$WD^2(\mathcal{P}) \geqslant max\left\{LB_{WD}^{(2c)}, LB_{WD}^{(2r)}, LB_{WD}^{(23)}\right\},$$

where

$$LB_{WD}^{(2c)} = \left(\frac{11}{8}\right)^s - \left(\frac{4}{3}\right)^s + \frac{1}{n^2}\left(\frac{5}{4}\right)^s \sum_{r=1}^{s} \frac{1}{5^r}\binom{s}{r} s_{n,r,2}\left(1 - \frac{s_{n,r,2}}{2^r}\right),$$

$$LB_{WD}^{(2r)} = -\left(\frac{4}{3}\right)^s + \frac{1}{n}\left(\frac{3}{2}\right)^s + \frac{n-1}{n}\left(\frac{5}{4}\right)^s\left(\frac{6}{5}\right)^{\lambda}, \tag{2.6.17}$$

and

$$LB_{WD}^{(23)} = -\left(\frac{4}{3}\right)^s + \frac{1}{n}\left(\frac{3}{2}\right)^s + \frac{1}{n^2}\left(\frac{5}{4}\right)^s\left[p\left(\frac{6}{5}\right)^w + q\left(\frac{6}{5}\right)^{w+1}\right]. \tag{2.6.18}$$

Here, $s_{n,r,2}$ and λ are defined in Theorem 2.6.3, and p, q, and w are defined in Theorem 2.6.4.

Proof Following Fang et al. (2003b), we know

$$\mathrm{WD}^2(\mathcal{P}) = \left(\frac{11}{8}\right)^s - \left(\frac{4}{3}\right)^s + \frac{1}{n^2}\sum_{r=1}^{s}\frac{1}{5^r}\binom{s}{r}B_r(\mathcal{P}) \tag{2.6.19}$$

$$= -\left(\frac{4}{3}\right)^s + \frac{1}{n^2}\left(\frac{5}{4}\right)^s\sum_{i=1}^{s}\sum_{j=1}^{s}\left(\frac{6}{5}\right)^{\delta_{ij}(\mathcal{P})}. \tag{2.6.20}$$

Following similar arguments in the proofs of Theorems 2.6.3 and 2.6.4, we can obtain the lower bounds $LB_{WD}^{(2c)}$, $LB_{WD}^{(2r)}$, and $LB_{WD}^{(23)}$. The theorem is proved.

Figure 2.10 shows that $LB_{WD}^{(23)}$ is better than $LB_{WD}^{(2c)}$ and $LB_{WD}^{(2r)}$ for small n and large s and $LB_{WD}^{(2c)}$ is better than others for large n and small s.

For three-level designs, Fang et al. (2002) gave the following lower bound of WD.

Theorem 2.6.7 *Let $\mathcal{P} \in \mathcal{D}(n; 3^s)$, then $WD^2(\mathcal{P}) \geqslant LB_{WD}^{(31)}$, where*

$$LB_{WD}^{(31)} = -\left(\frac{4}{3}\right)^s + \frac{1}{n^2}\left(\frac{23}{18}\right)^s\sum_{r=0}^{s}\binom{s}{r}\left(\frac{4}{23}\right)^r [ng_r + z_r(g_r+1)], \tag{2.6.21}$$

g_r is the largest integer contained in $n/3^r$ and $t_r = n - 3^r f_r$. In particular, if $n = 3^{s-k}$ $(1 \leqslant k \leqslant s-1)$, then $WD^2(\mathcal{P}) \geqslant LB_{WD}^{(32)}$, where

$$LB_{WD}^{(32)} = -\left(\frac{4}{3}\right)^s + \left(\frac{23}{18}\right)^s\sum_{r=0}^{s-k}\binom{s}{r}\left(\frac{4}{69}\right)^r + \frac{1}{n}\left(\frac{23}{18}\right)^s\sum_{r=s-k+1}^{s}\binom{s}{r}\left(\frac{4}{23}\right)^r.$$

Proof Note that

$$\mathrm{WD}^2(\mathcal{P}) = -\left(\frac{4}{3}\right)^s + \frac{1}{n^2}\boldsymbol{y}_{\mathcal{P}}^T Q_s \boldsymbol{y}_{\mathcal{P}}, \tag{2.6.22}$$

where Q_s is the s-fold Kronecker products of $\frac{2}{9}\boldsymbol{I}_3 + \frac{23}{18}\boldsymbol{J}_3$. Being along lines of the proof of Theorem 2.6.2, we can complete the proof.

Similarly, the lower bounds in Theorem 2.6.7 are often non-attainable and, therefore, conservative. In order to improve the lower bounds in Theorem 2.6.7, Fang et al. (2003b) and Zhang et al. (2015) gave new lower bounds of WD for U-type designs, respectively. We summarize them as the following theorem.

Theorem 2.6.8 *Let $\mathcal{P} \in \mathcal{U}(n; 3^s)$ and $\Delta_3 = -\left(\frac{4}{3}\right)^s + \frac{1}{n}\left(\frac{3}{2}\right)^s$, then*

$$WD^2(\mathcal{P}) \geqslant max\left\{LB_{WD}^{(3c)}, LB_{WD}^{(3r)}, LB_{WD}^{(33)}\right\},$$

where

$$LB_{WD}^{(3c)} = -\left(\frac{4}{3}\right)^s + \left(\frac{73}{54}\right)^s + \frac{1}{n^2}\left(\frac{23}{18}\right)^s \sum_{r=1}^{s}\left(\frac{4}{23}\right)^r \binom{s}{r} s_{n,r,3}\left(1 - \frac{s_{n,r,3}}{3^r}\right),$$

$$LB_{WD}^{(3r)} = \Delta_3 + \frac{n-1}{n}\left(\frac{23}{18}\right)^s \left(\frac{27}{23}\right)^\lambda \tag{2.6.23}$$

and

$$LB_{WD}^{(33)} = \Delta_3 + \frac{1}{n^2}\left(\frac{23}{18}\right)^s \left[p\left(\frac{27}{23}\right)^w + q\left(\frac{27}{23}\right)^{w+1}\right]. \tag{2.6.24}$$

Here, $s_{n,r,3}$ is the remainder at division of n by 3^r (n mod 3^r), $\lambda = s(n-3)/[3(n-1)]$, $p + q = n(n-1)$, $pw + q(w+1) = n(n-3)s/3$, and w is the largest integer contained in λ.

Proof From Fang et al. (2003b), we know

$$\mathrm{WD}^2(\mathcal{P}) = -\left(\frac{4}{3}\right)^s + \left(\frac{73}{54}\right)^s + \frac{1}{n^2}\left(\frac{23}{18}\right)^s \sum_{r=1}^{s}\left(\frac{4}{23}\right)^r \binom{s}{r} B_r(\mathcal{P}) \tag{2.6.25}$$

$$= -\left(\frac{4}{3}\right)^s + \frac{1}{n^2}\left(\frac{23}{18}\right)^s \sum_{i=1}^{s}\sum_{j=1}^{s}\left(\frac{27}{23}\right)^{\delta_{ij}(\mathcal{P})}. \tag{2.6.26}$$

Following similar arguments in the proofs of Theorems 2.6.3 and 2.6.4, we can obtain the lower bounds $LB_{WD}^{(3c)}$, $LB_{WD}^{(3r)}$, and $LB_{WD}^{(33)}$. The theorem is proved.

Figure 2.11 also shows that $LB_{WD}^{(33)}$ is better than $LB_{WD}^{(3c)}$ and $LB_{WD}^{(3r)}$ for small n and large s and $LB_{WD}^{(3c)}$ is better than others for large n and small s.

For q-level designs, where q is a positive integer, Fang et al. (2005) proposed a lower bound of WD for U-type designs. From the analytical expression of equation (2.3.6), it is easy to see that $\mathrm{WD}^2(\mathcal{P})$ is only a function of products of $\alpha_{ij}^k \equiv |x_{il} - x_{jl}|(1 - |x_{il} - x_{jl}|)$ ($i, j = 1, \ldots, n, i \neq j$ and $k = 1, \ldots, s$). For a U-type design $\mathcal{P} \in \mathcal{D}(n; q^s)$, when q is even, α-values can only take $q/2 + 1$ possible values, i.e., 0, $2(2q-2)/(4q^2), 4(2q-4)/(4q^2), \ldots, q^2/(4q^2)$; when q is odd, α-values can only take $(q+1)/2$ possible values, i.e., $0, 2(2q-2)/(4q^2), 4(2q-4)/(4q^2), \ldots, (q-1)(q+1)/(4q^2)$. Table 2.3 gives the distribution of α-values over the set $\{\alpha_{ij}^k : 1 \leqslant i < j \leqslant n, 1 \leqslant k \leqslant s\}$ for both even and odd q.

Theorem 2.6.9 *Let $\mathcal{P} \in \mathcal{U}(n; q^s)$, then when q is even, $WD^2(\mathcal{P}) \geqslant LB_{WD}^{(even)}$, where*

$$LB_{WD}^{(even)} = \Delta_3 + \frac{n-1}{n}\left(\frac{3}{2}\right)^{\frac{s(n-q)}{q(n-1)}} \left(\frac{5}{4}\right)^{\frac{sn}{q(n-1)}} \prod_{i=1}^{q/2-1}\left[\frac{3}{2} - \frac{2i(2q-2i)}{4q^2}\right]^{\frac{2sn}{q(n-1)}};$$

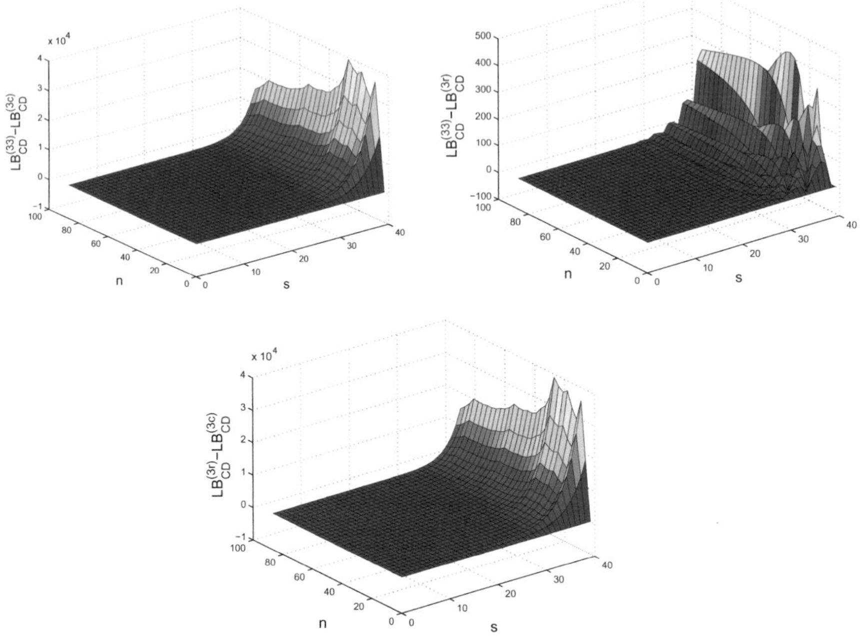

Fig. 2.11 Comparison among $LB_{WD}^{(3c)}$, $LB_{WD}^{(3r)}$, and $LB_{WD}^{(33)}$

when q is odd, $WD^2(\mathcal{P}) \geqslant LB_{WD}^{(odd)}$, where

$$LB_{WD}^{(odd)} = \Delta_3 + \frac{n-1}{n}\left(\frac{3}{2}\right)^{\frac{s(n-q)}{q(n-1)}}\left(\frac{5}{4}\right)^{\frac{sn}{q(n-1)}}\prod_{i=1}^{(q-1)/2}\left[\frac{3}{2} - \frac{2i(2q-2i)}{q^2}\right]^{\frac{2sn}{q(n-1)}}.$$

Proof By Eq. (2.3.6), to minimize $\mathrm{WD}^2(\mathcal{P})$ for a given design $\mathcal{P} \in \mathcal{U}(n; q^s)$ is equivalent to minimizing $\sum_{1\leqslant i\neq j\leqslant n}\prod_{k=1}^{s}(\frac{3}{2} - \alpha_{ij}^k)$ with respect to α_{ij}^k. From Table 2.3, we know that for given (n, q, s), the distributions of α-values are the same, so $\prod_{1\leqslant i\neq j\leqslant n}\prod_{k=1}^{s}(\frac{3}{2} - \alpha_{ij}^k)$ is a constant and $\frac{3}{2} - \alpha_{ij}^k > 0$. Based on the geometric and arithmetic mean inequality, $[WD(d(\gamma))]^2$ arrives at its minimum if all $\prod_{lk1}^{s}(\frac{3}{2} - \alpha_{ij}^k)$ are the same for $1 \leqslant i \neq j \leqslant n$. The expression of the lower bound of $[WD(d(\gamma))]^2$ is straightforward according to Table 2.3, which completes the proof.

When $q = 2, 3$, the lower bounds in Theorem 2.6.9 are equivalent to $LB_{WD}^{(2r)}$ in (2.6.17) and $LB_{WD}^{(3r)}$ in (2.6.23), respectively.

For mixed-level designs, Chatterjee et al. (2005) initiated a research on the derivation of lower bounds of WD for factorials with two and three mixed levels. Note that optimal asymmetrical factorials with two and three mixed levels are most demanded in practice, so an accurate lower bound for WD-value of this kind of asymmetrical factorials is valuable.

Table 2.3 Distributions of α-values of a $\mathcal{P} \in \mathcal{D}(n; q^s)$

q even		q odd	
α-values	Number	α-values	Number
0	$\frac{sn(n-q)}{2q}$	0	$\frac{sn(n-q)}{2q}$
$\frac{2(2q-2)}{4q^2}$	$\frac{sn^2}{q}$	$\frac{2(2q-2)}{4q^2}$	$\frac{sn^2}{q}$
...	...	...	...
$\frac{(q-2)(q+2)}{4q^2}$	$\frac{sn^2}{q}$	$\frac{(q-3)(q+3))}{4q^2}$	$\frac{sn^2}{q}$
$\frac{q^2}{4q^2}$	$\frac{sn(n-q)}{2q}$	$\frac{(q-1)(q+1)}{4q^2}$	$\frac{sn(n-q)}{2q}$

Theorem 2.6.10 *Let $\mathcal{P} \in \mathcal{D}(n; 2^{s_1} \times 3^{s_2})$, then*

$$WD^2(\mathcal{P}) \geqslant LB_{WD}^{(1)},$$

where

$$LB_{WD}^{(1)} = -\left(\frac{4}{3}\right)^{s_1+s_2} + \frac{1}{n^2}\left(\frac{5}{4}\right)^{s_1}\left(\frac{23}{18}\right)^{s_2}\sum_{i=0}^{s_1}\sum_{j=0}^{s_2}\binom{s_1}{i}\binom{s_2}{j}\left(\frac{1}{5}\right)^i\left(\frac{4}{23}\right)^j\theta_{ij}, \tag{2.6.27}$$

$\theta_{ij} = ng_{ij} + l_{ij}(g_{ij} + 1)$, *$g_{ij}$ is the largest integer contained in $n/(2^i 3^j)$, $l_{ij} = n - h_{ij}g_{ij}$, $0 \leqslant i \leqslant s_1$, $0 \leqslant j \leqslant s_2$.*

The proof of Theorem 2.6.10 is similar to Theorem 2.6.2. When $s_1 = 0$ or $s_2 = 0$, the lower bound in Theorem 2.6.10 is equivalent to $LB_{WD}^{(31)}$ in (2.6.21) or $LB_{WD}^{(21)}$ in (2.6.15).

For general mixed-level U-type designs, Zhou and Ning (2008) obtained some lower bounds of WD for U-type design $\mathcal{P} \in \mathcal{U}(n; q_1^{s_1} \times \cdots \times q_m^{s_m})$. Without loss of generality, we assume that the numbers of levels $q_1, \ldots, q_t$ are odd and $q_{t+1}, \ldots, q_m$ are even, where $0 \leqslant t \leqslant s$, $t = 0$ means all of the levels are even, and $t = s$ means the levels are all odd.

Theorem 2.6.11 *Let $\mathcal{P} \in \mathcal{U}(n; q_1^{s_1} \times \cdots \times q_m^{s_m})$, $s = s_1 + \cdots + s_m$, and $\Delta_4 = -\left(\frac{4}{3}\right)^s + \frac{1}{n}\left(\frac{3}{2}\right)^s$, then*

$$WD^2(\mathcal{P}) \geqslant LB_{WD}^{(3)},$$

where

$$LB_{WD}^{(3)} = \Delta_4 + \frac{n-1}{n}\left(\frac{3}{2}\right)^{\sum_{r=1}^m \frac{s_r(n-q_r)}{q_r(n-1)}}\left(\frac{5}{4}\right)^{\frac{n}{n-1}\sum_{r=t+1}^m \frac{s_r}{q_r}}$$
$$\times \prod_{r=1}^{t}\prod_{i=1}^{(q_r-1)/2}\left[\frac{3}{2} - \frac{2i(2q_r - 2i)}{4q_r^2}\right]^{\frac{2s_r n}{q_r(n-1)}}$$

$$\times \prod_{r=t+1}^{m} \prod_{i=1}^{q_r/2-1} \left[\frac{3}{2} - \frac{2i(2q_r - 2i)}{4q_r^2}\right]^{\frac{2s_r n}{q_r(n-1)}}.$$

The proof of Theorem 2.6.11 is similar to Theorem 2.6.9. When $q_1 = \cdots = q_m = q$, the lower bound in Theorem 2.6.11 is equivalent to the lower bound in Theorem 2.6.9.

2.6.3 Lower Bounds of Mixture Discrepancy

For two-level designs, Zhou et al. (2013) proposed a lower bound of MD as in the following theorem.

Theorem 2.6.12 *Let $\mathcal{P} \in \mathcal{U}(n; 2^s)$, then $MD^2(\mathcal{P}) \geqslant LB_{MD}^{(2:1)}$, where*

$$LB_{MD}^{(2:1)} = \left(\frac{19}{12}\right)^s - 2\left(\frac{305}{192}\right)^s + \left(\frac{39}{24}\right)^s + \frac{1}{n^2}\left(\frac{3}{2}\right)^s \sum_{r=1}^{s} \binom{s}{r} \frac{s_{n,r,2}}{6^r}\left(1 - \frac{s_{n,r,2}}{2^r}\right)$$

where $s_{n,r,2}$ is defined in Theorem 2.6.3.

Proof From Zhou et al. (2013), we know

$$\mathrm{MD}^2(\mathcal{P}) = \left(\frac{19}{12}\right)^s - 2\left(\frac{305}{192}\right)^s + \left(\frac{39}{24}\right)^s + \frac{1}{n^2}\left(\frac{3}{2}\right)^s \sum_{r=1}^{s} \frac{1}{6^r}\binom{s}{r} B_r(\mathcal{P}).$$

Following similar arguments in the proof of Theorem 2.6.3, we can obtain the lower bound $LB_{MD}^{(2:1)}$. The theorem is proved.

In order to reach the lower bound $LB_{MD}^{(2:1)}$, a two-level orthogonal design $\mathcal{P}$ is required. Clearly, this requirement does not valid in more general situations. Moreover, even when the two-level design $\mathcal{P}$ is orthogonal, the lower bound $LB_{MD}^{(2:1)}$ may not be tight. Therefore, Ke et al. (2015) gave the following improved lower bound.

Theorem 2.6.13 *Let $\mathcal{P} \in \mathcal{U}(n; 2^s)$ and λ be defined in Theorem 2.6.3. When λ is an integer, then $MD^2(\mathcal{P}) \geqslant LB_{MD}^{(2:2)}$, where*

$$LB_{MD}^{(2:2)} = \left(\frac{19}{12}\right)^s - 2\left(\frac{305}{192}\right)^s + \frac{1}{n}\left(\frac{7}{4}\right)^s + \frac{n}{n-1}\left(\frac{3}{2}\right)^s \left(\frac{7}{6}\right)^\lambda.$$

Proof From Ke et al. (2015), we know

$$\mathrm{MD}^2(\mathcal{P}) = \left(\frac{19}{12}\right)^s - 2\left(\frac{305}{192}\right)^s + \left(\frac{39}{24}\right)^s + \frac{1}{n^2}\left(\frac{3}{2}\right)^s \sum_{i=1}^{s}\sum_{j=1}^{s} \left(\frac{7}{6}\right)^{\delta_{ij}(\mathcal{P})}. \tag{2.6.28}$$

Following similar arguments in the proof of Theorem 2.6.3, we can obtain the lower bound $LB_{MD}^{(2:1)}$. The theorem is proved.

The lower bound $LB_{MD}^{(2:2)}$ can be achieved if and only if $\delta_{ik}(\mathcal{P}) = \lambda$, for all $i \neq k$, where $\delta_{ik}(\mathcal{P})$ appears in (2.5.2). When λ is not an integer, the lower bound $LB_{MD}^{(2:2)}$ is not achievable. Elsawah and Qin (2015) proposed the following lower bound, which is more useful and sharper than $LB_{MD}^{(2:1)}$ and $LB_{MD}^{(2:2)}$.

Theorem 2.6.14 *Let $\mathcal{P} \in \mathcal{U}(n; 2^s)$, then $MD^2(\mathcal{P}) \geqslant LB_{MD}^{(2:3)}$, where*

$$LB_{MD}^{(2:3)} = \left(\frac{19}{12}\right)^s - 2\left(\frac{305}{192}\right)^s + \frac{1}{n}\left(\frac{7}{4}\right)^s + \frac{1}{n^2}\left(\frac{3}{2}\right)^s\left(\frac{7}{6}\right)^w\left(p + \frac{7}{6}q\right),$$

where p, q, and w are defined in Theorem 2.6.4.

Lemma 2.6.2 (Elsawah and Qin 2015) *Suppose $\sum_{i=1}^n z_i = c$ and z_i's are nonnegative, and then for any integer l, we have*

$$\sum_{i=1}^n l^{z_i} \geqslant l^w(\alpha + \beta l),$$

where α and β are integers such that $\alpha + \beta = n$, $\alpha w + \beta(w + 1) = c$ and w is the largest integer contained in c/n.

Proof of Theorem 2.6.14 Note that for any design $\mathcal{P} \in \mathcal{D}(n; 2^s)$,

$$\sum_{i=1}^n \sum_{j(\neq i)=1}^n \delta_{ij}(\mathcal{P}) = ns(n-2)/2$$

is a constant. From Lemma 2.6.2, it is straightforward to show

$$\sum_{i=1}^s \sum_{j(\neq i)=1}^s \left(\frac{7}{6}\right)^{\delta_{ij}(\mathcal{P})} \geqslant \left(\frac{7}{6}\right)^w\left(p + \frac{7}{6}q\right). \tag{2.6.29}$$

Combining (2.6.28) and (2.6.28), the proof is completed.

Figure 2.11 shows that $LB_{MD}^{(2:3)}$ is better than $LB_{MD}^{(2:1)}$ for small n and large s and, meanwhile, better than $LB_{MD}^{(2:2)}$ for large n and small s, and $LB_{MD}^{(2:2)}$ is better than $LB_{MD}^{(2:1)}$ for small n and large s (Fig. 2.12).

For a three-level design $\mathcal{P} \in \mathcal{D}(n; 3^s)$, let φ be the largest integer contained in $2s/3$, $n_\varphi = (\varphi + 1)n - 2ns/3$ and $\varphi_0 = \frac{ln(\frac{2}{3n})+sln(\frac{9}{8})}{ln(\frac{42}{41})}$. Ke et al. (2015) obtained the following lower bound of $\text{MD}^2(\mathcal{P})$ for U-type design $\mathcal{P} \in \mathcal{U}(n; 3^s)$.

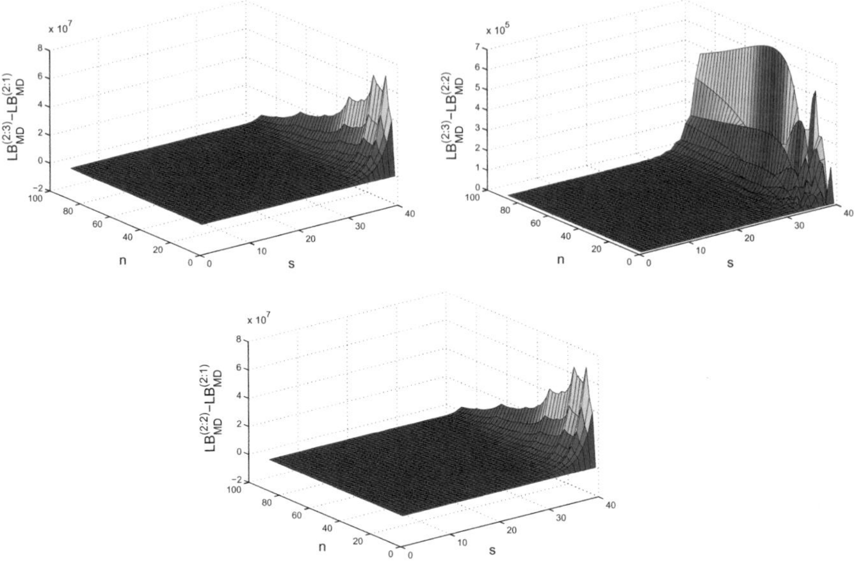

Fig. 2.12 Comparison among $LB_{MD}^{(2:1)}$, $LB_{MD}^{(2:2)}$, and $LB_{MD}^{(2:3)}$

Theorem 2.6.15 *For any* $\mathcal{P} \in \mathcal{U}(n; 3^s)$, *if* $\varphi_0 \geqslant \varphi$, *then* $MD^2(\mathcal{P}) \geqslant LB_{MD}^{(3:1)}$, *where*

$$LB_{MD}^{(3:1)} = \left(\frac{19}{12}\right)^s - \frac{2}{n}\left(\frac{5}{3}\right)^s \left[n_\varphi \left(\frac{14}{15}\right)^\varphi + (n - n_\varphi)\left(\frac{14}{15}\right)^{\varphi+1}\right] + \frac{1}{n^2}\left(\frac{15}{8}\right)^s \left[n_\varphi \left(\frac{41}{45}\right)^\varphi + (n - n_\varphi)\left(\frac{41}{45}\right)^{\varphi+1}\right] + \frac{n-1}{n}\left(\frac{15}{8}\right)^s e^{\widehat{\delta}}$$

and

$$\widehat{\delta} = \frac{2s(n-3)}{9(n-1)} ln\left(\frac{41}{45}\right) + \frac{4sn}{9(n-1)} ln\left(\frac{23}{27}\right) + \frac{2sn}{9(n-1)} ln\left(\frac{103}{135}\right).$$

Proof Define $\gamma_i = \left|\{i : x_{ik} \neq \frac{1}{2}\}\right|$, $\gamma_{ij}^{(1)} = \left|\{(i, j) : x_{ik} = x_{jk} = \frac{1}{6} \text{ or } \frac{5}{6}\}\right|$, $\gamma_{ij}^{(2)} = \left|\{(i, j) : (x_{ik}, x_{jk}) \in \{(\frac{1}{6}, \frac{1}{2}), (\frac{1}{2}, \frac{5}{6})\}\}\right|$, $\gamma_{ij}^{(3)} = \left|\{(i, j) : (x_{ik}, x_{jk}) = (\frac{1}{6}, \frac{5}{6})\}\right|$. From (2.3.13), we have

$$\mathrm{MD}^2(\mathcal{P}) = \left(\frac{19}{12}\right)^s - \frac{2}{n}\left(\frac{5}{3}\right)^s \sum_{i=1}^{n} \left(\frac{14}{15}\right)^{\gamma_i} + \frac{1}{n^2}\left(\frac{15}{8}\right)^s \sum_{i=1}^{n} \left(\frac{41}{45}\right)^{\gamma_i} + \frac{1}{n^2}\left(\frac{15}{8}\right)^s \sum_{i=1}^{s} \sum_{j(\neq i)=1}^{s} \left(\frac{41}{45}\right)^{\gamma_{ij}^{(1)}} \left(\frac{23}{27}\right)^{\gamma_{ij}^{(2)}} \left(\frac{103}{135}\right)^{\gamma_{ij}^{(3)}}. \quad (2.6.30)$$

Define $g(\gamma_1, \ldots, \gamma_n) = -\frac{2}{n}\left(\frac{5}{3}\right)^s \sum_{i=1}^n \left(\frac{14}{15}\right)^{\gamma_i} + \frac{1}{n^2}\left(\frac{15}{8}\right)^s \sum_{i=1}^n \left(\frac{41}{45}\right)^{\gamma_i}$. It is easy to check that if $\varphi_0 \geqslant \varphi$, then the lower bound of $g(\gamma_1, \ldots, \gamma_n)$ can be achieved, and

$$g(\gamma_1, \ldots, \gamma_n) = \geqslant -\frac{2}{n}\left(\frac{5}{3}\right)^s \left[n_\varphi \left(\frac{14}{15}\right)^\varphi + (n - n_\varphi)\left(\frac{14}{15}\right)^{\varphi+1}\right] + \frac{1}{n^2}\left(\frac{15}{8}\right)^s \left[n_\varphi \left(\frac{41}{45}\right)^\varphi + (n - n_\varphi)\left(\frac{41}{45}\right)^{\varphi+1}\right]. \quad (2.6.31)$$

Similar to the proof of $LB_{CD}^{(r)}$ in (2.6.3), we can obtain

$$\sum_{i=1}^{s} \sum_{j(\neq i)=1}^{s} \left(\frac{41}{45}\right)^{\gamma_{ij}^{(1)}} \left(\frac{23}{27}\right)^{\gamma_{ij}^{(2)}} \left(\frac{103}{135}\right)^{\gamma_{ij}^{(3)}} \geqslant n(n-1)e^{\widehat{\delta}}. \quad (2.6.32)$$

Combining (2.6.30), (2.6.31), and (2.6.32), the proof is completed.

For any design $\mathcal{P} \in \mathcal{D}(n; 3^s)$, define $f_3(x) = \frac{4}{41}\left(\frac{41}{24}\right)^s \left(\frac{45}{41}\right)^x - \frac{n}{7}\left(\frac{14}{9}\right)^s \left(\frac{15}{14}\right)^x, n = 3t$, $a_1 = ln(\frac{15}{8}) + 2ln(\frac{41}{24})$, $a_2 = ln(\frac{103}{72}) + 2ln(\frac{115}{72})$, $\zeta_1 = st(t-1)$, $\zeta_2 = 2st^2$, ψ is the largest integer contained in $s/3$, $p_3 + q_3 = n$, $p_3\psi + q_3(\psi + 1) = st$, and $\zeta = a_1\zeta_1 + a_2\zeta_2$. Elsawah and Qin (2015) gave the following another lower bound of $\mathrm{MD}^2(\mathcal{P})$ for U-type design $\mathcal{P} \in \mathcal{U}(n; 3^s)$.

Theorem 2.6.16 *For any $\mathcal{P} \in \mathcal{U}(n; 3^s)$, if $f_3(\psi) \geqslant f_3(0)$, then $MD^2(\mathcal{P}) \geqslant LB_{MD}^{(3:2)}$, where*

$$LB_{MD}^{(3:2)} = \left(\frac{19}{12}\right)^s - \frac{2}{n}\left(\frac{14}{9}\right)^s \left(\frac{15}{14}\right)^\psi \left(p_3 + \frac{15}{14}q_3\right) + \frac{1}{n^2}\left(\frac{41}{24}\right)^s \left(\frac{45}{41}\right)^\psi \left(p_3 + \frac{45}{41}q_3\right) + \frac{n-1}{n} e^{\frac{\zeta}{n(n-1)}}.$$

Proof Define $\varrho_{ij} = \left|\{(i,j) : (x_{ik}, x_{jk}) \in \{(\frac{1}{6}, \frac{1}{2}), (\frac{1}{2}, \frac{5}{6}), (\frac{1}{2}, \frac{1}{6}), (\frac{5}{6}, \frac{1}{2})\}\}\right|$, $\tau_{ij} = \left|\{(i,j) : x_{ik} = x_{jk}) \neq \frac{1}{2}\}\right|$, $\sigma_{ij} = \left|\{(i,j) : (x_{ik}, x_{jk}) \in \{(\frac{1}{6}, \frac{5}{6}), (\frac{5}{6}, \frac{1}{6})\}\}\right|$, $\lambda_{ij} = \left|\{(i,j) : x_{ik} = x_{jk}) = \frac{1}{2}\}\right|$. From (2.3.13), we have

$$\mathrm{MD}^2(\mathcal{P}) = \left(\frac{19}{12}\right)^s - \frac{2}{n}\left(\frac{14}{9}\right)^s \sum_{i=1}^n \left(\frac{15}{14}\right)^{\lambda_{ii}} + \frac{1}{n^2}\left(\frac{41}{24}\right)^s \sum_{i=1}^n \left(\frac{45}{41}\right)^{\lambda_{ii}} + \frac{1}{n^2}\sum_{i=1}^{s} \sum_{j(\neq i)=1}^{s} \left(\frac{15}{8}\right)^{\lambda_{ij}} \left(\frac{41}{24}\right)^{\tau_{ij}} \left(\frac{115}{72}\right)^{\varrho_{ij}} \left(\frac{103}{72}\right)^{\sigma_{ij}}. \quad (2.6.33)$$

Following similar arguments in the proof of Theorem 2.6.15, the theorem is proved.

Note that the lower bound in Theorem 2.6.16 is sharper than the lower bound in Theorem 2.6.15. When $s/3$ is an integer, i.e., $h = s/3$, then lower bound $LB_{MD}^{(3:2)}$

becomes $\left(\frac{19}{12}\right)^s - 2\left(\frac{14}{9}\right)^s\left(\frac{15}{14}\right)^h + \frac{1}{n}\left(\frac{41}{24}\right)^s\left(\frac{45}{41}\right)^h + \frac{n-1}{n}e^{\frac{\zeta}{n(n-1)}}$. Figure 2.11 shows that $LB_{MD}^{(3:2)}$ is better than $LB_{MD}^{(3:1)}$ for large s.

Similarly, for any design $\mathcal{P} \in \mathcal{D}(n; 4^s)$, define $f_4(x) = -\frac{144n}{1181}\left(\frac{1181}{768}\right)^s\left(\frac{1253}{1181}\right)^x + \frac{2}{27}\left(\frac{27}{16}\right)^s\left(\frac{29}{27}\right)^x$, $n = 4\theta$, $b_1 = ln(\frac{53}{32}) + ln(\frac{45}{32}) + 2ln(\frac{3}{2}) + 2ln(\frac{51}{32})$, $b_2 = ln(\frac{27}{16}) + 2ln(\frac{29}{16})$, $\eta_1 = 2s\theta^2$, $\eta_2 = 2s\theta(\theta-1)$, ω is the largest integer contained in $s/2$, $p_4 + q_4 = n$, $p_4\omega + q_4(\omega+1) = 2s\theta$, and $\eta = b_1\eta_1 + b_2\eta_2$. Elsawah and Qin (2015) proposed the following lower bound of $\mathrm{MD}^2(\mathcal{P})$ for $\mathcal{P} \in \mathcal{U}(n; 4^s)$.

Theorem 2.6.17 *For any U-type design $\mathcal{P} \in \mathcal{U}(n; 4^s)$, if $f_4(\omega) \geqslant f_4(0)$, then $MD^2(\mathcal{P}) \geqslant LB_{MD}^{(4:1)}$, where*

$$LB_{MD}^{(4:1)} = \left(\frac{19}{12}\right)^s - \frac{2}{n}\left(\frac{1181}{768}\right)^s\left(\frac{1253}{1181}\right)^\omega\left(p_4 + \frac{1253}{1181}q_4\right) + \frac{1}{n^2}\left(\frac{27}{16}\right)^s\left(\frac{29}{27}\right)^\omega\left(p_4 + \frac{29}{27}q_4\right) + \frac{n-1}{n}e^{\frac{\eta}{n(n-1)}}.$$

The proof is similar to that of Theorems 2.6.16 and 2.6.17. For details, ones can refer to Elsawah and Qin (2015). In particular, when $s/2$ is an integer, *i.e.*, $\omega = s/2$, then lower bound $LB_{MD}^{(4:1)}$ becomes $\left(\frac{19}{12}\right)^s - 2\left(\frac{1181}{768}\right)^s\left(\frac{1253}{1181}\right)^\omega + \frac{1}{n}\left(\frac{27}{16}\right)^s\left(\frac{29}{27}\right)^\omega + \frac{n-1}{n}e^{\frac{\eta}{n(n-1)}}$.

For any asymmetrical design $\mathcal{P} \in \mathcal{D}(n; 2^{s_1} \times 3^{s_2})$, let $a_3 = \ln\left(\frac{7}{6}\right)$, $\zeta_1^* = \frac{1}{9}s_2n(n-3)$, $\zeta_2^* = \frac{2}{9}s_2n^2$, $\zeta_3^* = \frac{1}{2}s_1n(n-2)$, μ be the largest integer contained in $s_2/3$, $p_5 + q_5 = n$, $p_5\mu + q_5(\mu+1) = \frac{ns_2}{3}$ and $\zeta^* = a_1\zeta_1^* + a_2\zeta_2^* + a_3\zeta_3^*$, where a_1 and a_2 are defined in Theorem 2.6.16. Define $f_{23}(x) = \frac{4}{41}\left(\frac{7}{4}\right)^{s_1}\left(\frac{41}{24}\right)^{s_2}\left(\frac{45}{41}\right)^x - \frac{n}{7}\left(\frac{305}{192}\right)^{s_1}\left(\frac{14}{9}\right)^{s_2}\left(\frac{15}{14}\right)^x$. Recently, Elsawah and Qin (2016) obtained the following lower bound of $\mathrm{MD}^2(\mathcal{P})$ for $\mathcal{P} \in \mathcal{U}(n; 2^{s_1} \times 3^{s_2})$.

Theorem 2.6.18 *For any U-type design $\mathcal{P} \in \mathcal{U}(n; 2^{s_1} \times 3^{s_2})$, if $f_{23}(\mu) \geqslant f_{23}(0)$, then $MD^2(\mathcal{P}) \geqslant LB_{MD}^{(23:1)}$, where*

$$LB_{MD}^{(23:1)} = \left(\frac{19}{12}\right)^{s_1+s_2} - \frac{2}{n}\left(\frac{305}{192}\right)^{s_1}\left(\frac{14}{9}\right)^{s_2}\left(\frac{15}{14}\right)^\mu\left(p_5 + \frac{15}{14}q_5\right) + \frac{1}{n^2}\left(\frac{7}{4}\right)^{s_1}\left(\frac{41}{24}\right)^{s_2}\left(\frac{45}{41}\right)^\mu\left(p_5 + \frac{45}{41}q_5\right) + \frac{n-1}{n}\left(\frac{3}{2}\right)^{s_1} e^{\frac{\zeta^*}{n(n-1)}}.$$

The proof is similar to that of Theorem 2.6.16. For details, ones can refer to Elsawah and Qin (2016). When $s_2/3$ is an integer, i.e., $\mu = s_2/3$, then lower bound $LB_{MD}^{(23:1)}$ becomes $\left(\frac{19}{12}\right)^{s_1+s_2} - 2\left(\frac{305}{192}\right)^{s_1}\left(\frac{14}{9}\right)^{s_2}\left(\frac{15}{14}\right)^\mu + \frac{1}{n}\left(\frac{7}{4}\right)^{s_1}\left(\frac{41}{24}\right)^{s_2}\left(\frac{45}{41}\right)^\mu + \frac{n-1}{n}\left(\frac{3}{2}\right)^{s_1} e^{\frac{\zeta^*}{n(n-1)}}$. When $s_1 = 0$ or $s_2 = 0$, the lower bound in Theorem 2.6.18 is equivalent to $LB_{MD}^{(3:2)}$ in Theorem 2.6.16 or $LB_{MD}^{(2:3)}$ in Theorem 2.6.14.

2.6.4 Lower Bounds of Discrete Discrepancy

For discrete discrepancy, Fang et al. (2003a) firstly gave the following lower bound of DD($\mathcal{P}$) for U-type design $\mathcal{P} \in \mathcal{U}(n; q_1 \times \cdots \times q_s)$.

Theorem 2.6.19 *Let* $\mathcal{P} \in \mathcal{U}(n; q_1 \times \cdots \times q_s)$, *then*

$$DD^2(\mathcal{P}) \geqslant LB_{DD}^{(1)},$$

where

$$LB_{DD}^{(1)} = -\prod_{j=1}^{s}\left[\frac{a+(q_j-1)b}{q_j}\right] + \frac{a^s}{n} + \frac{n-1}{n}b^s\left(\frac{a}{b}\right)^{\lambda},$$

and the lower bound $LB_{DD}^{(1)}$ *can be achieved if and only if* $\lambda = \sum_{j=1}^{s}(n/q_j - 1)/(n-1)$ *is a positive integer and* $\delta_{ik}(\mathcal{P}) = \lambda$, *for all* $i \neq k$.

Proof It is easy to check that for $\mathcal{P} \in \mathcal{D}(n; q_1 \times \cdots \times q_s)$, $\sum_{i=1}^{n}\sum_{j=1}^{n}\delta_{ij}(\mathcal{P}) = \sum_{i=1}^{s}\frac{n^2}{q_i}$, i.e., $\sum_{1 \leqslant i<j \leqslant n}\delta_{ij}(\mathcal{P}) = \sum_{i=1}^{s}\frac{n^2}{q_i} - ns$ is a constant. Let $\mathcal{P}^* \in \mathcal{D}(n; q_1 \times \cdots \times q_s)$, and $\delta_{ij}(\mathcal{P}^*) = \lambda$ for $i \neq j$. In view of theory of majorization, we know that $(\delta_{12}(\mathcal{P}), \ldots, \delta_{1n}(\mathcal{P}), \delta_{23}(\mathcal{P}), \ldots, \delta_{2n}(\mathcal{P}), \ldots, \delta_{(n-1)n}(\mathcal{P}))$ is majorized by $(\delta_{12}(\mathcal{P}^*), \ldots, \delta_{1n}(\mathcal{P}^*), \delta_{23}(\mathcal{P}^*), \ldots, \delta_{2n}(\mathcal{P}^*), \ldots, \delta_{(n-1)n}(\mathcal{P}^*))$. Noting that DD$^2(\mathcal{P})$ in (2.5.6) is a Schur-concave function of the vector $(\delta_{12}(\mathcal{P}), \ldots, \delta_{1n}(\mathcal{P}), \delta_{23}(\mathcal{P}), \ldots, \delta_{2n}(\mathcal{P}), \ldots, \delta_{(n-1)n}(\mathcal{P}))$, we have DD$^2(\mathcal{P}) \geqslant [DD(\mathcal{P}^*)]^2$, i.e., DD$^2(\mathcal{P}) \geqslant LB_{DD}^{(1)}$.

When $\lambda = \sum_{j=1}^{s}(n/q_j - 1)/(n-1)$ is not a positive integer, Qin and Fang (2004) obtained the following lower bound of $LD^2(\mathcal{P})$ and gave a necessary and sufficient condition for a design $\mathcal{P} \in \mathcal{U}(n; q_1 \times \cdots \times q_s)$ reaching to this lower bound.

Theorem 2.6.20 *Let* $\mathcal{P} \in \mathcal{U}(n; q_1 \times \cdots \times q_s)$, *then*

$$DD^2(\mathcal{P}) \geqslant LB_{DD}^{(2)},$$

where

$$LB_{DD}^{(2)} = -\prod_{j=1}^{s}\left[\frac{a+(q_j-1)b}{q_j}\right] + \frac{a^s}{n} + \frac{(n-1)[b(\gamma+1-\lambda)+a(\lambda-\gamma)b^s}{nb}\left(\frac{a}{b}\right)^{\gamma}$$

and γ *is the integer part of* λ. *The lower bound* $LB_{DD}^{(2)}$ *can be achieved if and only if for any* i*th run of* $\mathcal{P}$, *among the* $(n-1)$ *values of* $\delta_{ik}(\mathcal{P})$, *for* $i \neq k$, *there are* $(n-1)(\gamma+1-\lambda)$ *with the value* γ *and* $(n-1)(\lambda-\gamma)$ *with the value* $\gamma+1$.

Proof Let $\mathcal{P}^* \in \mathcal{D}(n; q_1 \times \cdots \times q_s)$, and among the $n(n-1)$ values of $\delta_{ij}(\mathcal{P}^*)$ for $i \neq j$, there are $n(n-1)(\gamma+1-\lambda)$ with the value γ and $n(n-1)(\lambda-\gamma)$ with the value $\gamma + 1$. It is easy to see that $(\delta_{12}(\mathcal{P}), \ldots, \delta_{1n}(\mathcal{P}), \delta_{23}(\mathcal{P}), \ldots, \delta_{2n}(\mathcal{P}), \ldots, \delta_{(n-1)n}(\mathcal{P}))$ is majorized by $(\delta_{12}(\mathcal{P}^*), \ldots, \delta_{1n}(\mathcal{P}^*), \delta_{23}(\mathcal{P}^*), \ldots, \delta_{2n}(\mathcal{P}^*), \ldots, \delta_{(n-1)n}(\mathcal{P}^*))$. Following the proof of Theorem 2.6.19, the proof is completed.

Note that the lower bound $LB_{DD}^{(2)}$ or $LB_{DD}^{(1)}$ is based on the Hamming distance for rows of $\mathcal{P}$. Hence, $LB_{DD}^{(2)}$ or $LB_{DD}^{(1)}$ is more useful for assessing nearly saturated orthogonal arrays or supersaturated designs.

When $q_1 = \cdots = q_s = q$, from the above theorem, we have the following result.

Theorem 2.6.21 *Let $\mathcal{P} \in \mathcal{U}(n; q^s)$, then*

$$DD^2(\mathcal{P}) \geqslant LB_{DD}^{(3)},$$

where

$$LB_{DD}^{(3)} = -\left[\frac{a + (q-1)b}{q}\right]^s + \frac{a^s}{n} + \frac{(n-1)[b(\gamma + 1 - \lambda) + a(\lambda - \gamma)b^s}{nb}\left(\frac{a}{b}\right)^\gamma,$$

$\lambda = s(n-q)/[q(n-1)]$ and γ is the integer part of λ.

Based on the column balance of $\mathcal{P}$, Qin and Li (2006) gave the following another lower bound of $DD(\mathcal{P})$.

Theorem 2.6.22 *Let $\mathcal{P} \in \mathcal{D}(n; q^s)$, then*

$$DD^2(\mathcal{P}) \geqslant LB_{DD}^{(4)},$$

where

$$LB_{DD}^{(4)} = \frac{b^s}{n^2}\sum_{i=1}^{s}\binom{s}{i}\left(\frac{a-b}{b}\right)^i R_{n,i,q}\left(1 - \frac{R_{n,i,q}}{q^i}\right),$$

$R_{n,i,q}$ is the residual of $n(\mathrm{mod}\ q^i)$.

Proof Note that

$$\mathrm{DD}^2(\mathcal{P}) = \frac{b^s}{n^2}\sum_{i=1}^{s}\sum_{j=1}^{s}\binom{s}{i}\left(\frac{a-b}{b}\right)^i B_i(\mathcal{P}).$$

Following the proof of Theorem 2.6.3, the proof is completed.

It is clear that $LB_{DD}^{(4)}$ is sharper than $LB_{DD}^{(3)}$, which is useful for assessing the uniformity of an orthogonal array.

For $1 \leqslant j \leqslant s$, let ω_j be the largest integer contained in n/q^j, $\theta_j = n - q^j\omega_j$ and $\theta_j^* = n\omega_j + \theta_j(1 + \omega_j)$. Chatterjee and Qin (2008) gave the following improved lower bound of $DD(\mathcal{P})$ for $\mathcal{P} \in \mathcal{D}(n; q^s)$.

Theorem 2.6.23 *Let $\mathcal{P} \in \mathcal{D}(n; q^s)$, then*

$$DD^2(\mathcal{P}) \geqslant LB_{DD}^{(5)},$$

where

$$LB_{DD}^{(5)} = -\left[\frac{a+(q-1)b}{q}\right]^s + b^s + \frac{b^s}{n^2}\sum_{i=1}^{s}\binom{s}{i}\left(\frac{a-b}{b}\right)^i \theta_j^*.$$

Proof It is easy to check that

$$\mathrm{DD}^2(\mathcal{P}) = -\left[\frac{a+(q-1)b}{q}\right]^s + \frac{1}{n^2}\boldsymbol{y}_{\mathcal{P}}^T \boldsymbol{M}_s \boldsymbol{y}_{\mathcal{P}},$$

where $\boldsymbol{M}_s$ is the s-fold Kronecker products of $(a-b)\boldsymbol{I}_s + b\boldsymbol{J}_s$. Following the proof of Theorem 2.6.2, the proof is completed.

Following Qin and Fang (2004), for any design $\mathcal{P} \in \mathcal{D}(n; q^s)$, we know that the following equations hold:
(I) When $q = 2$, $a = 5/4$, and $b = 1$,

$$\mathrm{DD}^2(\mathcal{P}) = \mathrm{CD}^2(\mathcal{P}) + 2\left(\frac{35}{32}\right)^s - \left(\frac{13}{12}\right)^s - \left(\frac{9}{8}\right)^s;$$

(II) When $q = 2$, $a = 3/2$, and $b = 5/4$,

$$\mathrm{DD}^2(\mathcal{P}) = \mathrm{WD}^2(\mathcal{P}) + \left(\frac{4}{3}\right)^s - \left(\frac{11}{8}\right)^s;$$

(III) When $q = 3$, $a = 3/2$, and $b = 23/18$,

$$\mathrm{DD}^2(\mathcal{P}) = \mathrm{WD}^2(\mathcal{P}) + \left(\frac{4}{3}\right)^s - \left(\frac{73}{54}\right)^s.$$

It is easy to check that when $q = 2$, if $a = 5/4$ and $b = 1$, then the lower bound $LB_{DD}^{(3)}$ in Theorem 2.6.21, the lower bound $LB_{DD}^{(4)}$ in Theorem 2.6.22, and the lower bound $LB_{DD}^{(5)}$ in Theorem 2.6.23 are, respectively, the lower bound $LB_{CD}^{(r)}$ in Theorem 2.6.3, the lower bound $LB_{CD}^{(c)}$ in Theorem 2.6.3, and the lower bound $LB_{CD}^{(2)}$ in Theorem 2.6.2; if $a = 3/2$ and $b = 5/4$, then the lower bound $LB_{DD}^{(3)}$ in Theorem 2.6.21, the lower bound $LB_{DD}^{(4)}$ in Theorem 2.6.22, and the lower bound $LB_{DD}^{(5)}$ in Theorem 2.6.23 are, respectively, the lower bound $LB_{WD}^{(2r)}$ in Theorem 2.6.6, the lower bound $LB_{WD}^{(2c)}$ in Theorem 2.6.6, and the lower bound $LB_{WD}^{(21)}$ in Theorem 2.6.5, and when $q = 3$, if $a = 3/2$ and $b = 23/18$, then the lower bound $LB_{DD}^{(3)}$ in Theorem 2.6.21, the lower bound $LB_{DD}^{(4)}$ in Theorem 2.6.22, and the lower bound

$LB^{(5)}_{DD}$ in Theorem 2.6.23 are equivalent to the lower bound $LB^{(3r)}_{WD}$ in Theorem 2.6.8, the lower bound $LB^{(3c)}_{WD}$ in Theorem 2.6.8, and the lower bound $LB^{(31)}_{WD}$ in Theorem 2.6.7, respectively.

2.6.5 Lower Bounds of Lee Discrepancy

For q-level U-type designs, where q is a positive integer, Zhou et al. (2008) firstly proposed a lower bound of LD as the following theorem.

Theorem 2.6.24 *Let $\mathcal{P} \in \mathcal{U}(n; q^s)$, then when q is even, $LD^2(\mathcal{P}) \geqslant LB^{(even)}_{LD}$, where*

$$LB^{(even)}_{LD} = \frac{1}{n} - \left(\frac{3}{4}\right)^s + \frac{n-1}{n}\left(\frac{1}{2}\right)^{\frac{ns}{q(n-1)}}\left[\prod_{i=1}^{q/2-1}\frac{q+2i}{2q}\right]^{\frac{2sn}{q(n-1)}};$$

when q is odd, $LD^2(\mathcal{P}) \geqslant LB^{(odd)}_{LD}$, where

$$LB^{(odd)}_{LD} = \frac{1}{n} - \left(\frac{3}{4} + \frac{1}{4q^2}\right)^s + \frac{n-1}{n}\left[\prod_{i=1}^{(q-1)/2}\frac{q+2i-1}{2q}\right]^{\frac{2sn}{q(n-1)}}.$$

The proof of Theorem 2.6.24 is similar to Theorem 2.6.9.

When $q = 2, 3$, the lower bounds in Theorem 2.6.24 are given in the following theorem.

Theorem 2.6.25 *Let $\mathcal{P} \in \mathcal{U}(n; q^s)$, then when $q = 2$, $LD^2(\mathcal{P}) \geqslant LB^{(21)}_{LD}$, where*

$$LB^{(21)}_{LD} = \frac{1}{n} - \left(\frac{3}{4}\right)^s + \frac{n-1}{n}\left(\frac{1}{2}\right)^{\frac{ns}{2(n-1)}};$$

when $q = 3$, $LD^2(\mathcal{P}) \geqslant LB^{(31)}_{LD}$, where

$$LB^{(31)}_{LD} = \frac{1}{n} - \left(\frac{7}{9}\right)^s + \frac{n-1}{n}\left(\frac{2}{3}\right)^{\frac{2sn}{3(n-1)}}.$$

When $q = 2, 3$, Zou et al. (2009) gave the following improved lower bounds, which are sharper than the lower bounds in Theorem 2.6.25.

Theorem 2.6.26 *Let $\mathcal{P} \in \mathcal{D}(n; q^s)$ be U-type, then when $q = 2$, $LD^2(\mathcal{P}) \geqslant LB^{(22)}_{LD}$, where*

$$LB^{(22)}_{LD} = -\left(\frac{3}{4}\right)^s + \frac{1}{2^s n^2}\sum_{r=0}^{s}\binom{s}{r}[nf_r + z_r(f_r+1)],$$

f_r and z_r are defined in Theorem 2.6.2; when $q = 3$, $LD^2(\mathcal{P}) \geqslant LB_{LD}^{(32)}$, where

$$LB_{LD}^{(32)} = -\left(\frac{7}{9}\right)^s + \frac{1}{3^s n^2}\sum_{r=0}^{s}\binom{s}{r}2^{s-r}[ng_r + t_r(g_r+1)],$$

g_r and t_r are defined in Theorem 2.6.7.

Proof Note that for $q = 2$,

$$\mathrm{LD}^2(\mathcal{P}) = -\left(\frac{3}{4}\right)^s + \frac{1}{n^2}\boldsymbol{y}_{\mathcal{P}}^T L_s^{(2)} \boldsymbol{y}_{\mathcal{P}}$$

and for $q = 3$,

$$\mathrm{LD}^2(\mathcal{P}) = -\left(\frac{7}{9}\right)^s + \frac{1}{n^2}\boldsymbol{y}_{\mathcal{P}}^T L_s^{(3)} \boldsymbol{y}_{\mathcal{P}},$$

where $L_s^{(2)}$ and $L_s^{(3)}$ are the s-fold Kronecker products of $\frac{1}{2}(\boldsymbol{I}_2 + \boldsymbol{J}_2)$ and $\frac{1}{3}\boldsymbol{I}_3 + \frac{2}{3}\boldsymbol{J}_3$, respectively. Being along the lines of the proof of Theorem 2.6.3, the proof is completed.

From Fig. 2.13, we know that $LB_{LD}^{(21)}$, $LB_{LD}^{(31)}$ is better than $LB_{LD}^{(22)}$, $LB_{LD}^{(32)}$, respectively, for small s (Fig. 2.14).

For general mixed-level U-type designs, Zhou et al. (2008) obtained some lower bounds of LD for $\mathcal{P} \in \mathcal{U}(n; q_1^{s_1} \times \cdots \times q_m^{s_m})$, where $q_1, \ldots, q_t$ are odd and $q_{t+1}, \ldots, q_m$ are even, where $0 \leqslant t \leqslant s$.

Theorem 2.6.27 *Let $\mathcal{P} \in \mathcal{U}(n; q_1^{s_1} \times \cdots \times q_m^{s_m})$, then*

$$LD^2(\mathcal{P}) \geqslant LB_{LD}^{(3)},$$

Fig. 2.13 Comparison between $LB_{MD}^{(3:1)}$ and $LB_{MD}^{(3:2)}$

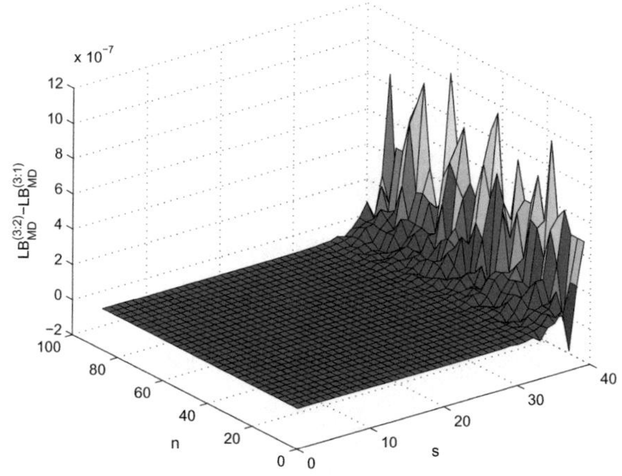

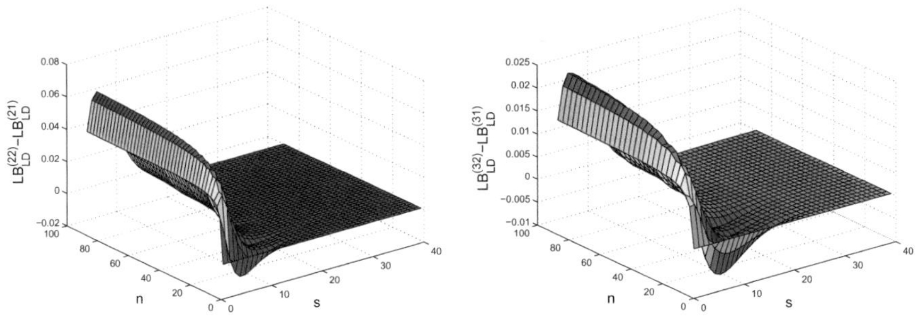

Fig. 2.14 Comparison between $LB_{LD}^{(21)}$ and $LB_{LD}^{(22)}$, between $LB_{LD}^{(31)}$ and $LB_{LD}^{(32)}$

where

$$LB_{LD}^{(3)} = \Delta_5 + \frac{n-1}{n}\left(\frac{1}{2}\right)^{\sum_{r=t+1}^{m}\frac{ns_r}{q_r(n-1)}} \prod_{r=1}^{t}\left[\prod_{j=1}^{(q_r-1)/2}\frac{q_r+2j-1}{2q_r}\right]^{\frac{2s_r n}{q_r(n-1)}} \times \prod_{r=t+1}^{m}\left[\prod_{j=1}^{q_r/2-1}\frac{q_r+2j}{2q_r}\right]^{\frac{2s_r n}{q_r(n-1)}},$$

and $\Delta_5 = \frac{1}{n} - \left(\frac{4}{3}\right)^{\sum_{j=t+1}^{m} s_j}\prod_{j=1}^{t}\left(\frac{3}{4}+\frac{1}{4q_j^2}\right)^{s_j}$.

The proof of Theorem 2.6.27 is similar to Theorem 2.6.9.

When $q_1 = \cdots = q_m = q$, the lower bound in Theorem 2.6.27 is equivalent to the lower bound in Theorem 2.6.24. When $q_1 = \cdots = q_t = 3$ and $q_{t+1} = \cdots = q_m = 2$, then we can get the following theorem.

Theorem 2.6.28 *Let $\mathcal{P} \in \mathcal{U}(n; 2^{s_1} \times 3^{s_2})$, then*

$$LD^2(\mathcal{P}) \geqslant \frac{1}{n} - \left(\frac{3}{4}\right)^{s_1}\left(\frac{7}{9}\right)^{s_2} + \frac{n-1}{n}\left(\frac{1}{2}\right)^{\frac{ns_1}{2(n-1)}}\left(\frac{2}{3}\right)^{\frac{2ns_2}{3(n-1)}}.$$

For asymmetrical factorials with two and three levels, Chatterjee et al. (2012b) also obtain the following lower bound, which is more useful and sharper than the lower bound in Theorem 2.6.28.

Theorem 2.6.29 *Let $\mathcal{P} \in \mathcal{D}(n; 2^{s_1} \times 3^{s_2})$, then*

$$LD^2(\mathcal{P}) \geqslant -\left(\frac{3}{4}\right)^{s_1}\left(\frac{7}{9}\right)^{s_2} + \frac{1}{n^2}\left(\frac{1}{2}\right)^{s_1}\left(\frac{2}{3}\right)^{s_2}\sum_{i=0}^{s_1}\sum_{j=0}^{s_2}\binom{s_1}{i}\binom{s_2}{j}\left(\frac{1}{2}\right)^{j}\theta_{ij},$$

where θ_{ij} is defined in Theorem 2.6.10.

The proof of Theorem 2.6.29 is similar to Theorem 2.6.2.

Exercises

2.1

For a one-factor experiment ($s = 1$), prove that the n-run design with minimum $\mathcal{L}_\infty$-star discrepancy is the following set of evenly spaced points:

$$\mathcal{P} = \left\{ \frac{1}{2n}, \frac{3}{2n}, \ldots, \frac{2n-1}{2n} \right\}.$$

2.2

Prove that a minimum star discrepancy one-run design takes the form $\mathcal{P} = \{(z, \ldots, z)\}$ for z satisfying $z^s + z - 1 = 0$.

2.3

Let $\mathcal{P} = \{\boldsymbol{x}_k = (x_{k1}, \ldots, x_{ks}), k = 1, \ldots, n\}$ be a set of n points on the unit cube $C^s = [0, 1]^s$. The wrap-around L_2-discrepancy can be calculated by

$$(WD(\mathcal{P}))^2 = -\left(\frac{4}{3}\right)^s + \frac{1}{n^2}\left(\frac{3}{2}\right)^s \sum_{k=1}^{n}\sum_{l=1}^{n}\left(\frac{5}{6}\right)^{d_H(k,l)},$$

where $d_H(k, l)$ is the Hamming distance between $\boldsymbol{x}_k$ and $\boldsymbol{x}_l$.

Prove that $WD_2(D_1) = WD_2(D_2)$ if D_1 and D_2 are equivalent. Two U-type designs are called equivalent if one can be obtained from the other by (i) exchanging rows or/and (ii) exchanging columns.

2.4

Compare the uniformity of the following two designs under WD, CD, and MD:

$$\boldsymbol{X}_1 = \begin{bmatrix} 3/4 & 3/4 & 3/4 \\ 3/4 & 3/4 & 3/4 \\ 3/4 & 1/4 & 3/4 \\ 3/4 & 3/4 & 1/4 \\ 1/4 & 3/4 & 1/4 \end{bmatrix}, \quad \boldsymbol{X}_2 = \begin{bmatrix} 1/4 & 1/4 & 1/4 \\ 1/4 & 3/4 & 3/4 \\ 3/4 & 1/4 & 1/4 \\ 3/4 & 3/4 & 1/4 \\ 1/4 & 3/4 & 1/4 \end{bmatrix}.$$

2.5

Write a MATLAB code for calculating $WD(\mathcal{P})$, $CD(\mathcal{P})$, $MD(\mathcal{P})$, where $\mathcal{P}$ is a design on the domain $[0, 1]^s$ with n runs and s-factors. Apply your own code to the two designs in Example 2.3.1.

2.6

Prove that the CD, WD, and MD defined in Sect. 2.3 can be expressed as function of their Hamming distances.

2.7

Consider the following four designs.

$\mathcal{P}_{5-1}$		$\mathcal{P}_{5-2}$		$\mathcal{P}_{5-3}$		$\mathcal{P}_{5-4}$	
1	2	1	2	1	5	1	1
2	4	2	5	2	1	2	5
3	1	3	3	3	4	3	4
4	3	4	1	4	2	4	3
5	5	5	4	5	3	5	2

Calculate the corresponding star discrepancy, CD, WD, MD, and LD for each design, and show your conclusion.

2.8

The inequality between arithmetic mean and geometric mean has been known and is useful for deriving some lower bounds. Prove this inequality stated as follows:

Let $a_1, \ldots, a_m$ be m nonnegative numbers, and then

$$\bar{a} \equiv \frac{1}{m}\sum_{i=1}^{m} a_i \geqslant \left[\prod_{i=1}^{m} a_j\right]^{1/m} \equiv \bar{a}_g,$$

where $\bar{a}$ is the arithmetic mean and $\bar{a}_g$ is the geometric mean of $a_1, \ldots, a_m$. The above equality holds if and only if all the a_i's are the same.

2.9

For easily understanding the so-called curse of dimensionality, let us study on the volume of the ball in R^n. A ball of radius r with the center at the origin in R^n can be expressed as

$$B_n(r) = \{\boldsymbol{x} : \boldsymbol{x}^T\boldsymbol{x} = x_1^2 + \cdots + x_n^2 \leqslant r^2\}.$$

It is known that the volume of $B_3(r)$ is $\frac{4}{3}\pi r^3$. In general,

$$\text{Vol}(B_n(r)) = \frac{\pi^{n/2}}{\Gamma(\frac{n}{2}+1)} r^n.$$

Intuitively, one may think that the volume of $B_n(r)$ becomes larger and larger. For the unit ball, its volume increases in the first five dimensions, but decreases as n tends to infinity. Study the behavior of the volume of $B_n(r)$ as n increases.

2.10

What is the Cauchy–Schwarz inequality in the linear inner product space? Give a proof. From the literature, find more applications of the Cauchy–Schwarz inequality.

2.11

Prove Theorems 2.6.27 and 2.6.29.

References

Aronszajn, N.: Theory of reproducing kernels. Trans. Amer. Math. Soc. **68**, 337–404 (1950)

Bundschuh, P., Zhu, Y.C.: A method for exact calculation of the discrepancy of low-dimensional finite point sets (I). In: Abhandlungen aus Math. Seminar, Univ.Hamburg, Bd. 63 (1993)

Chatterjee, K., Qin, H.: A new look on discrete discrepancy. Statist. Probab. Lett. **78**, 2988–2991 (2008)

Chatterjee, K., Fang, K.T., Qin, H.: Uniformity in factorial designs with mixed levels. J. Statist. Plann. Inference **128**, 593–607 (2005)

Chatterjee, K., Li, Z., Qin, H.: A new lower bound to centered and wrap-round l_2-discrepancies. Statist. Probab. Lett. **82**, 1367–1373 (2012a)

Chatterjee, K., Qin, H., Zou, N.: Lee discrepancy on two and three mixed level factorials. Sci. China Ser. A **55**, 663–670 (2012b)

Clerk, L.d.: A method for exact calculation of the star-discrepancy of plane sets applied to the sequences of hammersley. Mh. Math. **101**, 261–278 (1986)

D'Agostino, R.B., Stephens, M.A.: Goodness-of-Fit Techniques. Marcel Dekker, New York (1986)

Elsawah, A.M., Qin, H.: New lower bound for centered L_2-discrepancy of four-level U-type designs. Statist. Probab. Lett. **93**, 65–71 (2014)

Elsawah, A.M., Qin, H.: Mixture discrepancy on symmetric balanced designs. Statist. Probab. Lett. **104**, 123–132 (2015)

Elsawah, A.M., Qin, H.: Asymmetric uniform designs based on mixture discrepancy. J. App. Statist. **43**(12), 2280–2294 (2016)

Fang, K.T.: The uniform design: application of number-theoretic methods in experimental design. Acta Math. Appl. Sinica **3**, 363–372 (1980)

Fang, K.T., Ma, C.X.: Wrap-around L_2-discrepancy of random sampling, Latin hypercube and uniform designs. J. Complexity **17**, 608–624 (2001b)

Fang, K.T., Mukerjee, R.: A connection between uniformity and aberration in regular fractions of two-level factorials. Biometrika **87**, 193–198 (2000)

Fang, K.T., Wang, Y.: A note on uniform distribution and experiment design. Chinese Sci. Bull. **26**, 485–489 (1981)

Fang, K.T., Ma, C.X., Mukerjee, R.: Uniformity in fractional factorials. In: Fang, K.T., Hickernell, F.J., Niederreiter, H. (eds.) Monte Carlo and Quasi-Monte Carlo Methods, pp. 232–241. Springer (2002)

Fang, K.T., Ge, G.N., Liu, M.Q.: Uniform supersaturated design and its construction. Sci. China Ser. A **45**, 1080–1088 (2002)

Fang, K.T., Lin, D.K.J., Liu, M.Q.: Optimal mixed-level supersaturated design. Metrika **58**, 279–291 (2003a)

Fang, K.T., Lu, X., Winker, P.: Lower bounds for centered and wrap-around L_2-discrepancies and construction of uniform. J. Complexity **20**, 268–272 (2003b)

Fang, K.T., Tang, Y., Yin, J.X.: Lower bounds for wrap-around L_2-discrepancy and constructions of symmetrical uniform designs. J. Complexity **21**, 757–771 (2005)

Fang, K.T., Li, R., Sudjianto, A.: Design and Modeling for Computer Experiments. Chapman and Hall/CRC, New York (2006a)

Fang, K.T., Maringer, D., Tang, Y., Winker, P.: Lower bounds and stochastic optimization algorithms for uniform designs with three or four levels. Math. Comp. **75**, 859–878 (2006b)

Hickernell, F.J.: A generalized discrepancy and quadrature error bound. Math. Comp. **67**, 299–322 (1998a)

Hickernell, F.J.: Lattice rules: How well do they measure up? In: Hellekalek, P., Larcher, G. (eds.) Random and Quasi-Random Point Sets, pp. 106–166. Springer (1998b)

Hickernell, F.J., Liu, M.Q.: Uniform designs limit aliasing. Biometrika **89**, 893–904 (2002)

Hua, L.K., Wang, Y.: Applications of Number Theory to Numerical Analysis. Springer and Science Press, Berlin and Beijing (1981)

Ke, X., Zhang, R., Ye, H.J.: Two- and three-level lower bounds for mixture L_2-discrepancy and construction of uniform designs by threshold accepting. J. Complexity **31**, 741–753 (2015)
Koehler, J.R., Owen, A.B.: Computer experiments. In: Ghosh, S., R.Rao, C. (eds.) Handbook of Statistics, vol. 13, Elsevier Science B. V., Amsterdam, pp. 261–308 (1996)
Liu, M.Q., Hickernell, F.J.: $E(s^2)$-optimality and minimum discrepancy in 2-level supersaturated designs, Statist. Sinica **12**, 931–939 (2002)
McKay, M., Beckman, R., Conover, W.: A comparison of three methods for selecting values of input variables in the analysis of output from a computer code. Technometrics **21**, 239–245 (1979)
Niederreiter, H.: Application of diophantine approximations to numerical integration. In: Osgood, C.F. (ed.) Diophantine Approximation and Its Applications, pp. 129–199. Academic Press, New York (1973)
Niederreiter, H.: Random number generation and Quasi-Monte Carlo methods. In: SIAM CBMS-NSF Regional Conference. Applied Mathematics, Philadelphia (1992)
Owen, A.B.: A central limit theorem for Latin hypercube sampling. J. R. Statist. Soc. B **54**, 541–551 (1992)
Qin, H., Fang, K.T.: Discrete discrepancy in factorial designs. Metrika **60**, 59–72H (2004)
Qin, H., Li, D.: Connection between uniformity and orthogonality for symmetrical factorial designs. J. Stat. Plann. Inference **136**, 2770–2782 (2006)
Roth, R.M.: Introduction to Coding Theory. Cambridge University Press, Cambridge (2006)
Stein, M.L.: Large sample properties of simulations using latin hypercube sampling. Technometrics **29**, 143–151 (1987)
Tang, B.: Orthogonal array-based Latin hypercubes. J. Amer. Statat. Assoc. **88**, 1392–1397 (1993)
Warnock, T.T.: Computational investigations of low discrepancy point sets. In: Zaremba, S.K. (ed.) Applications of Number Theory to Numerical Analysis, pp. 319–343. Academic Press, New York (1972)
Weyl, H.: über die gleichverteilung der zahlem mod eins. Math. Ann. **77**, 313–352 (1916)
Winker, P., Fang, K.T.: Application of threshold accepting to the evaluation of the discrepancy of a set of points. SIAM Numer. Anal. **34**, 2038–2042 (1997)
Zhang, Q., Wang, Z., Hu, J., Qin, H.: A new lower bound for wrap-around L_2-discrepancy on two and three mixed level factorials. Statist. Probab. Lett. **96**, 133–140 (2015)
Zhou, Y.D., Ning, J.H.: Lower bounds of the wrap-around L_2-discrepancy and relationships between mlhd and uniform design with a large size. J. Stat. Plann. Inference **138**, 2330–2339 (2008)
Zhou, Y.D., Ning, J.H., Song, X.B.: Lee discrepancy and its applications in experimental designs. Statist. Probab. Lett. **78**, 1933–1942 (2008)
Zhou, Y.D., Fang, K.T., Ning, J.H.: Mixture discrepancy for quasi-random point sets. J. Complexity **29**, 283–301 (2013)
Zou, N., Ren, P., Qin, H.: A note on Lee discrepancy. Statist. Probab. Lett. **79**, 496–500 (2009)

Chapter 3
Construction of Uniform Designs—Deterministic Methods

In this chapter and the following chapters, the different construction methods for uniform design tables are given. Typically, there are three approaches of constructing uniform design tables:

(i) Quasi-Monte Carlo methods;
(ii) Combinatorial methods;
(iii) Numerical search.

The first two approaches are discussed in this chapter, and the last one will be presented in the next chapter. Uniform design tables have been collected on the Web site http://www.math.hkbu.edu.hk/UniformDesign/ and http://web.stat.nankai.edu.cn/cms-ud/.

The first section of this chapter provides a rather general explanation about many aspects of construction of uniform designs and gives a brief review on the construction of uniform designs for one-factor and multifactor experiments. It will be shown that the uniform design for a one-factor experiment can be obtained in closed form. For multifactor experiment, Sect. 3.2.1 shows that there is an essential complexity in constructing uniform designs. Quasi-Monte Carlo methods provide some useful construction methods for multifactor uniform designs. Among the quasi-Monte Carlo methods, the good lattice point method and its extensions are introduced in Sects. 3.3 and 3.4. Section 3.5 gives the linear permutation method, and Sect. 3.6 gives the combinatorial construction methods for uniform designs.

K.-T. Fang et al., *Theory and Application of Uniform Experimental Designs*, Lecture Notes in Statistics 221,
https://doi.org/10.1007/978-981-13-2041-5_3

3.1 Uniform Design Tables

For practical use, one needs uniform designs for various sizes. There are two ways to find the required uniform designs: (1) from the database where many uniform designs have obtained by the literature; (2) from a directly computational search with computer software help. In this section, some important issues for constructing uniform design tables are introduced. Some theoretical results for one-factor experiments are also given.

3.1.1 Background of Uniform Design Tables

For constructing uniform designs, it should determine several important issues such as experimental parameters, experimental domain, uniformity criterion, and candidate designs.

Experimental parameters. The parameters include the number of runs n, the number of factors s.

Experimental domain. There are several types of *experimental domain*, $\mathcal{X}$:

(a) A hypercube: $\mathcal{X} = [a_1, b_1] \times [a_2, b_2] \times \cdots \times [a_s, b_s]$ (see Cartesian product in (1.2.3)), and there is no constrain among the factors. Without loss of generality, one can use a linear transformation mapping the hypercube into the unit hypercube $[0, 1]^s$;
(b) A simplex: In *experiments with mixtures*, the experimental domain $\mathcal{X}$ is

$$T^s = \{(x_1, \ldots, x_s) : x_j \geqslant 0,\ j = 1, \ldots, s,\ x_1 + \cdots + x_s = 1\}, \quad (3.1.1)$$

or

$$T^s(\boldsymbol{a}, \boldsymbol{b}) = \left\{(x_1, \ldots, x_s) : 0 < a_i \leqslant x_i \leqslant b_i < 1,\ i = 1, \ldots, s,\ \sum_{i=1}^{s} x_i = 1\right\},$$

where $\boldsymbol{a} = (a_1, \ldots, a_s)$ and $\boldsymbol{b} = (b_1, \ldots, b_s)$.
(c) A set of lattice points: If the factors are categorical and whose levels are $q_1, \ldots, q_s$ for the s-factors, then the experimental domain becomes

$$\mathcal{X} = \{1, \ldots, q_1\} \times \cdots \times \{1, \ldots, q_s\}, \quad (3.1.2a)$$

or the domain with q levels per factor,

$$\mathcal{X} = \{1, \ldots, q\}^s. \quad (3.1.2b)$$

Note that all of these s-factor domains are the Cartesian product of one-factor domains.

Uniformity criterion When the experimental domain is C^s, there are several measures of uniformity discussed in the previous chapter. In this book, we shall choose the centered L_2-discrepancy (CD) in Sect. 2.3.2, the wrap-around L_2-discrepancy in Sect. 2.3.3, the mixture discrepancy in Sect. 2.3.5, and the discrete discrepancy or Lee discrepancy as defined in Sect. 2.5 for the finite-level domain in (3.1.2). When the experimental domain is a simplex, how to choose a measure of uniformity is not easy. Chapter 8 will discuss this problem.

Candidate designs Let $\mathcal{D}(n; C^s)$ be a set of n points on the experimental domain C^s. For a given discrepancy, denoted by D, a uniform design with n runs on the experimental domain C^s is a design on $\mathcal{D}(n; C^s)$ with the smallest discrepancy D. Here, $\mathcal{D}(n; C^s)$ is the set of candidate designs. Surely, the number of candidate designs is infinite and it is difficult to find the best one under a given uniformity criterion D.

For reducing the number of candidate designs, one may restrict the designs on lattice points. Assume that each factor has n levels, the number of total lattice points is n^s. Denote $D(n; n^s)$ be a design formed by n points from the n^s lattice points. Denote by $\mathcal{D}(n; n^s)$ the set of all $D(n; n^s)$. The number of candidate designs in $\mathcal{D}(n; n^s)$ is $\binom{n^s}{n}$ or n^{sn} if the repeated point is not permitted or permitted, respectively.

Obviously, the number of candidate designs in $\mathcal{D}(n; n^s)$ is too huge to afford even for moderate n and s. Then, it should reduce the number of candidate designs again. In the literature, it was shown that the U-type property is a reasonable requirement for uniform designs. A U-type design $U(n; n^s)$ is defined in Definition 1.3.6. It corresponds to a $n \times s$ matrix $\boldsymbol{X}$ such that each column is a permutation of the n levels, which are $\{\frac{2i-1}{2n}, i = 1, \ldots, n\}$ traditionally. Denote by $\mathcal{U}(n; n^s)$ the set of all $U(n; n^s)$. The number of candidate designs in $\mathcal{U}(n; n^s)$ is $(n!)^s$.

The number of levels is less than n for many practical applications. For s-factor cases each having levels $q_1, \ldots, q_s$, respectively, the total number of lattice points is $N = q_1 \ldots q_s$. From Definition 1.3.6, a candidate design is a U-type $U(n; q_1, \ldots, q_s)$. If $q_1 = \cdots = q_s = q$, the corresponding design is called as *symmetrical design*, otherwise *asymmetrical design*. Denote by $\mathcal{U}(n; q_1 \times \cdots \times q_s)$ the set of all $U(n; q_1 \times \cdots \times q_s)$. The number of candidate designs in $\mathcal{U}(n; q_1 \times \cdots \times q_s)$ is $\prod_{i=1}^s \left[\binom{n}{q_i}\binom{n-q_i}{q_i}\cdots\binom{2q_i}{q_i}\right]$, which is much smaller than that of $\mathcal{U}(n; n^s)$. In summary, the candidate designs can be reduced as follows.

$$\boxed{\mathcal{D}(n; C^s)} \Longrightarrow \boxed{\mathcal{D}(n; n^s)} \Longrightarrow \boxed{\mathcal{U}(n; n^s)} \Longrightarrow \boxed{\mathcal{U}(n; q_1 \times \cdots \times q_s)} \tag{3.1.3}$$

If some q_i's are equal, we denote the asymmetrical U-type design by $U(n; q_1^{s_1} \times \cdots \times q_m^{s_m})$, where $s = \sum_{i=1}^m s_i$. Denote by $\mathcal{U}(n; q_1^{s_1} \times \cdots \times q_m^{s_m})$ and $\mathcal{U}(n; q^s)$ the

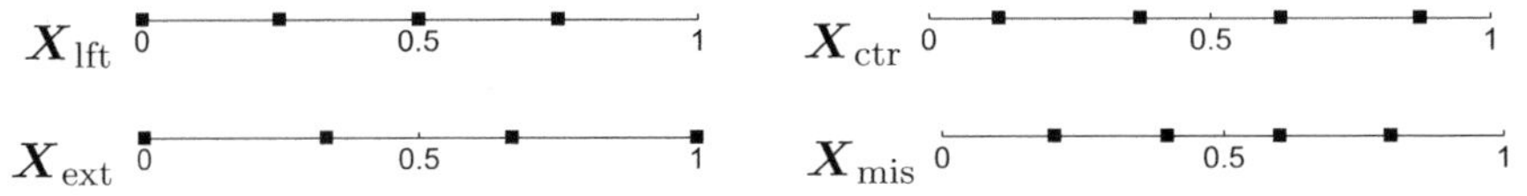

Fig. 3.1 Four different mappings of $\boldsymbol{U} = (1, 2, 3, 4)^T$ given by (3.1.4) for $q = 4$

set of all $U(n; q_1^{s_1} \times \cdots \times q_m^{s_m})$ and $U(n; q^s)$, respectively. For simplicity, sometimes we denote the set of candidate designs by $\mathcal{U}$ when the experimental domain is for a general case.

Design table. A design $\mathcal{P} = \{\boldsymbol{x}_1, \ldots, \boldsymbol{x}_n\} \subset [0, 1]^s$ can be denoted as a $n \times s$ matrix, $\boldsymbol{X}$, whose n rows, $\boldsymbol{x}_1^T, \ldots, \boldsymbol{x}_n^T$ correspond to the n runs of the design. To emphasize the matrix notation, $\boldsymbol{X}$ is called *design matrix*.

According to the discussion of the candidate in (3.1.3) for searching the uniform designs, one can limit the design space on $\mathcal{U}(n; q^s)$ or $\mathcal{U}(n; q_1 \times \cdots \times q_s)$ for symmetrical or asymmetrical designs, respectively. Then, for practical application, the experimental parameters can be described as follows.

Experimental parameters. The parameters include the number of runs n, the number of factors s, and the number of levels for each factor $q_1, \ldots, q_s$.

If the experimental domain is a hyperrectangle $\mathcal{X} = [a_1, b_1] \times, \ldots, [a_s, b_s]$. By a linear transformation, the domain $\mathcal{X}$ maps into C^s. Suppose that one wants to employ a uniform design table $U_n(q^s)$ for determination of design points. There are several ways to map one-factor domains $\{1, \ldots, q\}$ into $[0, 1]$ as follows:

$$\text{Left:}\quad x = (u-1)/q, \qquad u = 1, \ldots, q, \tag{3.1.4a}$$
$$\text{Centered:}\quad x = (u-1/2)/q, \qquad u = 1, \ldots, q, \tag{3.1.4b}$$
$$\text{Endpoints:}\quad x = (u-1)/(q-1), \qquad u = 1, \ldots, q, \tag{3.1.4c}$$
$$\text{Missing Endpoints:}\quad x = u/(q+1), \qquad u = 1, \ldots, q. \tag{3.1.4d}$$

Let $\boldsymbol{U} = (1, 2, 3, 4)^T$ be a design table for one-factor design with four runs having four levels. Figure 3.1 shows the corresponding design points on $[0, 1]$ by four transformations. The design $\boldsymbol{X}_{\text{lft}} = (0, 1/4, 1/2, 3/4)^T$ has evenly spaced points shifted to the left. The design $\boldsymbol{X}_{\text{ctr}} = (1/8, 3/8, 5/8, 7/8)^T$ has evenly spaced points, centered in the intervals $[(i-1)/4, i/4]$ for $i = 1, \ldots, 4$. The design $\boldsymbol{X}_{\text{ext}} = (0, 1/3, 2/3, 1)^T$ preserves even spacing of the points, but pushes the endpoints to the extremes of the experimental domain. The design $\boldsymbol{X}_{\text{mis}} = (1/5, 2/5, 3/5, 4/5)^T$ is like $\boldsymbol{X}_{\text{ext}}$ but without the extreme endpoints. In most practice, the centered mapping $x = (u - 1/2)/q, u = 1, \ldots, q$ has been widely used.

Example 3.1.1 To explain the relative advantages of the four different mappings in (3.1.4), consider the example of studying the period of a simple pendulum as a function of its amplitude. One plans to measure the period at several different amplitudes and model the dependency of the response (period) on the factor (amplitude). The experimental domain is [0°, 180°], with 0° corresponding to the pendulum starting straight down at rest and 180° corresponding to the pendulum starting straight up. Four designs with four runs each derived from $\boldsymbol{U} = (1, 2, 3, 4)^T$, the mapping (3.1.4), and a scaling to map [0, 1] to [0°, 180°] are given as follows:

$$\boldsymbol{X}_{\text{lft}} = (0^\circ,\ 45^\circ,\ 90^\circ,\ 135^\circ)^T\,, \tag{3.1.5a}$$

$$\boldsymbol{X}_{\text{ctr}} = (22.5^\circ,\ 67.5^\circ,\ 112.5^\circ,\ 157.5^\circ)^T \tag{3.1.5b}$$

$$\boldsymbol{X}_{\text{ext}} = (0^\circ,\ 60^\circ,\ 120^\circ,\ 180^\circ)^T\,, \tag{3.1.5c}$$

$$\boldsymbol{X}_{\text{mis}} = (36^\circ,\ 72^\circ,\ 108^\circ,\ 144^\circ)^T\,. \tag{3.1.5d}$$

Next subsection will show that the centered mapping (3.1.4b) is a more popular choice. By such mapping, a design matrix $\boldsymbol{U}$ for n runs and s factors is transformed column by column (factor by factor), to obtain a design $\boldsymbol{X}$ (or $\boldsymbol{X}_U$ for emphasizing $\boldsymbol{U}$) in the domain $[0, 1]^s$. A similar transformation can be applied to asymmetrical designs.

Definition 3.1.1 For any design $\boldsymbol{U} \in \mathcal{U}(n; q_1 \times \cdots \times q_s)$, by the mapping $f : u \to (2u - 1)/(2q_i)$, $u = 1, \ldots, q_i$ for the ith column, $i = 1, \ldots, s$, the resulting design matrix $\boldsymbol{X}$ is called the induced matrix of $\boldsymbol{U}$. We define $D(\boldsymbol{U}) = D(\boldsymbol{X})$.

The transformation mentioned in Definition 3.1.1 is one to one.

Example 3.1.2 Given a U-type symmetrical design, $U(12, 12^4)$ below

$$\boldsymbol{U} = \begin{pmatrix} 1 & 10 & 4 & 7 \\ 2 & 5 & 11 & 3 \\ 3 & 1 & 7 & 9 \\ 4 & 6 & 1 & 5 \\ 5 & 11 & 10 & 11 \\ 6 & 9 & 8 & 1 \\ 7 & 4 & 5 & 12 \\ 8 & 2 & 3 & 2 \\ 9 & 7 & 12 & 8 \\ 10 & 12 & 6 & 4 \\ 11 & 8 & 2 & 10 \\ 12 & 3 & 9 & 6 \end{pmatrix},$$

the corresponding induced matrix is

$$X = \begin{pmatrix} 1/24 & 19/24 & 7/24 & 13/24 \\ 3/24 & 9/24 & 21/24 & 5/24 \\ 5/24 & 1/24 & 13/24 & 17/24 \\ 7/24 & 11/24 & 1/24 & 9/24 \\ 9/24 & 21/24 & 19/24 & 21/24 \\ 11/24 & 17/24 & 15/24 & 1/24 \\ 13/24 & 7/24 & 9/24 & 23/24 \\ 15/24 & 3/24 & 5/24 & 3/24 \\ 17/24 & 13/24 & 23/24 & 15/24 \\ 19/24 & 23/24 & 11/24 & 7/24 \\ 21/24 & 15/24 & 3/24 & 19/24 \\ 23/24 & 5/24 & 17/24 & 11/24 \end{pmatrix} = \begin{pmatrix} 0.042 & 0.792 & 0.292 & 0.542 \\ 0.125 & 0.375 & 0.875 & 0.208 \\ 0.208 & 0.042 & 0.542 & 0.708 \\ 0.292 & 0.458 & 0.042 & 0.375 \\ 0.375 & 0.875 & 0.792 & 0.875 \\ 0.458 & 0.708 & 0.625 & 0.042 \\ 0.542 & 0.292 & 0.375 & 0.958 \\ 0.625 & 0.125 & 0.208 & 0.125 \\ 0.708 & 0.542 & 0.958 & 0.625 \\ 0.792 & 0.958 & 0.458 & 0.292 \\ 0.875 & 0.625 & 0.125 & 0.792 \\ 0.958 & 0.208 & 0.708 & 0.458 \end{pmatrix}.$$

The squared discrepancy values are $CD^2(\boldsymbol{X}) = 0.0114$, $WD^2(\boldsymbol{X}) = 0.0339$, and $MD^2(\boldsymbol{X}) = 0.0386$, respectively. Then, the discrepancy of $\boldsymbol{U}$ is equal to that of $\boldsymbol{X}$ according to Definition 3.1.1.

Definition 3.1.2 Given a candidate design space $\mathcal{U}$, a design $\mathcal{P}^*$ is called a *uniform design* if it minimizes the predetermined discrepancy measure D on $\mathcal{U}$, i.e.,

$$D(\mathcal{P}^*) = \min_{\mathcal{P}\in\mathcal{U}} D(\mathcal{P}). \tag{3.1.6}$$

Typically, the design space $\mathcal{U}$ is chosen as $\mathcal{D}(n; C^s)$ or the U-type designs $\mathcal{U}(n; q^s)$ or $\mathcal{U}(n; q_1 \times \cdots \times q_s)$. A uniform design on $\mathcal{U}(n; q^s)$ is denoted by $U_n(q^s)$, while a uniform design on $\mathcal{U}(n; q_1 \times \cdots \times q_s)$ is denoted by $U_n(q_1 \times \cdots \times q_s)$. A uniform design can be a tabular notation that is similar as the notation for orthogonal designs and is called *uniform design table*.

The minimization problem (3.1.6) always has many solutions when $s > 1$, because reordering the runs or the factors of a design keeps the discrepancy value unchanged due to the criterion C_1. Also, many discrepancies are invariant to reflections of the design points through the center of the domain. Designs with the same discrepancy are called *equivalent designs*. Efficient search algorithms for uniform designs should avoid as much as possible evaluating the discrepancies of designs that are equivalent to those already considered.

A design that approximately solves the minimization problem (3.1.6) is called a *low discrepancy design* or a *nearly uniform design*. It should be mentioned that many designs used in the literature are nearly uniform designs. Because the optimization problem is often extremely difficult to solve exactly, nearly uniform designs are typically accepted as uniform designs.

3.1.2 One-Factor Uniform Designs

The canonical experimental domain for one-factor experiments with a continuous range of levels is [0, 1]. In this subsection, it is shown that the uniform designs under centered L_2-discrepancy, wrap-around L_2-discrepancy, mixture discrepancy, and star discrepancy, can be analytically given.

Theorem 3.1.1 *Let the candidate space of one-factor designs be* $\mathcal{D}(n; [0,1])$. *The one-factor uniform designs under different uniformity criteria are as follows.*

(a) *(Fang and Wang 1994) The design*

$$\boldsymbol{X}^* = \left(\frac{1}{2n}, \frac{3}{2n}, \ldots, \frac{2n-1}{2n}\right)^T. \tag{3.1.7}$$

is the unique uniform design over [0,1] under the star discrepancy, whose value $D^*(X^*) = 1/(2n)$.

(b) *(Fang et al. 2002c) The design in* (3.1.7) *is also the unique uniform design over [0, 1] under the centered L_2-discrepancy and the squared CD-value* $CD^2(\boldsymbol{X}^*) = 1/(12n^2)$.

(c) *(Fang and Ma 2001a) The design*

$$\boldsymbol{X}_\delta = \left(\frac{\delta}{n}, \frac{1+\delta}{n}, \ldots, \frac{n-1+\delta}{n}\right)^T \tag{3.1.8}$$

is a uniform design over [0, 1] under the wrap-around L_2-discrepancy for any real number $\delta \in [0,1]$ *and the squared WD-value* $WD^2(\boldsymbol{X}_\delta) = 1/(6n^2)$.

(d) *(Zhou et al. 2013) The design in* (3.1.7) *is also the unique uniform design over [0, 1] under the mixture discrepancy and the squared MD-value* $MD^2(\boldsymbol{X}^*) = 1/(8n^2)$.

Proof We only prove the assertion (d). The proofs of the assertions (b) and (c) are similar, and the proof of Theorem 3.1.1(a) can be seen in Fang and Wang (1994), pp. 16–17. Let $\mathcal{P} = \{x_1, x_2, \ldots, x_n\}$ be a set on [0, 1]. Without loss of generality, suppose $x_1 \leqslant x_2 \leqslant \cdots \leqslant x_n$ and let $y_k = x_k - 1/2,\ k = 1, 2, \ldots, n$. Then, we have

$$\begin{aligned}
&\mathrm{MD}^2(\mathcal{P})\\
&= \frac{19}{12} - \frac{2}{n}\sum_{k=1}^{n}\left(\frac{5}{3} - \frac{|y_k|}{4} - \frac{|y_k|^2}{4}\right)\\
&+ \frac{1}{n^2}\sum_{k=1}^{n}\sum_{j=1}^{n}\left(\frac{15}{8} - \frac{|y_k|}{4} - \frac{|y_j|}{4} - \frac{3|y_k - y_j|}{4} - \frac{|y_k - y_j|^2}{2}\right)\\
&= \frac{19}{12} + \frac{1}{n^2}\sum_{k=1}^{n}\sum_{j=1}^{n}\left[\left(\frac{15}{8} - \frac{|y_k|}{4} - \frac{|y_j|}{4} - \frac{3|y_k - y_j|}{4} - \frac{|y_k - y_j|^2}{2}\right)\right.
\end{aligned}$$

$$
\begin{aligned}
&-\left(\frac{5}{3}-\frac{|y_k|}{4}-\frac{|y_k|^2}{4}\right)-\left(\frac{5}{3}-\frac{|y_j|}{4}-\frac{|y_j|^2}{4}\right)\Bigg]\\
&=\frac{1}{8}+\frac{1}{2n^2}\sum_{k=1}^{n}\sum_{j=1}^{n}\left((y_i-y_k)^2-|y_i-y_k|\right)+\frac{1}{2n}\left(\sum_{k=1}^{n}y_k^2-\frac{1}{n}\sum_{k>j}(y_k-y_j)\right)\\
&=\frac{1}{8}+\frac{1}{2n^2}\sum_{k=1}^{n}\sum_{j=1}^{n}\left(|y_i-y_k|-\frac{|k-j|}{n}\right)^2+\frac{1}{2}\left(\frac{1}{6n^2}-\frac{1}{6}\right)\\
&\quad+\frac{1}{2n}\sum_{k=1}^{n}\left(y_k-\frac{2k-1-n}{2n}\right)^2+\frac{1}{2}\left(\frac{1}{12n^2}-\frac{1}{12}\right)\\
&=\frac{1}{8n^2}+\frac{1}{2n^2}\sum_{k=1}^{n}\sum_{j=1}^{n}\left(|y_i-y_k|-\frac{|k-j|}{n}\right)^2+\frac{1}{n}\sum_{k=1}^{n}\left(y_k-\frac{2k-1-n}{2n}\right)^2\\
&\geqslant\frac{1}{8n^2}.
\end{aligned}
$$

Thus, $MD^2(\mathcal{P})$ achieves its minimum if and only if $y_k=\frac{2k-1-n}{2n}$, i.e., if and only if $x_k=\frac{2k-1}{2n}$, $k=1,2,\ldots,n$. The proof is completed.

Theorem 3.1.1(a), (b), and (d) provide a justification for the centered mapping in (3.1.4b) with $q=n$, i.e., under this mapping, the one-factor U-type design with n runs, $\boldsymbol{U}=(1,\ldots,n)^T$, minimizes the CD, MD, or star discrepancy. However, Theorem 3.1.1(c) shows that the wrap-around L_2-discrepancy has minimum discrepancy for a whole family of shifted points with equal spacing in between them. The design matrix $\boldsymbol{X}_\delta$ in this theorem corresponds to the centered mapping of the one-factor U-type design with n runs, $\boldsymbol{U}=(1,\ldots,n)^T$, for $\delta=1/2$. The left mapping corresponds to $\delta=0$. Theorem 3.1.1 also shows that when one chooses different discrepancies the corresponding uniform designs may be different.

The centered mapping is often chosen because the one-factor design $\boldsymbol{X}_{\text{ctr}}$ has the smallest centered L_2-discrepancy and mixture discrepancy as well as wrap-around L_2-discrepancy for all possible four-level designs. A disadvantage of the design arising from centered mapping is that the values of the levels may not be nice round numbers. This may be an inconvenience to the experimentalist. For example, $\boldsymbol{X}_{\text{ctr}}$ here has levels that are fractions of a degree, whereas the other designs above that are an integer number of degrees.

The design $\boldsymbol{X}_{\text{lft}}$ has minimum wrap-around L_2-discrepancy, but not minimum centered L_2-discrepancy or mixture discrepancy.

The design $\boldsymbol{X}_{\text{ext}}$ samples the extreme values of the factor. The advantage is that in fitting a model one need not extrapolate, only interpolate. Also, the values of the levels in applications may be nice round numbers. A disadvantage of $\boldsymbol{X}_{\text{ext}}$ is that in some applications the experiment cannot be performed for extreme values of the factors. In the pendulum Example 3.1.1, it is physically impossible to measure the period for amplitudes 0° and 180°, which are infinite.

The design $\boldsymbol{X}_{\text{mis}}$ is similar to $\boldsymbol{X}_{\text{ext}}$ in that in practice the levels tend to be nice round numbers. However, $\boldsymbol{X}_{\text{mis}}$ leaves out both extremes of the experimental domain to avoid the problem of having level-combinations where the experiment cannot be physically performed.

Designs $\boldsymbol{X}_{\text{lft}}$ and $\boldsymbol{X}_{\text{ctr}}$ are both special cases of an *equidistant design* with a shift δ/n, as given in (3.1.8). This is called a *shifted design*. The case $\delta = 0$ corresponds to $\boldsymbol{X}_{\text{lft}}$, and $\boldsymbol{X}_{\text{ctr}}$ corresponds to $\delta = 1/2$. Recall that the shifted design has minimum centered L_2-discrepancy and mixture discrepancy for a shift of $1/2$ and a minimum wrap-around L_2-discrepancy for any shift.

Choosing δ to be a uniform random number removes the bias in a design. This means, for example, that the expected value of the sample mean of a function sampled at the random design points equals the average value of the function. This is virtually never the case if the design is chosen deterministically. A disadvantage of an arbitrary shift δ is that the resulting design has levels that are not simple round numbers. This may be inconvenient for laboratory experiments, but should be no problem for computer experiments.

3.2 Uniform Designs with Multiple Factors

Whereas Theorem 3.1.1 gives the one-factor uniform design, there is no extension of this theorem to the case of more than one factor. In this subsection, the complexity of construction and representing method for designs with multiple factors are discussed.

3.2.1 Complexity of the Construction

Section 3.1.1 pointed out complexity for construction of uniform design with multiple factors. First of all, the canonical design space $\mathcal{D}(n; [0, 1]^s)$ is too complicated and the discrepancy is in general a multimodal function on the space of $\mathcal{D}(n; [0, 1]^s)$, which makes traditional optimization methods infeasible. To overcome this difficulty, the candidate set of designs considered is often chosen to be some well-structured subset of all possible designs. Flowchart (3.1.3) recommends to employ U-type designs, $\mathcal{U}(n; q_1 \times \cdots \times q_s)$ or $\mathcal{U}(n; q^s)$, as design space.

Definition 3.1.2 shows that to find a uniform design is an optimization problem

$$D(\mathcal{P}^*) = \min_{\mathcal{P} \in \mathcal{U}} D(\mathcal{P}), \tag{3.2.1}$$

i.e., (3.1.6), where $\mathcal{U}$ is a candidate design space. There are two approaches:

(A) *Theoretical Approach.* With reference to one-factor experiment, the last section provides uniform designs under different discrepancies based on theoretical approach. For multifactor experiments, there are some theoretical methods for generating uniform designs. In the next section, we shall introduce the good lattice point method (*glpm* for short) that was proposed by Korobov (1959) who is an expert in the number theory. He gave some theoretical justification to this method and a number of NT-nets for practical use. The *glpm* is the first theoretical method used for the construction of lower discrepancy sets of points and has a strong impact in quasi-Monte Carlo methods as well as in the construction of uniform design. Another approach is to establish connection between the combinatorial design and the uniform design. Section 3.6 will introduce the combinatorial design approach that applies the theory of the combinatorial design to construct uniform designs.

(B) *Numerical optimization approach.* So far, most of the existing uniform designs are obtained by numerical optimization approach. Note that the traditional optimization algorithms are useless for finding uniform designs. If we consider $\mathcal{D}(n; [0, 1]^s)$ as the design space, it is an optimization problem in R^{ns}. There are many local minimal points of the objective function D in (3.1.6) on the domain $\mathcal{U} = \mathcal{D}(n; [0, 1]^s)$. Even n and s are moderate, it is difficult to reach the global optimum design with the minimum discrepancy. For the two-dimensional case, the authors of this book found that it is not easy to find a uniform design with size $n > 10$ by computer software using the traditional optimization algorithm. Both of the *glpm* and optimization approach suggest to reduce the design space. As we mentioned in flowchart (3.1.3), the U-type designs are recommended.

Example 3.2.1 Let us demonstrate some comparisons between the uniform designs on the domain $\mathcal{D}(n; [0, 1]^2)$ and on $U(n, n^2)$ for $2 \leqslant n \leqslant 13$. Tables 3.1 and 3.2 give more justifications and display uniform designs under centered L_2-discrepancy for two factors with $n \leqslant 13$ runs on U-type $\mathcal{U}(n; n^2)$ and on $\mathcal{D}(n; [0, 1]^2)$, and denoted as $\boldsymbol{U}^*$ and $\boldsymbol{U}_0^*$, respectively. These uniform U-type designs $\boldsymbol{U}^*$ are displayed in Fig. 3.2. For simplicity of presentation, we give the design tables for the U-type designs $\boldsymbol{U}^*$, whose discrepancies are calculated from the corresponding induced matrices, $\boldsymbol{X}^*$, defined in Definition 3.1.1. Similarly, the discrepancies of $\boldsymbol{U}_0^*$ are also calculated from the corresponding induced matrices, $\boldsymbol{X}_0^* = (\boldsymbol{U}_0^* - 0.5)/n$. It is easy to see that two kinds of uniform designs and their centered L_2-discrepancies are close to each other. It is reasonable to restrict searching uniform designs in the design space $\mathcal{U}(n; n^s)$. Moreover, U-type designs have the advantage of evenly spaced levels, which is more convenient for the experimentalist.

3.2.2 Remarks

From now on, we choose U-type designs as our design space. Note that the U-type design space is a discrete point set in R^{ns}. As a consequence, there is no continuous concept as well as derivatives of the objective function, and the traditional optimization methods are useless. It needs new optimization technique for (3.2.1). The heuristic global optimization methods are recommended to find nearly uniform designs on the candidate design set. There are many heuristic global optimization

Table 3.1 Comparisons of two-factor uniform designs on different domains

Number of runs, n	U^* on $\mathcal{U}(n; n^2)$			U_0^* on $\mathcal{D}(n; [0,1]^2)$		
	u_{i1}	u_{i2}	Discrepancy	u_{i1}	u_{i2}	Discrepancy
2	1	2	6.228×10^{-2}	1.062	1.938	5.989×10^{-2}
	2	1		1.938	1.062	
3	1	2	2.958×10^{-2}	1.000	2.000	2.884×10^{-2}
	2	1		2.000	1.000	
	3	3		2.910	2.911	
4	1	3	1.626×10^{-2}	1.033	2.992	1.619×10^{-2}
	2	1		2.008	1.033	
	3	4		2.992	3.967	
	4	2		3.967	2.008	
5	1	4	1.105×10^{-2}	1.055	3.986	1.094×10^{-2}
	2	1		2.014	1.055	
	3	3		3.000	3.000	
	4	5		3.985	4.945	
	5	2		4.945	2.015	
6	1	4	7.628×10^{-3}	1.021	4.010	7.612×10^{-3}
	2	2		1.985	1.986	
	3	6		2.990	5.978	
	4	1		4.010	1.022	
	5	5		5.015	5.014	
	6	3		5.978	2.989	
7	1	3	5.824×10^{-3}	1.036	3.039	5.782×10^{-3}
	2	6		1.991	6.009	
	3	1		3.039	1.036	
	4	4		4.000	4.000	
	5	7		4.961	6.963	
	6	2		6.009	1.991	
	7	5		6.963	4.961	
8	1	3	4.475×10^{-3}	1.053	2.976	4.456×10^{-3}
	2	7		1.996	7.005	
	3	5		3.019	4.987	
	4	1		3.980	1.017	
	5	8		5.019	7.983	
	6	4		5.980	4.013	
	7	2		7.004	1.996	
	8	6		7.947	6.023	
9	1	4	3.583×10^{-3}	1.029	4.011	3.575×10^{-3}
	2	7		1.983	7.001	
	3	2		2.999	1.983	
	4	9		4.010	8.971	
	5	5		5.000	5.000	
	6	1		5.990	1.028	
	7	8		7.001	8.017	
	8	3		8.017	2.999	
	9	6		8.971	5.990	

Table 3.2 Comparisons of two-factor uniform designs on different domains

Number of runs, n	$\boldsymbol{U}^*$ on $\mathcal{U}(n; n^2)$			$\boldsymbol{U}_0^*$ on $\mathcal{D}(n; [0, 1]^s)$		
	u_{i1}	u_{i2}	Discrepancy	u_{i1}	u_{i2}	Discrepancy
10	1	8	2.953×10^{-3}	1.072	8.031	2.937×10^{-3}
	2	2		2.004	2.002	
	3	6		3.014	5.996	
	4	4		4.003	4.002	
	5	10		4.974	9.986	
	6	1		6.026	1.011	
	7	7		6.997	6.999	
	8	5		7.984	5.005	
	9	9		8.996	8.996	
	10	3		9.928	2.970	
11	1	7	2.467×10^{-3}	1.022	6.973	2.453×10^{-3}
	2	2		2.006	2.007	
	3	10		2.983	10.011	
	4	4		4.011	4.010	
	5	9		5.046	8.970	
	6	5		6.000	5.000	
	7	1		7.048	1.024	
	8	11		8.022	10.949	
	9	6		8.999	5.962	
	10	3		10.013	2.990	
	11	8		10.950	8.022	
12	1	6	2.076×10^{-3}	1.018	5.973	2.065×10^{-3}
	2	10		1.994	10.055	
	3	2		2.946	1.993	
	4	8		3.997	8.014	
	5	4		4.987	3.997	
	6	12		5.972	11.988	
	7	1		7.027	1.012	
	8	9		8.014	9.004	
	9	5		9.004	4.987	
	10	11		10.054	11.008	
	11	3		11.007	2.945	
	12	7		11.988	7.028	
13	1	6	1.747×10^{-3}	1.016	6.007	1.744×10^{-3}
	2	11		1.995	11.024	
	3	2		2.975	1.994	
	4	9		4.002	8.979	
	5	4		5.021	4.002	
	6	13		6.007	12.980	
	7	7		7.000	7.000	
	8	1		7.992	1.017	
	9	10		8.979	9.998	
	10	5		9.997	5.020	
	11	12		11.024	12.005	
	12	3		12.005	2.976	
	13	8		12.979	7.992	

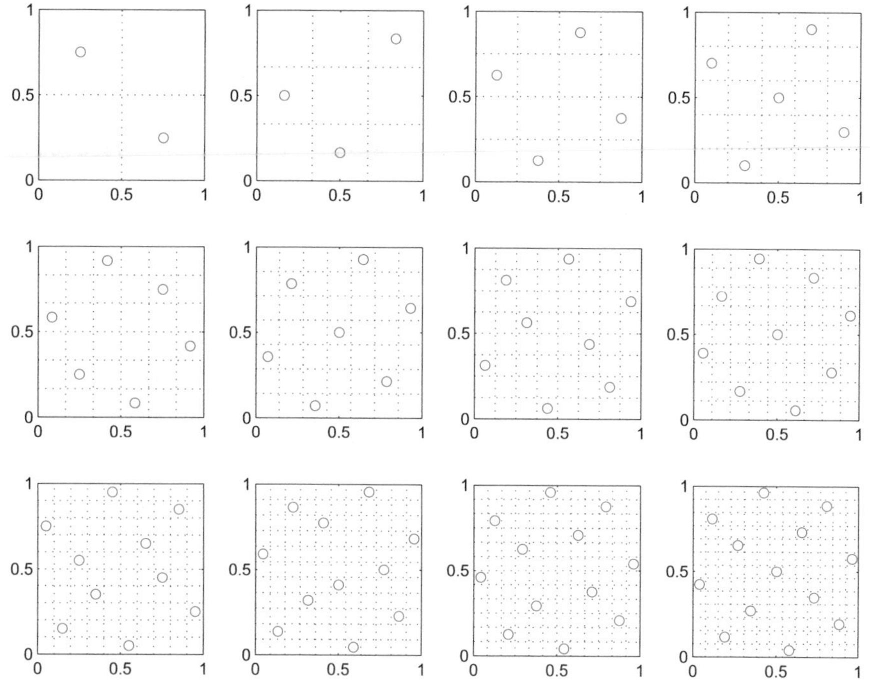

Fig. 3.2 Two-factor uniform U-type designs

algorithms, such as the threshold-accepting algorithm, simulated annealing, genetic algorithms. Following such strategies means that the design found may not have an absolute minimum discrepancy. However, this seems a small price to pay for making the problem of constructing designs tractable. Designs constructed in this manner are still called uniform designs even though they may not attain the absolute minimum discrepancy.

There are several important considerations involved in both theoretical and numerical optimization approaches.

i. One must identify good candidate sets of designs to consider. U-type designs and certain subsets of U-type designs are suitable choices.
ii. Reordering the runs and/or the factors of a design keeps the discrepancy unchanged. Also, many discrepancies are invariant to reflections of the design points through the center of the domain. The designs in the candidate set must be coded so that equivalent designs can be recognized.
iii. A good search algorithm must be developed to search a better design on the candidate set. There is no best algorithm for all the cases. To choose a good algorithm depends on the size of the candidate set. It can be an exhaustive search for a small candidate set. For larger candidate sets, one may use a random search, such as the threshold-accepting algorithm or an evolutionary algorithm. Some-

times good designs are obtained by the projection of low discrepancy designs or selecting some points from them.

iv. Although a uniform design commonly has as many levels per factor as runs, the number of runs per factor is sometimes chosen to be a smaller more manageable number in physical experiments. This complicates the coding process and search algorithm somewhat.

The following remarks are useful for understanding some problems/treatments in the construction of uniform designs.

Remark 1 Many different types of discrepancies are described in Chap. 2. Fortunately, the choice of discrepancy does not affect the choice of design in a dramatic way. Designs with small discrepancy relative to other designs for one kernel tend to have relatively small discrepancy for other kernels as well. In this chapter, the centered L_2-discrepancy is often used to discriminate between designs.

Although the particular choice of discrepancy does not influence the construction of designs too much, there is an important trade-off between having uniformity when considering a few factors at a time versus many factors at a time. The scaling of the parameters in the definition of the kernel can give higher weight to the former or the latter case. The designs constructed here are based on a common assumption that model terms involving the interaction of a few factors are more important than those involving many factors.

Remark 2 Laboratory experiments typically require only a modest number of runs, say in the tens, whereas some computer experiments can require thousands or millions of runs. Although the candidate sets for both types of experimental designs may be similarly constructed, the method for choosing the most uniform design is typically different. For smaller numbers of runs, it is worth to try to find a design with the absolute minimum discrepancy as location for each run is important. For large numbers of runs, the measure of uniformity is typically chosen to be efficient to compute, and one may narrow the search to a small subset of the original candidate set of designs.

Remark 3 It is easily known that the uniform design based on U-type designs may not be the uniform design in $\mathcal{D}(n; [0, 1]^s)$. Usually, the minimum discrepancy over U-type designs is very close to the discrepancy of the uniform design in $\mathcal{D}(n; [0, 1]^s)$.

The cardinality in the U-type design set $\mathcal{U}(n; n^s)$ is $(n!)^s$. Without loss of generality, let the first column of design be $(1, \ldots, n)^T$. Then, the cardinality decreases to $(n!)^{s-1}$. If $(n!)^{s-1}$ is affordable, one can use the enumeration method to find the uniform U-type design; otherwise, one needs to further reduce the design space. One approach is to reduce the complexity of the *design space*. Historically, the *glpm* is the first method in this approach. Next section will give an introduction to *glpm* and some new development.

3.3 Good Lattice Point Method and Its Modifications

The good lattice point method has been widely used in quasi-Monte Carlo methods. The *glpm* was proposed by Korobov (1959) for numerical evaluation of multivariate integrals. Further developments were discussed in Hua and Wang (1981), Shaw (1988), and Fang and Wang (1994). In this section, we start with the introduction of *glpm* and then its modifications.

3.3.1 Good Lattice Point Method

Let $\boldsymbol{X}_1$ and $\boldsymbol{X}_2$ be two designs. If $\boldsymbol{X}_1$ can be obtained from the permutation of rows and/or columns of $\boldsymbol{X}_2$, then $\boldsymbol{X}_1$ and $\boldsymbol{X}_2$ are equivalent to each other. For given n and s, there are at most $(n!)^{s-1}$ non-inequivalent $U(n; n^s)$ designs in $\mathcal{U}(n; n^s)$ if we account equivalent U-type designs to be one. However, even for a moderate number of runs, factors, and levels per factor, the number of possible designs can be astronomical. We have to further reduce the complexity of the design space again. Fang (1980) and Fang and Wang (1981) pioneered the application of good lattice point method for the construction of uniform designs.

The purpose of the *glpm* is to generate a proper subset of $\mathcal{U}(n; n^s)$. The key procedure of the *glpm* is to generate the first row of a U-type design, called as *generating vector*. Denote this row by $\boldsymbol{h} = (h_1, \ldots, h_s)$. Then, the jth column of the design can be obtained by

$$\boldsymbol{h}_j = (h_j, 2h_j, \ldots, nh_j)^T \ (\widetilde{\text{mod}}\, n), \tag{3.3.1}$$

where operator $\widetilde{\text{mod}}$ is the special modulo operation defined in Sect. 1.2. This modification of the modulo operator will be used in the part related to the *glpm*. Obviously, each $\boldsymbol{h}$ should satisfy:

(a) h_i is a positive integer and $h_i < n, i = 1, \ldots, s$.
(b) Each $\boldsymbol{h}_j$ is a permutation of $\{1, 2, \ldots, n\}$.
(c) h's are distinct, so we can assume that $1 \leqslant h_1 < h_2 \cdots < h_s < n$.
(d) The matrix of $[\boldsymbol{h}_1, \ldots, \boldsymbol{h}_s]$ has a rank of s.

Note that $\boldsymbol{h}_j = (h_j, 2h_j, \ldots, nh_j)^T \ (\widetilde{\text{mod}}\ n)$ may be formed by a permutation of $\{1, 2, \ldots, n\}$ or not. For example, when $n = 8$, $h_2 = 2$, and $h_3 = 3$ we have two generated columns as follows: $(h_2, 2h_2, \ldots, 8h_2) = (2, 4, 6, 8, 2, 4, 6, 8)$ and $(h_3, 2h_3, \ldots, 8h_3) = (3, 6, 1, 4, 7, 2, 5, 8)$, respectively. The second one is formed by a permutation of $\{1, 2, \ldots, 8\}$ while the first one is not. The necessary and sufficient condition to ensure the condition (b) is that the greatest common divisor of n and h is one, i.e., $\gcd(n, h) = 1$. Thus, we can choose $h_1 = 1$ in most cases.

Let m be the number of h_j's satisfying the above conditions (a)–(c). From the number theory, it is known that $m = \phi(n)$, where $\phi(\cdot)$ is the Euler function (see Hua and Wang 1981) that is defined as follows. For each positive integer n, there is a unique prime decomposition $n = p_1^{r_1} \dots p_t^{r_t}$, where $p_1, \dots, p_t$ are different primes and $r_1, \dots, r_t$ are positive integers. Then, the Euler function is given by

$$\phi(n) = n\left(1 - \frac{1}{p_1}\right)\cdots\left(1 - \frac{1}{p_t}\right). \tag{3.3.2}$$

When n is a prime, it is easy to see $\phi(n) = n - 1$. For example, $\phi(31) = 30$ as 31 is a prime. The prime decomposition of 30 is $30 = 2 \times 3 \times 5$ and $\phi(30) = 30(1 - 1/2)(1 - 1/3)(1 - 1/5) = 8$. When $s > m = \phi(n)$, we cannot obtain a U-type design of $U(n; n^s)$ by the *glpm*. When $s < m$, there are possible $\binom{m}{s}$ U-type designs of $U(n; n^s)$. But some of these U-type designs may have a rank less than s. Denote the candidate set of positive integers

$$\mathcal{H}_n = \{h : h < n, \text{the greatest common divisor of } n \text{ and } h \text{ is one}\}.$$

If $h \in \mathcal{H}_n$, then $n - h \in \mathcal{H}_n$. It is easy to see that the columns, denoted by $\boldsymbol{h}_h$ and $\boldsymbol{h}_{n-h}$, generated by h and $n - h$, have $\boldsymbol{h}_h + \boldsymbol{h}_{n-h} = (n, \dots, n, 2n)'$. Moreover, if $\tilde{h} \in \mathcal{H}_n$ and $\tilde{h}$ is not equal to h or $n - h$, then $\boldsymbol{h}_h + \boldsymbol{h}_{n-h} = \boldsymbol{h}_{\tilde{h}} + \boldsymbol{h}_{n-\tilde{h}}$. Therefore, the elements $\{h, n - h, \tilde{h}, n - \tilde{h}\}$ cannot be all in a generating vector; otherwise, the design matrix will be singular. Therefore, the number of possible columns generated by *glpm* is limited to

$$k(n) = \phi(n)/2 + 1, \tag{3.3.3}$$

which means that at most one pair $\{h, n - h\}$ is permitted in the generating vector.

Let $1 = h_1 < h_2 < \cdots < h_{\phi(n)/2}$ in $\mathcal{H}_n$. Among the $\phi(n)/2$ pairs $\{h_i, n - h_i\}$, $i = 1, \dots, \phi(n)/2$, we randomly choose one pair $\{h_i, n - h_i\}$ and randomly choose one element in each of the other $\phi(n)/2 - 1$ pairs to form the $k(n)$ elements in the candidate of generating vector such that the corresponding design has full column rank. There are $2^{\phi(n)/2-1}\phi(n)/2$ such choices, and for each choice we can get $\binom{k(n)}{s}$ U-type designs. Then, there are at most $\phi(n)2^{\phi(n)/2-2}\binom{k(n)}{s}$ U-type designs that form the design space, denoted by $\mathcal{G}_{n,s}$. The cardinality of $\mathcal{G}_{n,s}$ is much less than the cardinality of the space $\mathcal{U}(n; n^s)$. This is an advantage of the *glpm*. A design with the lowest discrepancy on this space is a nearly uniform design. Then, the algorithm of *glpm* is given as follows:

Algorithm 3.3.1 (*Good Lattice Point Method*)

Step 1. Find the candidate set of positive integers $\mathcal{H}_n$. If $s \leqslant k(n)$, go to Step 2; otherwise, we fail to generate a nearly uniform design by the method.

Step 2. Let $\boldsymbol{h}^{(t)} = (h_{t1}, \dots, h_{ts})$ be the tth set of s distinct elements of $\mathcal{H}_n$ and generate a $n \times s$ matrix $\boldsymbol{U}^{(t)} = (u_{ij}^{(t)})$ where $u_{ij}^{(t)} = ih_{tj} \ (\widetilde{\text{mod}}\ n)$. Denote $\boldsymbol{U}^{(t)}$ by $\boldsymbol{U}(n, \boldsymbol{h}^{(t)})$, where $\boldsymbol{h}^{(t)}$ is the generating vector of the $\boldsymbol{U}^{(t)}$. Denote by $\mathcal{G}_{n,s}$ the set of all such matrices $\boldsymbol{U}(n, \boldsymbol{h}^{(t)})$ with rank s.

Step 3. Find a generating vector $\boldsymbol{h}^*$ such that its corresponding $\boldsymbol{U}(n, \boldsymbol{h}^*)$ has the smallest pre-decided discrepancy over the set $\mathcal{G}_{n,s}$. This $\boldsymbol{U}(n, \boldsymbol{h}^*)$ is a (nearly) uniform design $U_n(n^s)$.

A design $\boldsymbol{U}$ generated by the *glpm* is also called a good lattice point set. In fact, we can further reduce the computation complexity of the *glpm* by the following facts:

(a) For any integer h if there exists h^* such that $hh^* = 1 \pmod n$, then h^* is called as *inverse of* h (mod n) and may be denoted as h^{-1}. If h^{-1} exists, it can be easy to show that $h^{-1} \in \mathcal{H}_n$ if and only if $h \in \mathcal{H}_n$.

(b) For each generating vector $\boldsymbol{h} = (h_1, \ldots, h_s)$, where $h_1 < \cdots < h_s$, if $h_1 \neq 1$, the first column $\boldsymbol{h}_1 = (h_1, 2h_1, \ldots, nh_1) \pmod n$ is formed by a permutation of $\{1, 2, \ldots, n\}$. Then, we can do the row permutation such that the first column becomes $(1, 2, \ldots, n)^T$. Then, the permuted design is another glp set with generating vector $\tilde{\boldsymbol{h}} = h_1^{-1}(h_1, h_2, \ldots, h_s) \equiv (1, \tilde{h}_2, \ldots, \tilde{h}_s) \pmod n$. Since the row permutation does not change the discrepancy value, the two glp sets are equivalent to each other. Then, let $h_1 = 1$ and we only need to determine the values of $\{h_2, \ldots, h_s\}$. Therefore, we only need to compare $\binom{\phi(n)-1}{s-1}$ glp sets.

(c) **Fibonacci sequence**. For two-factor experiments, there exists an analytic construction for lattice designs with low discrepancy. Consider the Fibonacci sequence, $1, 1, 2, 3, 5, 8, \ldots$, defined by $F_0 = F_1 = 1$, and $F_m = F_{m-1} + F_{m-2}, m = 2, 3, \ldots$. The choice $n = F_m$ and $\boldsymbol{h} = (1, F_{m-1})^T$ gives a low discrepancy lattice design as $m \to \infty$ (Hua and Wang 1981, Section 4.8).

Table 3.3 gives the "best" generating vectors for $4 \leqslant n \leqslant 31$ and $s \leqslant 5$ that is modified from P. 98 of Fang and Ma (2001b), according to the limitation of number of factors in (3.3.3) and the reader can easily to find the corresponding uniform design tables.

3.3.2 *The Leave-One-Out glpm*

In some cases such as $s > k(n)$, the candidate set belonging to $\mathcal{H}_n$ is not enough to generate a *glp* set with full column rank. However, if $k(n+1)$ is larger than s, one can consider the leave one out *glpm*. Let us look at the following illustrative example.

Example 3.3.1 For $n = 12$, the cardinality of the set of $\mathcal{H}_{12} = \{1, 5, 7, 11\}$ is 4, and $k(12) = 3$ by (3.3.3). It is impossible to construct a nearly uniform design $U_{12}(12^s), s > 3$ by the *glpm* in Algorithm 3.3.1. However, note that

$$\mathcal{H}_{13} = \{1, 2, 3, 4, 5, 6, 7, 8, 9, 10, 11, 12\}.$$

Select s elements from $\mathcal{H}_{13}$, and obtain a U-type designs $U(13; 13^s)$ by the *glpm*, and its last row is a vector $(13, \ldots, 13)$. Deleting this row, the remaining $12 \times s$ matrix gives a U-type $U(12; 12^s)$. Comparing all such designs, we can obtain a nearly uniform design $U_{12}(12^s)$. This modification may be called leave-one-out *glpm*.

Table 3.3 The generating vectors under CD

n	$s = 2$	$s = 3$	$s = 4$	$s = 5$
4	(1 3)			
5	(1 2)	(1 2 3)		
6	(1 5)			
7	(1 3)	(1 2 3)	(1 2 3 5)	
8	(1 5)	(1 3 5)		
9	(1 4)	(1 4 7)	(1 2 4 7)	
10	(1 3)	(1 3 7)		
11	(1 7)	(1 5 7)	(1 2 5 7)	(1 2 3 5 7)
12	(1 5)	(1 5 7)		
13	(1 5)	(1 4 6)	(1 4 5 11)	(1 3 4 5 11)
14	(1 9)	(1 9 11)	(1 3 5 13)	
15	(1 11)	(1 4 7)	(1 4 7 13)	(1 2 4 7 13)
16	(1 7)	(1 5 9)	(1 5 9 13)	(1 3 5 9 13)
17	(1 10)	(1 4 10)	(1 4 5 14)	(1 4 10 14 15)
18	(1 7)	(1 7 13)	(1 5 7 13)	
19	(1 8)	(1 6 8)	(1 6 8 14)	(1 6 8 14 15)
20	(1 9)	(1 9 13)	(1 9 13 17)	(1 3 7 11 19)
21	(1 13)	(1 4 5)	(1 5 8 19)	(1 4 10 13 16)
22	(1 13)	(1 5 13)	(1 5 7 13)	(1 3 5 7 13)
23	(1 9)	(1 7 18)	(1 7 18 20)	(1 4 7 17 18)
24	(1 17)	(1 11 17)	(1 11 17 19)	(1 5 7 13 23)
25	(1 11)	(1 6 16)	(1 6 11 16)	(1 6 11 16 21)
26	(1 11)	(1 11 17)	(1 5 11 17)	(1 3 5 11 17)
27	(1 16)	(1 8 10)	(1 8 20 22)	(1 8 20 22 23)
28	(1 11)	(1 9 11)	(1 9 11 15)	(1 9 11 15 23)
29	(1 18)	(1 9 17)	(1 8 17 18)	(1 7 16 20 24)
30	(1 19)	(1 17 19)	(1 17 19 23)	(1 7 11 13 29)
31	(1 22)	(1 18 24)	(1 6 14 22)	(1 6 13 20 27)

The procedure of the leave-one-out *glpm* is as follows.

Algorithm 3.3.2 (*Leave-one-out glpm*)

Step 1. If $s \leqslant \phi(n+1)/2$, go to Step 2; otherwise, we fail to generate a nearly uniform design by the method.

Step 2. To find $\mathcal{H}_{n+1}$ for given n, denote its elements by $\boldsymbol{h}^* = (h_1^*, \ldots, h_m^*)$. By use of the method in (3.3.1) to obtain a $(n+1) \times m$ matrix $\boldsymbol{U}(n+1, \boldsymbol{h}^*)$.

Step 3. Delete the last row of $\boldsymbol{U}(n+1, \boldsymbol{h}^*)$ to form a $n \times m$ matrix $\boldsymbol{H}$. Select s columns of $\boldsymbol{H}$ to form a U-type design $U(n; n^s)$ with rank s. Denote by $\mathcal{L}_{n,s}$ the set of all such U-type designs.

Step 4. Choose the one with the smallest discrepancy in $\mathcal{L}_{n,s}$ as a nearly uniform design $U_n(n^s)$.

It is easily known that the $(n+1)$th row of $\boldsymbol{U}(n+1, \boldsymbol{h}^*)$ is $(n+1, \ldots, n+1)$. After deleting this row, each column of $\boldsymbol{H}$ is a permutation of $(1, \ldots, n)^T$. If there are two elements h_{1i} and h_{1j} in the first row of $\boldsymbol{H}$ such that $h_{1i} + h_{1j} = n + 1$, then the ith column $\boldsymbol{h}_i$ and the jth column $\boldsymbol{h}_j$ in $\boldsymbol{H}$ are fully negatively correlated since $\boldsymbol{h}_i + \boldsymbol{h}_j = (n+1)\boldsymbol{1}_n$. Then, if $s > \phi(n+1)/2$, there exist at least two columns in $\boldsymbol{H}$ that are fully negatively correlated (for more detail, see Yuan et al. 2017). This is the reason that we judge the condition $s \leqslant \phi(n+1)/2$ in the Step 1 of Algorithm 3.3.2. When $\phi(n+1) > \phi(n)$, for example n is an even, the number of possible designs by leave-one-out *glpm* may be larger than that of the *glpm*. For example, let $n = 36, s = 4$. The number of possible *glp sets* is at most $\binom{\phi(36)-1}{4-1} = 165$. The corresponding generating vector of the *glpm* is $(1, 7, 11, 17)$, and its squared CD-value is 0.0029. On the other hand, the number of possible designs by leave-one-out *glpm* is at most $\binom{\phi(36+1)-1}{4-1} = 6545$, which is much larger than the former. Then, the resulted design by leave-one-out *glpm* may be more uniform. In the case with $n = 36, s = 4$, the generating vector of the leave-one-out *glpm* is $(1, 6, 27, 29)$ for $n+1$ level, and the corresponding squared CD-value is 0.0023.

Example 3.3.2 (*Example* 3.3.1 *continues*) Consider the construction of the uniform design $U_{12}(12^5)$ by the leave-one-out *glpm*. From the discussion of the procedure of *glpm* in the last subsection, let the element $h_1 = 1$ in the generating vector. Then, we choose other four elements in the generating vector from the set $\{2, 3, \ldots, 12\}$. There exist $\binom{11}{4} = 330$ different generating vectors. Among them, the generating vector (1 3 4 5 7) is the best one. The matrix $\boldsymbol{H}$ and the nearly uniform design $U_{12}(12^5)$ are as follows.

$\boldsymbol{H}$												$U_{12}(12^5)$				
1	2	3	4	5	6	7	8	9	10	11	12	1	2	3	4	5
2	4	6	8	10	12	1	3	5	7	9	11	2	4	6	8	10
3	6	9	12	2	5	8	11	1	4	7	10	3	6	9	12	2
4	8	12	3	7	11	2	6	10	1	5	9	4	8	12	3	7
5	10	2	7	12	4	9	1	6	11	3	8	5	10	2	7	12
6	12	5	11	4	10	3	9	2	8	1	7	6	12	5	11	4
7	1	8	2	9	3	10	4	11	5	12	6	7	1	8	2	9
8	3	11	6	1	9	4	12	7	2	10	5	8	3	11	6	1
9	5	1	10	6	2	11	7	3	12	8	4	9	5	1	10	6
10	7	4	1	11	8	5	2	12	9	6	3	10	7	4	1	11
11	9	7	5	3	1	12	10	8	6	4	2	11	9	7	5	3
12	11	10	9	8	7	6	5	4	3	2	1	12	11	10	9	8

Table 3.4 shows the CD-value of two methods and related generating vector in bracket for the case $n = 12$. It is clear that the leave-one-out *glpm* can improve the original *glpm*, i.e., the designs constructed by the leave-one-out *glpm* have less discrepancy value. Table 3.5 gives the generating vectors for $4 \leqslant n \leqslant 31$ and $s \leqslant 5$ by the leave-one-out *glpm* which is modified from P. 99 of Fang and Ma (2001b)

Table 3.4 CD-value and generating vector of two methods for $n = 12$

Method	$s = 2$	$s = 3$	$s = 4$	$s = 5$
glmp	0.0506	0.1112		
	(1, 5)	(1, 5, 7)		
Leave one out	0.0458	0.0782	0.1211	0.1656
glmp	(1, 5)	(1, 3, 4)	(1, 2, 3, 5)	(1, 2, 3, 4, 5)

Table 3.5 The generating vectors for $4 \leqslant n \leqslant 31$ and $s \leqslant 5$ by the leave-one-out *glpm*

n	$s = 2$	$s = 3$	$s = 4$	$s = 5$
4	(1 2)			
6	(1 2)	(1 2 3)		
7	(1 3)			
8	(1 2)	(1 2 4)		
9	(1 3)	(1 3 7)	(1 3 7 9)	
10	(1 3)	(1 2 3)	(1 2 3 4)	(1 2 3 4 5)
11	(1 5)			
12	(1 5)	(1 3 4)	(1 2 3 5)	(1 2 3 4 5)
13	(1 3)	(1 3 5)		
14	(1 4)	(1 2 4)	(1 2 4 7)	
15	(1 7)	(1 3 5)	(1 3 5 7)	
16	(1 5)	(1 3 5)	(1 3 4 5)	(1 2 3 5 8)
17	(1 5)	(1 5 7)		
18	(1 7)	(1 7 8)	(1 3 4 5)	(1 2 5 6 8)
19	(1 9)	(1 3 7)	(1 3 7 9)	
20	(1 8)	(1 4 5)	(1 2 5 8)	(1 2 4 5 8)
21	(1 5)	(1 3 5)	(1 3 5 7)	(1 3 5 7 9)
22	(1 7)	(1 4 10)	(1 4 5 7)	(1 3 4 5 7)
23	(1 7)	(1 5 7)	(1 5 7 11)	
24	(1 7)	(1 4 11)	(1 4 6 9)	(1 4 6 9 11)
25	(1 7)	(1 3 7)	(1 3 5 7)	(1 3 5 7 9)
26	(1 8)	(1 8 10)	(1 4 5 7)	(1 2 5 7 8)
27	(1 5)	(1 3 5)	(1 3 5 11)	(1 3 5 9 11)
28	(1 12)	(1 8 12)	(1 8 9 12)	(1 4 5 7 13)
29	(1 11)	(1 7 11)	(1 7 11 13)	
30	(1 12)	(1 7 9)	(1 4 13 14)	(1 4 5 6 14)
31	(1 7)	(1 7 9)	(1 7 9 15)	(1 3 5 11 13)

according to the limitation of the number of factors. Fang and Wang (1981), Fang and Li (1995) and Fang and Ma (2001b) showed that many nearly uniform designs obtained by this way have lower discrepancy than the corresponding designs generated by directly using Algorithm 3.3.1 for many n.

3.3.3 Good Lattice Point with Power Generator

The good lattice point method with the best generating vector can be time-consuming even for moderate n and s. Korobov (1959) proposed generating vectors of the form $\boldsymbol{h} = (1, h, h^2, \ldots, h^{s-1}) \pmod n$, where $\gcd(n, h^j)=1\ (j = 1, \ldots, s-1)$ and $h < n$. He showed that the discrepancy is asymptotically small for good choices of h and prime n, and this form has been used by Fang (1980), Wang and Fang (1981), Fang and Hickernell (1995) in searching low discrepancy designs. Thus, the *glpm* with a power generator is recommend, which has a very lower computation complexity. Denote the set constructed by good lattice point method with a power generator be *pglp set*. For given positive integer pair (n, s), we can generate a (nearly) uniform design $U_n(n^s)$ by the following algorithm:

Algorithm 3.3.3 (*Good Lattice Point Method with Power Generator*)

Step 1. Find the candidate set of positive integers

$$\mathcal{A}_{n,s} = \{a : a < n,\ \gcd(a, n) = 1, \text{and}\, a, a^2, \ldots, a^s \widetilde{(\text{mod}\, n)} \text{ are distinct}\}.$$

If the set $\mathcal{A}_{n,s}$ is nonempty, go to Step 2; otherwise, we fail to generate a nearly uniform design by the method.

Step 2. From each $a \in \mathcal{A}_{n,s}$, construct a U-type design $\boldsymbol{U}^a = (u_{ij}^a)$ as follows:

$$u_{ij}^a = ia^{j-1} \widetilde{(\text{mod}\, n)}, i = 1, \ldots, n;\ j = 1, \ldots, s.$$

Step 3. Find a $a_* \in \mathcal{A}_{n,s}$ such that $\boldsymbol{U}^{a_*}$ has the smallest pre-decided discrepancy over all possible $\boldsymbol{U}^a$'s with respect to a. This $\boldsymbol{U}^{a_*}$ is a (nearly) uniform design $U_n(n^s)$.

For each positive integer $a < n$, if the greatest common divisor of n and a is one, then the greatest common divisor of n and a^i is also one for any nonnegative integer i, and the jth column of $\boldsymbol{U}^a$ is a permutation of $\{1, \ldots, n\}$. According to the limitation that a and n are coprime in the definition $\mathcal{A}_{n,s}$, the cardinality of $\mathcal{A}_{n,s}$, denoted by $|\mathcal{A}_{n,s}|$, should not be larger than $\phi(n)$. For a prime n, it is shown that $|\mathcal{A}_{n,s}| = \phi(\phi(n)) = \phi(n-1)$ when $s = n-1$ and $|\mathcal{A}_{n,s}|$ falls in $[\phi(n-1), n-1]$ when $s < n-1$.

For example, $\phi(31) = 30$ and $\phi(30) = 8$. So the cardinality of $\mathcal{A}_{31,5}$ is in $[\phi(\phi(31)), 31-1] = [8, 30]$. The numerical calculation shows that the cardinality of $\mathcal{A}_{31,5}$ equals to 26. In step 3, we need to compare only 26 U-type design candidates if we choose $n = 31$. Many uniform designs were generated by the above method, for example, Fang (1980) and Fang and Ma (2001b). The leave-one-out method can be applied to the *glpm* with power generator. Fang and Li (1995) showed that many nearly uniform designs obtained by this way have lower discrepancy than the corresponding designs generated by directly using Algorithm 3.3.3.

In the literature, Hua and Wang (1981), for example, showed that the best order of the star discrepancy of U-type design generated by the *glpm* is $O(n^{-1}(\log\ n)^{s-1})$

and the best order of the star discrepancy of U-type design generated by the *glpm* with power generating vector is

$$O(n^{-1}(\log\ n)^{s-1} \log(\log\ n))$$

which is slightly worse than $O(n^{-1}(\log\ n)^{s-1})$. This gives a justification to the *glpm* with power generator.

3.4 The Cutting Method

Suppose that we want to obtain a uniform design $U_n(n^s)$, where n is not a prime. For small s, the design can be generated by the *glpm*, but it may have a poor uniformity; while for large s, the good lattice point method fails to generate such a design, as the maximum number of factors is less than $\phi(n)/2 + 1$ if the uniform design is generated by the *glpm*.

Ma and Fang (2004) suggested the *cutting method* to generate $U_n(n^s)$ for any n by cutting a larger uniform design. The key idea of the cutting method is as follows. Let $\boldsymbol{U}_p$ be a uniform design $U_p(p^s)$, where $n < p$ or $n \ll p$ and p or $p+1$ is a prime, and let $\mathcal{P}_p$ be its induced design. Let $\mathcal{Q}$ be a proper subset of $[0, 1]^u$ such that there are n points of $\mathcal{P}_p$ fell on $\mathcal{Q}$, and let $\mathcal{P}_n$ denote these n points. From the theory of quasi-Monte Carlo methods, it is clear that the points in $\mathcal{P}_n$ are uniformly scattered over $\mathcal{Q}$. Particularly, we can choose a suitable rectangle in C^s such that there are exact n points of $\mathcal{P}_p$ falling in this rectangle. These n points will form a (nearly) uniform design by some linear transformations. Figure 3.3 provides an illustrative example, where $p = 47$, $n = 8$, and $s = 2$. Choosing suitable rectangle such that the number of runs in the rectangle equals $n = 8$, and we reset the levels of these runs and obtain a U-type design with n levels. One can move the rectangle to find the design with best uniformity. It should be mentioned that the rectangle has wrap-around property. In fact, the cutting method can be carried out directly on $\boldsymbol{U}_p$ as given below.

Algorithm 3.4.1 (*Cutting Method*)

Step 1. **Initial design**. For given (n, s), find a $U_p(p^s)$, where $p \gg n$ and p or $p+1$ is a prime, denote it by $\boldsymbol{U}_p = (u_{ij})$, and call it the *initial design*.

Step 2. **Row sorting**. For $l = 1, \ldots, s$, reorder rows of $\boldsymbol{U}_p$ according to its lth column such that the elements in this column are ordered from 1 to p, and denote the reordered matrix by $\boldsymbol{U}_p^{(l)} = (u_{kj}^{(l)})$.

Step 3. **Cutting**. For $m = 1, \ldots, p$, let $\boldsymbol{U}_p^{(l,m)} = (u_{kj}^{(l,m)})$, where

$$u_{kj}^{(l,m)} = \begin{cases} u_{k+m-n-1,j}^{(l)}, & m > n, k = 1, \ldots, n, \\ u_{kj}^{(l)}, & m \leqslant n, k = 1, \ldots, m-1, \quad j = 1, \ldots, s. \\ u_{k+p-n,j}^{(l)}, & m \leqslant n, k = m, \ldots, n, \end{cases}$$

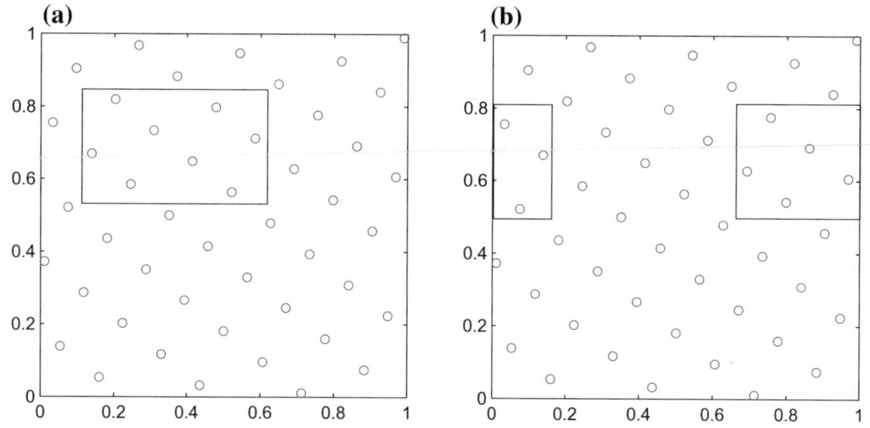

Fig. 3.3 Illustration of cutting method

Step 4. **New design space**. Relabel elements of each column of $\boldsymbol{U}_p^{(l,m)}$ by $1, 2, \ldots, n$ according to the magnitude of these elements. The resulted matrix becomes a U-type design $U(n, n^s)$ and is denoted by $\boldsymbol{U}^{(l,m)}$. We have ps such U-type designs.

Step 5. **Output design**. For a given measure of uniformity D, compare ps designs $\boldsymbol{U}^{(l,m)}$ obtained in the previous step and choose the one with the smallest D-value. That one is a nearly uniform design $U_n(n^s)$.

In Step 1, the *glpm* is often used to obtain a good initial design if its complexity is affordable; otherwise, the *glpm* with power generator is considered since it is the fastest way to give a good initial U-type design. Now, let us see an example for illustration.

Example 3.4.1 (*Construction of* $U_8(8^3)$) Consider the nearly uniform design $U_{47}(47^3)$ obtained by the *glpm* as an initial design. The generating vector is $(1, 18, 26)$, and the initial design is

$$\boldsymbol{U}_{47} = \left(\begin{array}{cccccccccccccccccccccccc} 1 & 2 & 3 & 4 & 5 & 6 & 7 & 8 & 9 & 10 & 11 & 12 & 13 & 14 & 15 & 16 & 17 & 18 & 19 & 20 & 21 & 22 & 23 & 24 \\ 18 & 36 & 7 & 25 & 43 & 14 & 32 & 3 & 21 & 39 & 10 & 28 & 46 & 17 & 35 & 6 & 24 & 42 & 13 & 31 & 2 & 20 & 38 & 9 \\ 26 & 5 & 31 & 10 & 36 & 15 & 41 & 20 & 46 & 25 & 4 & 30 & 9 & 35 & 14 & 40 & 19 & 45 & 24 & 3 & 29 & 8 & 34 & 13 \end{array}\right.$$

$$\left.\begin{array}{ccccccccccccccccccccccc} 25 & 26 & 27 & 28 & 29 & 30 & 31 & 32 & 33 & 34 & 35 & 36 & 37 & 38 & 39 & 40 & 41 & 42 & 43 & 44 & 45 & 46 & 47 \\ 27 & 45 & 16 & 34 & 5 & 23 & 41 & 12 & 30 & 1 & 19 & 37 & 8 & 26 & 44 & 15 & 33 & 4 & 22 & 40 & 11 & 29 & 47 \\ 39 & 18 & 44 & 23 & 2 & 28 & 7 & 33 & 12 & 38 & 17 & 43 & 22 & 1 & 27 & 6 & 32 & 11 & 37 & 16 & 42 & 21 & 47 \end{array}\right)^T.$$

The centered L_2-discrepancy of $\boldsymbol{U}_{47}$ is 6.57×10^{-4}. Consider to generate $U_8(8^3)$ from the $\boldsymbol{U}_{47}$. One can choose eight successive runs according to its first, second, and third factors. By this consideration, we have 47 choices of eight successive runs according to each factor; thus, there are $141 = 47 * 3$ such choices, i.e., $\boldsymbol{U}_p^{(l,m)}$ for $l = 1, 2, 3$ and $m = 1, \ldots, 47$, each of which results a U-type design $\boldsymbol{U}^{(l,m)}$ after relabeling the elements in each column. Among these 141 U-type designs, the one with the smallest CD is $\boldsymbol{U}^{(3,9)}$, i.e., $l = 3, m = 9$. First reorder the initial design according to its third column, we obtain

$$\boldsymbol{U}_{47}^{(3)} = \left(\begin{matrix} 38 & 29 & 20 & 11 & 2 & 40 & 31 & 22 & 13 & 4 & 42 & 33 & 24 & 15 & 6 & 44 & 35 & 26 & 17 & 8 & 46 & 37 & 28 & 19 \\ 26 & 5 & 31 & 10 & 36 & 15 & 41 & 20 & 46 & 25 & 4 & 30 & 9 & 35 & 14 & 40 & 19 & 45 & 24 & 3 & 29 & 8 & 34 & 13 \\ 1 & 2 & 3 & 4 & 5 & 6 & 7 & 8 & 9 & 10 & 11 & 12 & 13 & 14 & 15 & 16 & 17 & 18 & 19 & 20 & 21 & 22 & 23 & 24 \end{matrix}\right.$$

$$\left.\begin{matrix} 10 & 1 & 39 & 30 & 21 & 12 & 3 & 41 & 32 & 23 & 14 & 5 & 43 & 34 & 25 & 16 & 7 & 45 & 36 & 27 & 18 & 9 & 47 \\ 39 & 18 & 44 & 23 & 2 & 28 & 7 & 33 & 12 & 38 & 17 & 43 & 22 & 1 & 27 & 6 & 32 & 11 & 37 & 16 & 42 & 21 & 47 \\ 25 & 26 & 27 & 28 & 29 & 30 & 31 & 32 & 33 & 34 & 35 & 36 & 37 & 38 & 39 & 40 & 41 & 42 & 43 & 44 & 45 & 46 & 47 \end{matrix}\right)^T .$$

then we get

$$\boldsymbol{U}_p^{(3,9)} = \begin{pmatrix} 38 & 29 & 20 & 11 & 2 & 40 & 31 & 22 \\ 26 & 5 & 31 & 10 & 36 & 15 & 41 & 20 \\ 1 & 2 & 3 & 4 & 5 & 6 & 7 & 8 \end{pmatrix},$$

and transfer it into a U-type design

$$\boldsymbol{U}^{(3,9)} = \begin{pmatrix} 7 & 5 & 3 & 2 & 1 & 8 & 6 & 4 \\ 5 & 1 & 6 & 2 & 7 & 3 & 8 & 4 \\ 1 & 2 & 3 & 4 & 5 & 6 & 7 & 8 \end{pmatrix},$$

which is a nearly uniform design $U_8(8^3)$. The CD-value of $\boldsymbol{U}^{(3,9)}$ is 0.0129. For $n = 8, s = 3$, the CD of the design obtained by the *glpm* is 0.0132. The $U_8(8^3)$ obtained by the cutting method is better than that by *glpm*.

The cutting method has several advantages: (i) Given one initial design $U_p(p^s)$, one can find many nearly uniform designs $U_n(n^s)$ for $n < p$; (ii) the designs obtained by the cutting method may have better uniformity than those directly generated by the good lattice point method; (iii) the performance of the cutting method does not depend on specific measure of uniformity. Interested readers are referred to Ma and Fang (2004) for more examples and comparisons for the cutting method.

3.5 Linear Level Permutation Method

The design space generated by the *glpm* is a subset of U-type designs $\mathcal{U}(n, q^s)$ for given (n, s, q). It may find some ways to improve its performance of uniformity. Tang et al. (2012) showed that the level permutation technique can improve the space-filling property of regular designs. Zhou and Xu (2015) showed that the linear level permutation of good lattice point set can also improve its space-filling property. As a special type of uniform designs, good lattice point sets can also be used in computer experiments, which are becoming increasingly popular for researchers to investigate complex systems (Sacks et al. 1989; Santner et al. 2003; Fang et al. 2006a). Large designs are needed for computer experiments; for example, Morris (1991) considered many simulation models involving hundreds of factors. Good lattice point sets can be used for large design. In this section, the linear level permutation technique is considered for improving space-filling property of *glp* sets under discrepancy and maximin distance.

Johnson et al. (1990) proposed to use maxmin or minimax distance to measure the space-filling property for a design. For each $\boldsymbol{U} \in \mathcal{D}(n; q^s)$, denote the q levels be $0, 1, \ldots, q-1$. Define $d_p(\boldsymbol{x}, \boldsymbol{y}) = \sum_{i=1}^{s} |x_i - y_i|^p, p \geqslant 1$, as the *$L_p$-distance* of any two rows $\boldsymbol{x} = (x_1, \ldots, x_s)$ and $\boldsymbol{y} = (y_1, \ldots, y_s)$ in $\boldsymbol{U}$. When $p = 1$, the distance is the *rectangular distance*, and when $p > 1$, the L_p-distance is the pth power of the traditional L_p-norm. Define the L_p-distance of the design $\boldsymbol{U}$ to be

$$d_p(\boldsymbol{U}) = \min\{d_p(\boldsymbol{x}, \boldsymbol{y}) : \boldsymbol{x} \neq \boldsymbol{y}, \boldsymbol{x}, \boldsymbol{y} \text{ are rows of } \boldsymbol{U}\}. \tag{3.5.1}$$

Johnson et al. (1990) suggested to use $\max d_p(\boldsymbol{U})$, the maximin distance criterion, to measure the space-filling property for $\boldsymbol{U}$.

Definition 3.5.1 For an integer $q \geqslant 2$, a design $\boldsymbol{U}$ with s factors each having q levels and q^{s-k} runs is called as a *regular q^{s-k} design* if $\boldsymbol{U}$ can be constructed as the product of a $q^{(s-k)}$ full factorial design $\boldsymbol{M}$ on $Z_q = \{0, 1, \ldots, q-1\}$ and a generator matrix $\boldsymbol{G}$, i.e.,

$$\boldsymbol{U} = \boldsymbol{M}\boldsymbol{G} \pmod{q}. \tag{3.5.2}$$

Here, we define the regular q^{s-k} design by this way in order to consider distances between different runs in a regular design. For example, consider $q = 3, s = 4, k = 2$, and $\boldsymbol{M}$, $\boldsymbol{G}$, and $\boldsymbol{U}$ are as follows,

$$\boldsymbol{M} = \begin{pmatrix} 0\,0 \\ 0\,1 \\ 0\,2 \\ 1\,0 \\ 1\,1 \\ 1\,2 \\ 2\,0 \\ 2\,1 \\ 2\,2 \end{pmatrix}, \boldsymbol{G} = \begin{pmatrix} 1\,0\,1\,1 \\ 0\,1\,1\,2 \end{pmatrix}, \boldsymbol{U} = \begin{pmatrix} 0\,0\,0\,0 \\ 0\,1\,1\,2 \\ 0\,2\,2\,1 \\ 1\,0\,1\,1 \\ 1\,1\,2\,0 \\ 1\,2\,0\,2 \\ 2\,0\,2\,2 \\ 2\,1\,0\,1 \\ 2\,2\,1\,0 \end{pmatrix}.$$

As shown in Tang et al. (2012) and Zhou and Xu (2014), permuting levels for one or more factors do not change the essence of orthogonality of a design, but it can change its geometrical structure and space-filling properties. For regular designs, we consider linear level permutations over $Z_q = \{0, 1, \ldots, q-1\}$ which have a simple form. Given a regular q^{s-k} design $\boldsymbol{U}$ and an $s \times 1$ vector $\boldsymbol{u}$, let

$$\boldsymbol{U}_{\boldsymbol{u}} = \boldsymbol{U} + \boldsymbol{u} = \{\boldsymbol{x} + \boldsymbol{u} \pmod{q} : \boldsymbol{x} \text{ for each row of } \boldsymbol{U}\} \tag{3.5.3}$$

be a new design after a linear level permutation of $\boldsymbol{U}$ over Z_q, where

$$\boldsymbol{x} + \boldsymbol{u} \pmod q = \{x_1 + u_1 \pmod q, \ldots, x_s + u_s \pmod q\}.$$

When considering all possible linear level permutations, it is sufficient to consider permuting the k-dependent columns and keep the $(s-k)$-independent columns unchanged. Here, the independent columns means that there is no column which can be obtained from the linear combination of other columns by modulo q. Without loss of generality, we can assume that the first $(s-k)$ columns of a regular q^{s-k} design are independent columns, i.e., the first $(s-k)$ columns form a full factorial. For a regular design, we have the following result regarding the L_p-distance.

Theorem 3.5.1 *For a regular q^{s-k} design $\boldsymbol{U}$ over Z_q, a linear level permutation does not decrease the L_p-distance, i.e., $d_p(\boldsymbol{U_u}) \geqslant d_p(\boldsymbol{U})$, where $\boldsymbol{U_u}$ is defined in (3.5.3).*

The proof of Theorem 3.5.1 can be seen in Zhou and Xu (2015). Theorem 3.5.1 shows that a linear level permutation may improve the space-filling property under the maximin distance criterion. There exist designs such that any linear level permutation neither increases nor decrease the L_p-distance. For example, it can be checked that for any regular 3^{s-1} design, $s \leqslant 10$, any linear level permutation does not change its L_p-distance.

According to the definition of regular design in (3.5.2), a good lattice point set $\boldsymbol{U}$ can be constructed from a regular design by replacing 0 with n. Specifically, let $\boldsymbol{U}_0$ be a regular design whose generator matrix is $\boldsymbol{h}$ and the independent column is $M = (0, 1, \ldots, n-1)^T$, i.e.,

$$\boldsymbol{U}_0 = M\boldsymbol{h} \pmod n. \tag{3.5.4}$$

We obtain a good lattice point set $\boldsymbol{U}$ by replacing the first row $(0, \ldots, 0)$ in $\boldsymbol{U}_0$ with the row $(n, \ldots, n)$. Thus, good lattice point sets can be treated as a special class of regular designs, although the elements of a good lattice point set are defined over the set $\{1, \ldots, n\}$, which differs from the definition of a regular design in Definition 3.5.1, where elements are from $\{0, 1, \ldots, n-1\}$. Section 1.2 is referred for more clarification.

The replacement of 0 with n changes the geometrical structure and statistical properties of the design; nevertheless, the following result shows that the L_p-distance remains the same.

Corollary 3.5.1 *The good lattice point set $\boldsymbol{U}$ and the regular design $\boldsymbol{U}_0$ in (3.5.4) have the same L_p-distance.*

Proof For any nonzero row $\boldsymbol{x} = (x_1, \ldots, x_n) \in \boldsymbol{U}_0$, there exists an integer $c \in \{1, \ldots, n-1\}$ such that $\boldsymbol{x} = c\boldsymbol{h} \pmod n$. Then, $\boldsymbol{z} = (n-c)\boldsymbol{h} = (n - x_1, \ldots, n - x_n) \pmod n$ is a row in $\boldsymbol{U}$. The L_p-distance between $\boldsymbol{x}$ and the row vector $(0, \ldots, 0)$ is equal to the L_p-distance between $\boldsymbol{z}$ and the row vector $(n, \ldots, n)$. Since the other

$n-1$ design points are same in $\boldsymbol{U}$ and $\boldsymbol{U}_0$ and the $n-1$ L_p-distances between the point $(0,\ldots,0)$ and the other $n-1$ points have same distribution as the $n-1$ L_p-distances between the point $(n,\ldots,n)$ and the other $n-1$ points, then replacing 0 with n in $\boldsymbol{U}_0$ does not change the L_p-distance. The proof is finished.

Consider the linear level permutation (3.5.3) for a good lattice point set but replace 0 with n in the permuted design. From Theorem 3.5.1, any linear level permutation of $\boldsymbol{U}_0$ in (3.5.4) does not decrease the L_1-distance. Similar to Theorem 3.5.1, we have the following result for good lattice point sets.

Proposition 3.5.1 *Let $\boldsymbol{U}$ be a $n \times s$ good lattice point set. Any linear level permutation of $\boldsymbol{U}$ does not decrease the L_1-distance, i.e., $d_1(\boldsymbol{U}_u) \geqslant d_1(\boldsymbol{U})$, where $\boldsymbol{U}_u$ is defined in (3.5.3) with 0 replaced by n.*

Example 3.5.1 Let $n=7$ and $s=6$. Using generator vector $\boldsymbol{h}=(1,\ldots,6)$, we obtain a good lattice point set

$$\boldsymbol{U} = \begin{pmatrix} 1\,2\,3\,4\,5\,6 \\ 2\,4\,6\,1\,3\,5 \\ 3\,6\,2\,5\,1\,4 \\ 4\,1\,5\,2\,6\,3 \\ 5\,3\,1\,6\,4\,2 \\ 6\,5\,4\,3\,2\,1 \\ 7\,7\,7\,7\,7\,7 \end{pmatrix}$$

with $d_1(\boldsymbol{U})=12$. Without loss of generality, consider all linear level permutations of the last five dependent columns of U. Among the 16,807 permuted designs, 16,167 designs have the same L_1-distance as $\boldsymbol{U}$, and 640 designs have L_1-distance 13. For example, using the linear permuting vector $\boldsymbol{u}=(0,4,1,5,2,6)$, we obtain the design

$$\boldsymbol{U}_u = \begin{pmatrix} 1\,6\,4\,2\,7\,5 \\ 2\,1\,7\,6\,5\,4 \\ 3\,3\,3\,3\,3\,3 \\ 4\,5\,6\,7\,1\,2 \\ 5\,7\,2\,4\,6\,1 \\ 6\,2\,5\,1\,4\,7 \\ 7\,4\,1\,5\,2\,6 \end{pmatrix}$$

with $d_1(\boldsymbol{U}_u)=13$. Note that U_u is equivalent to the design $\boldsymbol{U}+3 \pmod 7$, i.e., they have same design points. Moreover, $\boldsymbol{U}+5 \pmod 7$ also has L_1-distance 13.

Proposition 3.5.1 shows that a linear level permutation can increase the L_1-distance. Usually, the improvement of maximin distance derives the improvement of uniformity, i.e., the discrepancy may be decreased. However, finding the best linear level permutation is not an easy task when n and s are large. Example 3.5.1 suggests that we can look at some special linear level permutations. Specifically, we consider simple linear level permutations $\boldsymbol{U}+i \pmod{n}$ for $i=1,\ldots,n-1$, which may lead to better space-filling designs.

Example 3.5.2 Let $n=37$ and $s=36$. The L_1-distance and centered L_2-discrepancy of the corresponding *glp* set $\boldsymbol{U}$ are 342 and 1200.89, respectively. Consider the simple linear level permutations $\boldsymbol{U}+i \pmod{n}$ for $i=1,\ldots,n-1$. Among the permuted designs, denote $\boldsymbol{U}_1^*$ and $\boldsymbol{U}_2^*$ be the design with smallest L_1-distance and the design with smallest CD-value, respectively. Then, the L_1-distance and CD-value of U_1^*, respectively, are 408 and 203.9, and the L_1-distance and CD-value of U_2^*, respectively, are 342 and 61.48. Then, the CD-value is decreased significantly, and the simple linear level permutation improves the space-filling property. On the other hand, one can generate $K=10^4$ permuting vectors randomly and get K linear permuted designs. Among the K designs, the L_1-distance and CD-value of U_1^*, respectively, are 376 and 79.15, and the L_1-distance and CD-value of U_2^*, respectively, are 358 and 63.34. Sum up the above results

$$\begin{aligned}
&D_p(\boldsymbol{U})=342,\ CD(\boldsymbol{U})=1200.89,\\
&D_p(\boldsymbol{U}_1^*)=408,\ CD(\boldsymbol{U}_1^*)=203.9,\\
&D_p(\boldsymbol{U}_2^*)=342,\ CD(\boldsymbol{U}_2^*)=61.48,\\
&D_p(\boldsymbol{U}_1^*)=376,\ CD(\boldsymbol{U}_1^*)=79.15,\ \text{among } K \text{ designs}\\
&D_p(\boldsymbol{U}_2^*)=358,\ CD(\boldsymbol{U}_2^*)=63.34,\ \text{among } K \text{ designs}.
\end{aligned}$$

Clearly, the space-filling property including uniformity and maximin distance is also improved.

Example 3.5.3 Let $n=89$ and $s=24$. Consider to generate the initial design $\boldsymbol{U}$ by the *glpm* with power generator. The generating element is $a=33$ and the corresponding squared CD-value is 3.4302. Using the permuting vector

$$\boldsymbol{u}=(0\ 20\ 25\ 64\ 45\ 7\ 82\ 8\ 9\ 31\ 87\ 1\ 12\ 62\ 13\ 49\ 37\ 10\ 55\ 33\ 32\ 52\ 41\ 81),$$

one obtains a design $\boldsymbol{U}_{\boldsymbol{u}}$ with squared CD-value 1.7146, which is much less than that of $\boldsymbol{U}$. Then, the linear level permutation can improve the uniformity.

Examples 3.5.1–3.5.3 show that the simple linear level permutation $\boldsymbol{U}+i \pmod{n}$ is useful. Moreover, with the increase in the number of randomly permuted vectors K, one can find design with better space-filling property.

Good lattice point sets are easy to generate, and we have shown that a linear level permutation can increase the minimum distance, which often improves the uniformity. An interesting question is to find good or best linear level permutations that maximize the minimum distance. For a small design, we can perform all possible linear level permutations. Complete search becomes infeasible when the numbers of runs and factors are large. One simple method is to randomly perform some linear level permutations and choose the best one. Another is to use a stochastic algorithm such as simulation annealing or threshold accepting to find an optimal linear level permutation, which is shown in the next chapter.

3.6 Combinatorial Construction Methods

In this section, we mainly employ the discrete discrepancy introduced in Sect. 2.5.1 as the uniformity measure. If we can find a design that can achieve the lower bound of the discrete discrepancy, this design is a uniform design under the discrete discrepancy. Some tight lower bounds for the discrete discrepancy have been presented in Sect. 2.6.4; thus, we only need to develop theories for finding some methods for constructing designs that can achieve the lower bounds.

Sections 3.6.1 and 3.6.2 focus on the construction of uniform design from combinatorial designs. Connection between uniform designs and uniformly resolvable designs is derived, and several construction approaches are introduced in these two subsections. Section 3.6.3 discusses the construction from saturated orthogonal arrays. The last section covers some further results, including results on projection uniformity and nearly U-type designs.

3.6.1 Connection Between Uniform Designs and Uniformly Resolvable Designs

Application of combinatorial designs in the area of statistical design of experiments and in the theory of error-correcting codes is quite rich. Moreover, in recent years, this application is found in experimental and theoretical computer science, communications, cryptography, and networking. In this regard, mention may be made to Colbourn et al. (1999). In this and subsequent subsections, we focus on a new application of combinatorial design theory in experimental design theory, i.e., constructing uniform designs from uniformly resolvable incomplete block designs. First, let us introduce some knowledge related to uniformly resolvable designs and show their relation with uniform designs.

A *block* is a set of relatively homogeneous experimental conditions. For example, an experiment in a chemical process may require two equipment. There perhaps exists a systematic difference between the equipment. If we are not specially interested in the effect of two equipment, the equipment can be regarded as a nuisance variable. In this experiment, each equipment would form a block, because the variability of the experimental conditions within equipment would be expected to be smaller than the variability between equipment.

Example 3.6.1 In a medical experiment, the experimenter wants to compare five drugs. For each drug, ten mice are used and one needs 50 mice. We should concern with the variability of the mice in this experiment. It is better that 50 mice are from the same parents as there are less variability among mice in the same family. It is more difficult to do so. Instead, one may consider to choose ten families of mice from each one can choose five mice. Here, every five mice from the same family form a block. There are ten blocks in this experiment.

Definition 3.6.1 (*Block design*) Suppose n treatments are arranged into b blocks, such that the jth block contains k_j experimental units and the ith treatment appears s_i times in the entire design, $i = 1, \ldots, n$; $j = 1, \ldots, b$.

1. A block design that does not have all the treatments in each block is called *incomplete*.
2. A block design is said to be *equireplicate* if $s_i = s$ for all i.
3. A block design is said to be *proper* if $k_j = k$ for all j.
4. A block design is said to be *binary* if every treatment appears in each block at most once.

Note that the notations of n and s have been previously used as the number of runs and factors, and this is consistent as we can see from our subsequent discussion. Equireplicate, proper and binary incomplete block designs have received much attention among block designs, and the most widely used one is the *balanced incomplete block design* (BIBD, for short).

Definition 3.6.2 An equireplicate, proper, and binary incomplete block design is called a balanced incomplete block design, denoted by BIBD(n, b, s, k, λ), if it satisfies that every pair of treatments occurs altogether in exact λ blocks.

It is easy to see that the five parameters satisfy the following two relations:

$$ns = bk \qquad \text{and} \qquad \lambda(n-1) = s(k-1). \tag{3.6.1}$$

Hence, we can write a BIBD with the three parameters n, k, λ as BIBD(n, k, λ).

Definition 3.6.3 A block design is said to be *resolvable* if its blocks can be partitioned into *parallel classes*, each of which consists of a set of blocks that partition all the treatments. A parallel class is *uniform* if every block in the parallel class is of the same size.

It is obvious that a resolvable block design is also equireplicate and binary. A resolvable BIBD(n, k, λ) is denoted by RBIBD(n, k, λ).

Let $\mathcal{A}$ be a subset containing the different values of the block sizes, $\mathcal{R}$ be a multiset with $|\mathcal{R}| = |\mathcal{A}|$, where $|\mathcal{A}|$ denotes the cardinality of the set $\mathcal{A}$.

Definition 3.6.4 Suppose that for each $k \in \mathcal{A}$ there corresponds a positive $r_k \in \mathcal{R}$ such that there are exactly r_k parallel classes of block size k. We use URBD$(n, \mathcal{A}, \mathcal{R})$ to denote a resolvable incomplete block design with uniform parallel classes. A URBD$(n, \mathcal{A}, \mathcal{R})$ with the property that every pair of treatments occurs in exactly λ blocks is called a *uniformly resolvable design* (URD) and denoted by URD$(n, \mathcal{A}, \lambda, \mathcal{R})$.

For a URD with $\mathcal{A} = \{k_1, \ldots, k_l\}$, and $\mathcal{R} = \{r_1, \ldots, r_l\}$, it is obviously that

$$\lambda(n-1) = \sum_{i=1}^{l} r_i(k_i - 1). \tag{3.6.2}$$

For a thorough discussion of block designs and the general background on design theory, the reader may refer to Caliński and Kageyama (2000) and Beth et al. (1999). For the self-completeness of the contents and for reading convenience, in the following there are some overlapping examples and discussion with Section 3.5 of the book by Fang et al. (2006a). Let us see an example of URD.

Example 3.6.2 (*A* URD(6, {3, 2}, 1, {1, 3})) Suppose that $n = 6$ treatments are arranged into $b = 11$ blocks of size k_j from $\mathcal{A} = \{3, 2\}$ (i.e., $k_1 = 3, k_2 = 2$), such that each treatment appears in exactly four parallel classes, which are denoted by P_1, P_2, P_3 and P_4, and shown below (in each parallel class, a $\{\cdots\}$ represents a block).

$$\begin{aligned} P_1 &= \{\{1, 2, 3\}, \{4, 5, 6\}\}; \\ P_2 &= \{\{1, 4\}, \{2, 5\}, \{3, 6\}\}; \\ P_3 &= \{\{3, 5\}, \{1, 6\}, \{2, 4\}\}; \\ P_4 &= \{\{2, 6\}, \{3, 4\}, \{1, 5\}\}. \end{aligned}$$

Note that every pair of treatments occurs in exactly $\lambda = 1$ block, so it is a URD(6, {3, 2}, 1, {1, 3}).

We now establish the relationship between U-type designs and URDs. To illustrate this, suppose that there exists a URD$(n, \mathcal{A}, \lambda, \mathcal{R})$. For convenience, let $1, \ldots, n$ denote the n treatments, let $\mathcal{A} = \{n/q_1, \ldots, n/q_s\}$, $\mathcal{R} = \{1, \ldots, 1\}$, and P_j denote the parallel class of block size n/q_j, $j = 1, \ldots, s$. Then, a U-type design $\boldsymbol{U} \in \mathcal{U}(n; q_1, \ldots, q_s)$ can be constructed as follows.

Algorithm 3.6.1 (*URD–UD*)

Step 1. Assign a natural order $1, \ldots, q_j$ to the q_j blocks in each parallel class P_j $(j = 1, \ldots, s)$.

Step 2. For each P_j, construct a q_j-level column $\boldsymbol{u}^j = (u_{ij})$ as follows: Set $u_{ij} = u$, if treatment i is contained in the uth block of P_j, $u = 1, \ldots, q_j$.

Step 3. Combine the s columns constructed from P_j for $j = 1, \ldots, s$ to form a U-type design $\boldsymbol{U} \in \mathcal{U}(n; q_1, \ldots, q_s)$.

From this method, it is easy to note that the number of coincidences between any two distinct rows $\boldsymbol{u}_i$ and $\boldsymbol{u}_j$ of $\boldsymbol{U}$ is just the number of blocks in which the pair of treatments i and j appears together; thus, it is a constant λ, where $\lambda = (\sum_{j=1}^{s} \frac{n}{q_j} - s)/(n-1)$ from (3.6.2). According to Theorem 2.6.19, the resulting design $\boldsymbol{U}$ is a uniform design $U_n(q_1, \ldots, q_s)$ under the discrete discrepancy criterion. Theorem 3.6.1 gives a summary of the above explanation.

Now, let us take the URD(6, {3, 2}, 1, {1, 3}) given in Example 3.6.2 for an illustration.

Example 3.6.3 (*An* $U_6(2^1 3^3)$) From the four parallel classes of the URD(6, {3, 2}, 1, {1, 3}) given in Example 3.6.2, four columns of $\boldsymbol{U}$ can be constructed as follows. In P_1, there are two blocks {1, 2, 3} and {4, 5, 6}. We put "1" in the cells located in rows 1, 2, and 3 of the first column of $\boldsymbol{U}$ and "2" in the cells located in rows 4, 5, and 6 of that column; thus, we obtain one two-level column of $\boldsymbol{U}$. Similarly, from P_2, we put "1" in the cells located in rows 1 and 4 of the second column of $\boldsymbol{U}$, "2" and "3" in the cells located in rows 2, 5 and 3, 6 of that column, respectively. One three-level column of $\boldsymbol{U}$ is thus generated. In this way, four columns are then constructed from these four parallel classes, which form a U-type design $\boldsymbol{U}$ as follows:

$$\boldsymbol{U} = \begin{pmatrix} 1 & 1 & 2 & 3 \\ 1 & 2 & 3 & 1 \\ 1 & 3 & 1 & 2 \\ 2 & 1 & 3 & 2 \\ 2 & 2 & 1 & 3 \\ 2 & 3 & 2 & 1 \end{pmatrix},$$

and it is easy to check that $\boldsymbol{U}$ is a $U_6(2^1 3^3)$.

On the contrary, given a uniform design with constant number of coincidences between any two rows, we can construct a URD. Let $\boldsymbol{U}$ be such a uniform design $U_n(q_1, \ldots, q_s)$, the construction method can be carried out as follows:

Algorithm 3.6.2

Step 1. For each column $\boldsymbol{u}^j$ of $\boldsymbol{U}$, construct a parallel class consisting of q_j disjoint blocks each of size n/q_j as follows: If the ith element of $\boldsymbol{u}^j$ takes the uth level, then let the uth block contain i, $i = 1, \ldots, n$; $u = 1, \ldots, q_j$.

Step 2. For $j = 1, \ldots, s$, joining the s parallel classes constructed in *Step 1* together results a URBD$(n, \mathcal{A}, \mathcal{R})$, where $\mathcal{A} = \{n/q_1, \ldots, n/q_s\}$ and $\mathcal{R} = \{1, \ldots, 1\}$.

Because $\boldsymbol{U}$ has a constant number of coincidences between any two rows, it is easy to verify that the resulting URBD$(n, \mathcal{A}, \mathcal{R})$ is a URD$(n, \mathcal{A}, \lambda, \mathcal{R})$ with $\lambda = (\sum_{j=1}^{s} n/q_j - s)/(n-1)$.

Note that Liu and Fang (2005) studied the uniformity of a certain kind of resolvable incomplete block designs and proposed a method for constructing such uniform block designs. In fact, their uniform block designs are just the URDs here, and the above method is the one they proposed.

Formally, we have the following theorem which plays an important role in the above construction methods.

Theorem 3.6.1 *Under the discrete discrepancy, there exists a* $U_n(q_1^{r_1} \ldots q_l^{r_l})$ *design* $\boldsymbol{U}$ *with* $\delta_{ij}(\boldsymbol{U}) = \delta$ *for all* $1 \leqslant i \neq j \leqslant n$ *if and only if there exists a* URD$(n, \mathcal{A}, \delta, \mathcal{R})$, *where* $\mathcal{A} = \{n/q_1, \ldots, n/q_l\}$, $\mathcal{R} = \{r_1, \ldots, r_l\}$, *and* $\delta = [\sum_{j=1}^{l} r_j(n/q_j - 1)]/(n-1)$.

The equivalence between a URD$(n, \mathcal{A}, \delta, \mathcal{R})$ and a $U_n(q_1^{r_1} \ldots q_l^{r_l})$ with constant number of coincidences between any two rows has been illustrated in Example 3.6.3. It is easy to see that there are some advantages of this approach to the construction of uniform designs:

1. We can find many uniform designs without any computational search.
2. The method can construct both symmetric and asymmetric uniform designs.
3. The method can be employed to generate supersaturated uniform designs. The latter will be introduced in Chap. 7.

In a factorial design, two columns are called *fully aliased* if one column can be obtained from another by permuting levels. For given (n, s, q), we also hope that any two columns of the resulting uniform design are not fully aliased, because one cannot use two fully aliased columns to accommodate two different factors. Hence when constructing a uniform design in $\mathcal{U}(n; q^s)$, we need to limit the number of any of the q^2 level-combinations between any two distinct columns and to avoid fully aliased columns. We shall always keep this remark in the construction of uniform designs, especially via combinatorial designs.

Before closing this subsection, let us now illustrate another example to construct a uniform design from a URD.

Example 3.6.4 Suppose we have a URD$(12, \{2, 3\}, 1, \{5, 3\})$ with the treatment set $\mathcal{V} = \{1, 2, 3, 4, 5, 6, 7, 8, 9, 10, 11, 12\}$ and the eight parallel classes below:

$$P_1 = \{\{1, 2\}, \{3, 4\}, \{5, 6\}, \{7, 8\}, \{9, 10\}, \{11, 12\}\};$$
$$P_2 = \{\{1, 3\}, \{2, 4\}, \{5, 7\}, \{6, 8\}, \{9, 11\}, \{10, 12\}\};$$
$$P_3 = \{\{1, 4\}, \{2, 3\}, \{5, 8\}, \{6, 7\}, \{9, 12\}, \{10, 11\}\};$$
$$P_4 = \{\{1, 7\}, \{2, 8\}, \{3, 12\}, \{4, 9\}, \{5, 10\}, \{6, 11\}\};$$
$$P_5 = \{\{3, 5\}, \{4, 6\}, \{1, 10\}, \{2, 11\}, \{7, 12\}, \{8, 9\}\};$$
$$P_6 = \{\{1, 5, 9\}, \{2, 6, 10\}, \{3, 7, 11\}, \{4, 8, 12\}\};$$
$$P_7 = \{\{1, 6, 12\}, \{2, 7, 9\}, \{3, 8, 10\}, \{4, 5, 11\}\};$$
$$P_8 = \{\{1, 8, 11\}, \{2, 5, 12\}, \{3, 6, 9\}, \{4, 7, 10\}\}.$$

Thorough Algorithm 3.6.1, a $U_{12}(6^5 4^3)$, which is shown in Table 3.6, can be constructed from this URD. On the contrary, we can also form the corresponding URD$(12, \{2, 3\}, 1, \{5, 3\})$ from this $U_{12}(6^5 4^3)$ following Algorithm 3.6.2.

3.6.2 Construction Approaches via Combinatorics

Algorithm 3.6.1 and Theorem 3.6.1 ensure that we can construct uniform designs from URDs, and the latter have been studied extensively in combinatorial design the-

Table 3.6 An $U_{12}(6^5 4^3)$

Row	1	2	3	4	5	6	7	8
1	1	1	1	1	3	1	1	1
2	1	2	2	2	4	2	2	2
3	2	1	2	3	1	3	3	3
4	2	2	1	4	2	4	4	4
5	3	3	3	5	1	1	4	2
6	3	4	4	6	2	2	1	3
7	4	3	4	1	5	3	2	4
8	4	4	3	2	6	4	3	1
9	5	5	5	4	6	1	2	3
10	5	6	6	5	3	2	3	4
11	6	5	6	6	4	3	4	1
12	6	6	5	3	5	4	1	2

ory. This subsection will introduce several kinds of combinatorial designs (most of which can reduce to URDs) and employ them to construct uniform designs through Algorithm 3.6.1. Examples will be given to illustrate these block designs. A–E introduce the approaches for constructing symmetrical uniform designs, and F–H for asymmetrical cases. Besides these approaches discussed in the following subsections, there are some miscellaneous known results on the existence of URDs; readers can refer to Fang et al. (2001, 2004a) and the references therein for these results.

A. Construction via Resolvable Balanced Incomplete Block Designs

From the definition of RBIBD introduced in the above subsection, we know that an RBIBD(n, k, λ) is in fact a special URD$(n, \mathcal{A}, \lambda, \mathcal{R})$ with $\mathcal{A} = \{k\}$ and $\mathcal{R} = \{s\}$, where $s = \lambda(n-1)/(k-1)$ from (3.6.1). Thus, RBIBDs can be employed to construct symmetrical uniform designs, $U_n((\frac{n}{k})^s)$'s; see Fang et al. (2002b, 2003a, 2004a) for the details. These works generalized the results of Nguyen (1996) and Liu and Zhang (2000). Their results can be summarized as the following theorem.

Theorem 3.6.2 *Suppose that the discrete discrepancy is employed as the measure of uniformity. The following $U_n(q^s)$ can be constructed by Algorithm 3.6.1, where n, m, k are positive integers and their values depend on the corresponding situation.*

(a) *If $n = 2m$ is even, then a $U_n(m^{k(n-1)})$ exists, where k ia a positive integer.*
(b) *If $n = 6m + 3$, then a $U_n((2m+1)^{\frac{n-1}{2}})$ exists.*
(c) *If $n = 3m$ and $n \neq 6$, then a $U_n(m^{n-1})$ exists.*
(d) *If $n = 12m + 4$, then a $U_n((3m+1)^{\frac{n-1}{3}})$ exists.*
(e) *If $n = 4m$, then a $U_n(m^{n-1})$ exists.*
(f) *If $n = 6m$ and $n \neq 174, 240$, then a $U_n(m^{n-1})$ exists.*
(g) *If $n = 6m$, then a $U_n(m^{2(n-1)})$ exists.*
(h) *If $n = 20m + 5$ and $n \neq 45, 225, 345, 465, 645$, then a $U_n((m+1)^{\frac{n-1}{4}})$ exists.*
(i) *If $n = 5m$ and $n \neq 10, 15, 70, 90, 135, 160, 190, 195$, then a $U_n(m^{n-1})$ exists.*

Table 3.7 An RBIBD(10, 2, 1)

	P_1	P_2	P_3	P_4	P_5	P_6	P_7	P_8	P_9
b_1^j	{1, 10}	{2, 10}	{4, 9}	{3, 7}	{2, 8}	{5, 7}	{5, 6}	{1, 7}	{1, 6}
b_2^j	{8, 9}	{5, 8}	{3, 10}	{4, 10}	{6, 9}	{2, 4}	{3, 4}	{2, 5}	{2, 7}
b_3^j	{4, 5}	{3, 6}	{7, 8}	{1, 2}	{5, 10}	{1, 9}	{1, 8}	{4, 6}	{4, 8}
b_4^j	{6, 7}	{7, 9}	{2, 6}	{5, 9}	{1, 3}	{3, 8}	{7, 10}	{3, 9}	{3, 5}
b_5^j	{2, 3}	{1, 4}	{1, 5}	{6, 8}	{4, 7}	{6, 10}	{2, 9}	{8, 10}	{9, 10}

b_i^j: ith block in the jth parallel class.

Table 3.8 A $U_{10}(5^9)$ derived from Table 3.7

Row	1	2	3	4	5	6	7	8	9
1	1	5	5	3	4	3	3	1	1
2	5	1	4	3	1	2	5	2	2
3	5	3	2	1	4	4	2	4	4
4	3	5	1	2	5	2	2	3	3
5	3	2	5	4	3	1	1	2	4
6	4	3	4	5	2	5	1	3	1
7	4	4	3	1	5	1	4	1	2
8	2	2	3	5	1	4	3	5	3
9	2	4	1	4	2	3	5	4	5
10	1	1	2	2	3	5	4	5	5

Tables 3.7 and 3.8 show us an example of RBIBD and the corresponding uniform design.

B. Construction via Room Squares

The "room square" is an important concept used in combinatorial design theory. For a comprehensive introduction, reference can be made to Colbourn and Dinita (1996). Fang et al. (2002a) applied it to the construction of symmetrical supersaturated designs, which are in fact uniform designs. Now, let us give the following definition.

Definition 3.6.5 Let $\mathcal{V}$ be a set of n elements (*treatments*). A *room square* of side $n-1$ (on treatment set $\mathcal{V}$) is an $(n-1)\times(n-1)$ array, $\mathcal{F}$, which satisfies the following properties:

1. Every cell of $\mathcal{F}$ either is empty or contains an unordered pair of treatments from $\mathcal{V}$.
2. Each treatment of $\mathcal{V}$ occurs once in each row and column of $\mathcal{F}$.
3. Every unordered pair of treatments occurs in precisely one cell of $\mathcal{F}$.

Table 3.9 illustrates a room square of side 7. From this table and Definition 3.6.5, it can be easily observed that if we take the unordered pair of treatments as blocks,

Table 3.9 A room square of side 7

81			26		57	34
45	82			37		61
72	56	83			41	
	13	67	84			52
63		24	71	85		
	74		35	12	86	
		15		46	23	87

Table 3.10 Uniform designs derived from Table 3.9

Row	1	2	3	4	5	6	7	8	9	10	11	12	13	14
1	1	4	4	1	3	3	1	1	3	4	3	3	2	2
2	2	2	1	4	2	3	3	3	1	3	1	3	4	3
3	4	3	3	1	1	2	3	4	3	1	4	1	4	1
4	4	1	4	3	2	1	2	2	4	3	2	4	2	1
5	3	1	2	4	4	2	1	2	2	4	4	2	1	3
6	2	4	2	2	1	4	2	4	2	2	1	4	3	2
7	3	3	1	2	3	1	4	3	4	2	3	1	1	4
8	1	2	3	3	4	4	4	1	1	1	2	2	3	4

then the blocks from each row of a room square of side $n-1$ form a parallel class, and the resulting $n-1$ parallel classes form an RBIBD$(n, 2, 1)$. Similarly, another RBIBD$(n, 2, 1)$ can be obtained when regarding each column of the room square as a parallel class. And joining these two RBIBD together, we have an RBIBD$(n, 2, 2)$. Thus, we have

Theorem 3.6.3 *Given a room square of side $n-1$, two* RBIBD$(n, 2, 1)$*'s and one* RBIBD$(n, 2, 2)$ *can be obtained, and hence, two $U_n((\frac{n}{2})^{n-1})$'s and one $U_n((\frac{n}{2})^{2n-2})$ can be constructed through Algorithm 3.6.1, and there are no fully aliased columns in these uniform designs.*

Table 3.10 illustrates two uniform designs $U_8(4^7)$ and one uniform design $U_8(4^{14})$ (listed as columns $1, \ldots, 7$, columns $8, \ldots, 14$ and columns $1, \ldots, 14$, respectively) constructed from the room square in Table 3.9. More results on this approach can be found in Fang et al. (2002a).

C. Construction via Resolvable Packing Designs

The resolvable packing design has been extensively studied in combinatorial design theory. Fang et al. (2004c) established a strong link between resolvable packing designs and supersaturated designs and employed such designs to construct supersaturated designs. In fact, the supersaturated designs constructed by their method are uniform designs under the discrete discrepancy. Now, let us introduce some concepts related to packing designs.

Table 3.11 A resolvable 2-(6, 2, 1) packing design

$\mathcal{V}$	$\{1, 2, 3, 4, 5, 6\}$				
$\mathcal{B}$	P_1	P_2	P_3	P_4	P_5
b_1^j	$\{1, 2\}$	$\{1, 3\}$	$\{1, 4\}$	$\{1, 5\}$	$\{1, 6\}$
b_2^j	$\{3, 4\}$	$\{2, 5\}$	$\{2, 6\}$	$\{2, 4\}$	$\{2, 3\}$
b_3^j	$\{5, 6\}$	$\{4, 6\}$	$\{3, 5\}$	$\{3, 6\}$	$\{4, 5\}$

b_i^j: ith block in the jth parallel class.

Table 3.12 $U_6(3^5)$ derived from Table 3.11

Row	1	2	3	4	5
1	1	1	1	1	1
2	1	2	2	2	2
3	2	1	3	3	2
4	2	3	1	2	3
5	3	2	3	1	3
6	3	3	2	3	1

Definition 3.6.6 Let $n \geqslant p \geqslant t$. A t-$(n, p, 1)$ *packing design* is a pair $(\mathcal{V}, \mathcal{B})$, where $\mathcal{V}$ is a set of n elements (*treatments*) and $\mathcal{B}$ is a collection of p-element subsets of $\mathcal{V}$ (*blocks*), such that every t-element subset of $\mathcal{V}$ occurs in at most one block of $\mathcal{B}$. The *packing number* $N(n, p, t)$ is the maximum number of blocks in any t-$(n, p, 1)$ packing design. And a t-$(n, p, 1)$ packing design $(\mathcal{V}, \mathcal{B})$ is *optimal* if $|\mathcal{B}| = N(n, p, t)$.

For more discussions about packing designs, refer to Stinson (1996). In particular, Stinson (1996) showed that

Lemma 3.6.1 (Theorem 33.5 of Stinson 1996) *The packing number has the following upper limit*

$$N(n, p, t) \leqslant \left\lfloor \frac{n}{p} \left\lfloor \frac{n-1}{p-1} \cdots \left\lfloor \frac{n-t+1}{p-t+1} \right\rfloor \right\rfloor \right\rfloor,$$

where $\lfloor x \rfloor$ *denotes the integer part of* x.

From Definition 3.6.6, it is easy to see that, given a *resolvable packing design*, a U-type design can be constructed following Algorithm 3.6.1. Table 3.11 provides us a resolvable 2-(6, 2, 1) packing design $(\mathcal{V}, \mathcal{B})$, where $\mathcal{V} = \{1, 2, 3, 4, 5, 6\}$, $\mathcal{B}$ is partitioned into five parallel classes, each of which consists of three disjoint blocks of size 2, and every unordered pair of elements occurs in *exactly one* block of $\mathcal{B}$. Note that this design is an *optimal* 2-(6, 2, 1) packing design, because $\mathcal{B}$ contains all the different blocks of size 2 and adding one more block, e.g. $\{k, l\}$, to $\mathcal{B}$ will cause pair $\{k, l\}$ appearing in two blocks of $\mathcal{B}$. Table 3.12 gives us the U-type design constructed from this resolvable optimal packing design. In fact, it is a uniform design $U_6(3^5)$.

However in general, the property of *constant number of coincidences* between any two rows of the resulting U-type designs is not ensured, as we are uncertain whether every pair of treatments occurs in the same number of blocks of the packing design. A natural question arises about the uniformity of the U-type deigns constructed from resolvable packing designs and other properties of such U-type designs. To answer these questions, let us first give a definition of largest frequency.

Definition 3.6.7 Let $\mathcal{U}(n; q^s; r)$ be a subset of $\mathcal{U}(n; q^s)$ such that for each design in $\mathcal{U}(n; q^s; r)$, any of the q^2 level-combinations in any two columns appears at most r times. The number r is called the *largest frequency* of the design.

From Definitions 3.6.6 and 3.6.7, Fang et al. (2004c) obtained the following connection between designs in $\mathcal{U}(n; q^s; r)$ and resolvable packing designs.

Theorem 3.6.4 *The existence of a design in $\mathcal{U}(n; q^s; r)$, where $2 \leqslant p = n/q \leqslant q$ and $1 \leqslant r < p$, is equivalent to the existence of a resolvable $(r+1)$-$(n, p, 1)$ packing design with s parallel classes.*

With this connection and the known result on the upper bound of the packing number $N(n, p, t)$ in Lemma 3.6.1, Fang et al. (2004c) gave the upper bound of the number of columns, s, of the resulting design in $\mathcal{U}(n; q^s; r)$.

Theorem 3.6.5 *For given (n, q, r) satisfying $2 \leqslant p = n/q \leqslant q$ and $1 \leqslant r < p$, the upper bound of s is given by*

$$s \leqslant \left\lfloor \frac{n-1}{p-1} \left\lfloor \frac{n-2}{p-2} \cdots \left\lfloor \frac{n-r}{p-r} \right\rfloor \right\rfloor \right\rfloor,$$

where $\lfloor x \rfloor$ denotes the integer part of x.

Note that for any design in $\mathcal{U}(n; q^s; 1)$, the number of coincidences between any two rows is *at most one*. Then from Theorems 2.6.20, 3.6.4, and 3.6.5, we have

Theorem 3.6.6 *For $2 \leqslant p = n/q \leqslant q$ and $s \leqslant \lfloor (n-1)/(p-1) \rfloor$, any design $\boldsymbol{U} \in \mathcal{U}(n; q^s; 1)$ is a $U_n(q^s)$ and can be constructed from resolvable 2-$(n, p, 1)$ packing designs.*

This theorem tells us that any design in $\mathcal{U}(n; q^s; 1)$ is uniform regardless of whether s achieves the upper bound given in Theorems 3.6.5 or not. When s achieves the upper bound, the corresponding resolvable packing design is optimal.

D. Construction via Large Sets of Kirkman Triple Systems

The *large set of Kirkman triple systems* is an important concept in combinatorial design theory (see Stinson 1991) which can be regarded as a kind of resolvable optimal packing design and can be used to construct uniform designs in $\mathcal{U}(n; q^s; 2)$.

Definition 3.6.8 A *Steiner triple system* of order n, denoted by STS(n), is a pair $(\mathcal{V}, \mathcal{B})$, where $\mathcal{V}$ is a set containing n elements (*treatments*) and $\mathcal{B}$ is a collection of

Table 3.13 An LKTS(9)

$\mathcal{V}$	$\{1, 2, 3, 4, 5, 6, 7, 8, 9\}$			
	P_1^i	P_2^i	P_3^i	P_4^i
$\mathcal{B}_1$	{{124}{356}{789}}	{{257}{468}{139}}	{{347}{158}{269}}	{{167}{238}{459}}
$\mathcal{B}_2$	{{235}{467}{189}}	{{361}{578}{249}}	{{451}{268}{379}}	{{271}{348}{569}}
$\mathcal{B}_3$	{{346}{571}{289}}	{{472}{618}{359}}	{{562}{378}{419}}	{{312}{458}{679}}
$\mathcal{B}_4$	{{457}{612}{389}}	{{513}{728}{469}}	{{673}{418}{529}}	{{423}{568}{719}}
$\mathcal{B}_5$	{{561}{723}{489}}	{{624}{138}{579}}	{{714}{528}{639}}	{{534}{678}{129}}
$\mathcal{B}_6$	{{672}{134}{589}}	{{735}{248}{619}}	{{125}{638}{749}}	{{645}{718}{239}}
$\mathcal{B}_7$	{{713}{245}{689}}	{{146}{358}{729}}	{{236}{748}{159}}	{{756}{128}{349}}

Note: {124} represents the block {1, 2, 4} etc.

three-element subsets of $\mathcal{V}$, called *triples* or *blocks*, such that every unordered pair of $\mathcal{V}$ appears in exactly one block. If $\mathcal{B}$ is resolvable, we call the STS(n) a *Kirkman triple system*, which is denoted by KTS(n). A *large set* of KTS(n), denoted by LKTS(n), is a collection of $(n-2)$ pairwise disjoint KTS(n)'s on the same set $\mathcal{V}$.

Note that from this definition, a KTS(n) is in fact a resolvable optimal 2-$(n, 3, 1)$ packing design. In such a design, there are exactly $n(n-1)/6$ triples which contain all the $\binom{n}{2}$ unordered pairs, and all the triples are partitioned into $(n-1)/2$ parallel classes. As an LKTS(n) contains all the $\binom{n}{3} = (n-2)[n(n-1)/6]$ different triples of $\mathcal{V}$, it is a resolvable optimal 3-$(n, 3, 1)$ packing design. It can be also observed that a KTS(n) is an RBIBD$(n, 3, 1)$, and an LKTS(n) is an RBIBD$(n, 3, n-2)$, which is formed by the $(n-2)$ RBIBD$(n, 3, 1)$'s. In this case, $q = n/3$ is a positive integer. Hence, from the discussions in the above subsections, we have

Theorem 3.6.7 *Given an* LKTS(n) $(n \geqslant 9)$, *let* $q = n/3$ *and* $m = (n-1)/2$, *then*

1. $(n-2)$ $U_n(q^m)$*'s in* $\mathcal{U}(n; q^m; 1)$ *and one* $U_n(q^{(n-2)m})$ *in* $\mathcal{U}(n; q^{(n-2)m}; 2)$ *can be constructed through Algorithm 3.6.1.*
2. *The number of columns in the* $U_n(q^{(n-2)m})$ *or any of the* $U_n(q^m)$*'s attains the upper bound given in Theorem 3.6.5.*
3. *For* $1 < l < n-2$, *a* $U_n(q^{lm})$ *in* $\mathcal{U}(n; q^{lm}; 2)$ *can be formed by any* l *of the* $U_n(q^m)$*'s, and here* 2 *is the smallest value of* r *for* $\mathcal{U}(n; q^s; r)$, *for given* $(n, q, s) = (n, n/3, lm)$.

Tables 3.13 and 3.14 provide us an example of an LKTS(9) and the corresponding uniform designs. There are seven parts each having four columns in the $U_9(3^{28})$ given in Table 3.14. This design has the following properties:

(a) Each part is an $L_9(3^4)$, i.e., $r = 1$.
(b) For any $l = 2, \ldots, 7$, the design formed by the first $4l$ columns is a $U_9(3^{4l})$ with the largest frequency $r = 2$.

Table 3.14 $U_9(3^{4s})(1 \leqslant s \leqslant 7)$ derived from Table 3.13

Row	1	2	3	4	5	6	7
1	1 3 2 1	3 1 1 1	2 2 3 1	2 1 2 3	1 2 1 3	2 3 1 2	1 1 3 2
2	1 1 3 2	1 3 2 1	3 1 1 1	2 2 3 1	2 1 2 3	1 2 1 3	2 3 1 2
3	2 3 1 2	1 1 3 2	1 3 2 1	3 1 1 1	2 2 3 1	2 1 2 3	1 2 1 3
4	1 2 1 3	2 3 1 2	1 1 3 2	1 3 2 1	3 1 1 1	2 2 3 1	2 1 2 3
5	2 1 2 3	1 2 1 3	2 3 1 2	1 1 3 2	1 3 2 1	3 1 1 1	2 2 3 1
6	2 2 3 1	2 1 2 3	1 2 1 3	2 3 1 2	1 1 3 2	1 3 2 1	3 1 1 1
7	3 1 1 1	2 2 3 1	2 1 2 3	1 2 1 3	2 3 1 2	1 1 3 2	1 3 2 1
8	3 2 2 2	3 2 2 2	3 2 2 2	3 2 2 2	3 2 2 2	3 2 2 2	3 2 2 2
9	3 3 3 3	3 3 3 3	3 3 3 3	3 3 3 3	3 3 3 3	3 3 3 3	3 3 3 3

(c) For given $(n, q, r) = (9, 3, 2)$, the largest value of s is 28, which is attained by this $U_9(3^{28})$.
(d) For given $(n, q, s) = (9, 3, 4l), l = 2, \ldots, 7$, the smallest value of r is 2 which is attained by the $U_9(3^{4l})$.

Note that these designs are all uniform designs. This is a magic result! Fang et al. (2004c) further showed that under some other criteria, e.g. $E(f_{NOD})$ that will be defined in Sect. 7.2, these designs have a slightly better performance than that obtained by Yamada et al. (1999) and Fang et al. (2000) through numerical searches. More detailed results on the designs derived from the resolvable packing designs and LKTS can be found in Fang et al. (2004c).

E. Construction via Super-Simple Resolvable *t*-designs

Super-simple resolvable t-designs can also be used to generate uniform designs in $\mathcal{U}(n; q^s; r)$. Let us introduce some concepts of resolvable t-designs.

Definition 3.6.9 A *t-design*, denoted by $S_\lambda(t, k, n)$, is an ordered pair $(\mathcal{V}, \mathcal{B})$, where $\mathcal{V}$ is a set of n elements (*treatments*) and $\mathcal{B}$ is a collection of k-subsets of $\mathcal{V}$, called *blocks*, such that every t-subset of $\mathcal{V}$ is contained in exactly λ blocks of $\mathcal{B}$. When $t = 2$, a $S_\lambda(t, k, n)$ is just a *balanced incomplete block design*. An $S_\lambda(t, k, n)$ is called *simple* if it contains no repeated blocks. A simple $S_\lambda(t, k, n)$ is called *super-simple* if no two blocks have more than two points in common. A resolvable $S_\lambda(t, k, n)$ is denoted by $RS_\lambda(t, k, n)$.

The reader can refer to Beth et al. (1999) for more discussions about super-simple $S_\lambda(t, k, n)$. From this definition, we can see that an $S_1(2, k, n)$, an $S_1(3, 4, n)$ and a simple $S_\lambda(2, 3, n)$ are all super-simple. Also, based on what have been discussed in Sect. 3.6.1, Fang et al. (2004b) established the following link.

Theorem 3.6.8 *The existence of a design in* $\mathcal{U}(n; q^s)$ *with* λ *coincidences among any t distinct rows is equivalent to the existence of an* $RS_\lambda(t, n/q, n)$ *with s parallel classes.*

Furthermore, if we use a super-simple $RS_\lambda(2, k, n)$ to construct a $U(n; q^s)$ design, it is easy to see that the largest frequency of the resulting $U(n; q^s)$ is 2, i.e., this design is in $\mathcal{U}(n; q^s; 2)$. Fang et al. (2004b) obtained the following result.

Theorem 3.6.9

1. *The existence of a design in $\mathcal{U}(n; q^s; 2)$ is equivalent to the existence of a super-simple $RS_\lambda(2, n/q, n)$ with s parallel classes, where $\lambda = s(n - q)/(q(n - 1))$.*
2. *Given a super-simple $RS_\lambda(2, k, n)$, the design in $\mathcal{U}(n; q^s; 2)$ constructed through Algorithm 3.6.1 is a uniform design and has no fully aliased columns.*

Example 3.6.5 shows us a super-simple $RS_2(2, 3, 12)$ and the uniform design generated from it for illustration. More results can be found in Fang et al. (2004b), for example, a $RS_2(2, 3, 9)$, a $RS_1(3, 4, 16)$ and the corresponding $U_9(3^8; 2)$, $U_{16}(4^{35}; 2)$, respectively.

Example 3.6.5 For $n = 12$, $\mathcal{V} = \{1, 2, 3, 4, 5, 6, 7, 8, 9, 10, 11, 12\}$ and one given parallel class P_1, which contains four blocks $b_1^1 = \{1, 2, 4\}, b_2^1 = \{3, 6, 8\}, b_3^1 = \{5, 9, 10\}$, and $b_4^1 = \{7, 11, 12\}$. Applying the permutation group generated by $(1, 2, 3, 4, 5, 6, 7, 8, 9, 10, 11)(12)$ to this parallel class, then 11 parallel classes are obtained each containing four blocks. For example, based on the permutation group generated by $(1, 2, 3, 4, 5, 6, 7, 8, 9, 10, 11)(12)$, adding one to each element of P_1 results the parallel classes $P_2 = \{\{2, 3, 5\}, \{4, 7, 9\}, \{6, 10, 11\}, \{8, 1, 12\}\}$. Every unordered pair of $\mathcal{V}$ appears in exactly two blocks, and any triple of $\mathcal{V}$ appears in at most one parallel class. So a super-simple $RS_2(2, 3, 12)$ is formed which is shown in Table 3.15. From this super-simple $RS_2(2, 3, 12)$, a uniform design $U_{12}(4^{11}; 2)$ is obtained by Algorithm 3.6.1 and is listed in Table 3.15 also. Extension of this example to the general case is straightforward.

F. Construction via Resolvable Group Divisible Designs

Fang et al. (2001) employed *resolvable group divisible designs* (RGDDs) for constructing both symmetrical and asymmetrical (mixed-level) uniform designs.

Definition 3.6.10 Let k and g be positive integers, and let n be a multiple of g. A *group divisible design* of *index* one, *order* n, and *type* $g^{n/g}$, denoted by k-GDD($g^{n/g}$), is a triple $(\mathcal{V}, \mathcal{G}, \mathcal{B})$, where

1. $\mathcal{V}$ is a set of n treatments.
2. $\mathcal{G}$ is a partition of $\mathcal{V}$ into *groups* of size g.
3. $\mathcal{B}$ is a family of blocks of $\mathcal{V}$, such that each block is of size k.
4. Every pair of treatments occurs in exactly one block or group, but not both.

If the blocks of a GDD can be partitioned into parallel classes, the GDD is then called a resolvable GDD, denoted by k-RGDD($g^{n/g}$). Obviously when $g = 1$, a k-RGDD is in fact a RBIBD$(n, k, 1)$.

Example 3.6.6 Start from the URD$(6, \{3, 2\}, 1, \{1, 3\})$ in Example 3.6.2 and take $\mathcal{V} = \{1, 2, 3, 4, 5, 6\}$, $\mathcal{G} = P_1$ and $\mathcal{B} = \bigcup_{i=2}^{4} P_i$, we can get a 2-RGDD($3^2$).

Table 3.15 A $U_{12}(4^{11}; 2)$ and the corresponding simple $RS_2(2, 3, 12)$

Row	1	2	3	4	5	6	7	8	9	10	11
1	1	4	3	3	2	4	2	3	1	2	1
2	1	1	4	3	3	2	4	2	3	1	2
3	2	1	1	4	3	3	2	4	2	3	1
4	1	2	1	1	4	3	3	2	4	2	3
5	3	1	2	1	1	4	3	3	2	4	2
6	2	3	1	2	1	1	4	3	3	2	4
7	4	2	3	1	2	1	1	4	3	3	2
8	2	4	2	3	1	2	1	1	4	3	3
9	3	2	4	2	3	1	2	1	1	4	3
10	3	3	2	4	2	3	1	2	1	1	4
11	4	3	3	2	4	2	3	1	2	1	1
12	4	4	4	4	4	4	4	4	4	4	4
$\mathcal{B}$	P_1	P_2	P_3	P_4	P_5	P_6	P_7	P_8	P_9	P_{10}	P_{11}
b_1^j	$\{124\}$	$\{235\}$	$\{346\}$	$\{457\}$	$\{568\}$	$\{679\}$	$\{78t_0\}$	$\{89t_1\}$	$\{9t_01\}$	$\{t_0t_12\}$	$\{t_113\}$
b_2^j	$\{368\}$	$\{479\}$	$\{58t_0\}$	$\{69t_1\}$	$\{7t_01\}$	$\{8t_12\}$	$\{913\}$	$\{t_024\}$	$\{t_135\}$	$\{146\}$	$\{257\}$
b_3^j	$\{59t_0\}$	$\{6t_0t_1\}$	$\{7t_11\}$	$\{812\}$	$\{923\}$	$\{t_034\}$	$\{t_145\}$	$\{156\}$	$\{267\}$	$\{378\}$	$\{489\}$
b_4^j	$\{7t_1t_2\}$	$\{81t_2\}$	$\{92t_2\}$	$\{t_03t_2\}$	$\{t_14t_2\}$	$\{15t_2\}$	$\{26t_2\}$	$\{37t_2\}$	$\{48t_2\}$	$\{59t_2\}$	$\{6t_0t_2\}$

b_i^j: ith block in the jth parallel class; t_i: $10 + i$, for $i = 0, 1, 2$.

Suppose $(\mathcal{V}, \mathcal{G}, \mathcal{B})$ is a k-RGDD$(g^{n/g})$. Let $\mathcal{B}^* = \mathcal{G} \bigcup \mathcal{B}$, then $\mathcal{B}^*$ can be regarded as a block design, where the groups of $\mathcal{G}$ just form a uniform parallel class with block size g. From the definition of GDD, we can see that every pair of elements of $\mathcal{V}$ occurs in exactly one block of $\mathcal{B}^*$. Thus from the resolvability and uniformity of the original RGDD, the block design $\mathcal{B}^*$ is a URD and can be used to generate a uniform design. Fang et al. (2001) gave the following theorem and a lot of new uniform/supersaturated designs.

Theorem 3.6.10 *If there exists a* k-RGDD$(g^{n/g})$, *then there exists a* URD$(n, \{g, k\}, 1, \{1, r_k\})$, *where* $r_k = (n - g)/(k - 1)$; *as a result, there exists a* $U_n((\frac{n}{g})^1 (\frac{n}{k})^{\frac{n-g}{k-1}})$.

G. Construction via Latin Squares

Latin squares are playing an important role in experimental design over a very long time. Here, we employ Latin squares to construct k-RGDD(g^k) designs as well as uniform designs. Now, let us review the definition of Latin square.

Definition 3.6.11 A $g \times g$ matrix with g symbols as its elements is called a *Latin square* of order g, denoted by $L = (L_{ij})$, if each symbol appears in each row as well as each column once and only once. We call this property a *Latin property*. Two Latin squares are said to be *orthogonal* if their superposition yields g^2 different ordered pair. A set of Latin squares is called a set of *pairwise orthogonal Latin squares* if any pair of which are orthogonal.

Table 3.16 Two orthogonal Latin squares of order 4

No. 1				No. 2			
8	6	7	5	9	11	12	10
5	7	6	8	11	9	10	12
6	8	5	7	12	10	9	11
7	5	8	6	10	12	11	9

Let $N(g)$ denote the maximum number of pairwise orthogonal Latin squares of order g. For the known results on $N(g)$, the reader can refer to Colbourn and Dinita (1996). Fang et al. (2001) showed that

Theorem 3.6.11 *If $N(g) \geqslant k-1$, then there exists a k-RGDD(g^k); thus, we have a $U_{kg}(k^1 g^g)$ with $n = kg$ runs and $g+1$ factors one having k levels and the other g factors having g levels.*

Let us see an example of constructing an RGDD and a uniform design from pairwise orthogonal Latin squares.

Example 3.6.7 Take the two orthogonal Latin squares of order 4 shown in Table 3.16 as an example; we may construct a 3-RGDD(4^3) $(\mathcal{V}, \mathcal{G}, \mathcal{B})$ as follows. Here, $\mathcal{V} = \{1, 2, 3, 4, 5, 6, 7, 8, 9, 10, 11, 12\}$, $\mathcal{G} = \{\{1, 2, 3, 4\}, \{5, 6, 7, 8\}, \{9, 10, 11, 12\}\}$, $\mathcal{B} = \bigcup_{i=1}^{4} P_i$ and

$$P_1 = \{\{1, 8, 9\}, \{2, 6, 11\}, \{3, 7, 12\}, \{4, 5, 10\}\};$$
$$P_2 = \{\{1, 5, 11\}, \{2, 7, 9\}, \{3, 6, 10\}, \{4, 8, 12\}\};$$
$$P_3 = \{\{1, 6, 12\}, \{2, 8, 10\}, \{3, 5, 9\}, \{4, 7, 11\}\};$$
$$P_4 = \{\{1, 7, 10\}, \{2, 5, 12\}, \{3, 8, 11\}, \{4, 6, 9\}\}.$$

From this RGDD, a $U_{12}(3^1 4^4)$ can be constructed as shown in Table 3.17.

Table 3.17 A $U_{12}(3^1 4^4)$

Row	1	2	3	4	5
1	1	1	1	1	1
2	1	2	2	2	2
3	1	3	3	3	3
4	1	4	4	4	4
5	2	4	1	3	2
6	2	2	3	1	4
7	2	3	2	4	1
8	2	1	4	2	3
9	3	1	2	3	4
10	3	4	3	2	1
11	3	2	1	4	3
12	3	3	4	1	2

H. Construction via Resolvable Partially Pairwise Balanced Designs

Fang et al. (2006b) proposed to employ the *resolvable partially pairwise balanced design* for constructing uniform designs with the property that the Hamming distances between any two rows differ by at most one. First, let us review some related terminologies.

Definition 3.6.12 Let $\mathcal{V}$ be a set of n treatments and $\mathcal{B}$ be a family of blocks of $\mathcal{V}$ with sizes from a set $\mathcal{A}$, such that every pair of distinct treatments occurs exactly in λ_1 or λ_2 blocks. This design is called a *partially pairwise balanced design* of order n, with block size set $\mathcal{A}$, and is denoted by $(n, \mathcal{A}; \lambda_1, \lambda_2)$-PPBD. A resolvable PPBD with mutually distinct and *uniform* parallel classes is denoted by $\widetilde{\text{R}}$PPBD (cf. Definition 3.6.3).

When $\lambda_1 = \lambda_2$, an $\widetilde{\text{R}}$PPBD is a URD with mutually distinct parallel classes. It should be emphasized that for given parameters n, $\mathcal{A} = \{k_1, \ldots, k_l\}$, λ_1 and λ_2, the number of parallel classes of an $(n, \mathcal{A}; \lambda_1, \lambda_2)$-$\widetilde{\text{R}}$PPBD is not a constant, but depends on its construction. In view of this, when $\widetilde{\text{R}}$PPBD contains r_i parallel classes of block size k_i $(i = 1, \ldots, l)$, we say that it is of *class type* $k_1^{r_1} \ldots k_l^{r_l}$.

Example 3.6.8 Suppose $\mathcal{V} = \{1, \ldots, 8\}$. Let $\lambda_1 = 2, \lambda_2 = 1$, and $\mathcal{A} = \{4, 2\}$. Then, the following blocks form a (8, {4, 2}; 2, 1)-$\widetilde{\text{R}}$PPBD with class type $4^4 2^1$.

$$
\begin{array}{c}
\{1,3,5,7\}, \quad \{2,4,6,8\},\\
\{1,3,6,8\}, \quad \{2,4,5,7\},\\
\{1,4,5,8\}, \quad \{2,3,6,7\},\\
\{1,4,6,7\}, \quad \{2,3,5,8\},\\
\{1,2\}, \ \{3,4\}, \ \{5,6\}, \ \{7,8\}.
\end{array}
$$

Following Algorithm 3.6.1, given an $\widetilde{\text{R}}$PPBD, a U-type design can be generated, and it can be easily seen that the resulting design has its numbers of coincidences between any two rows taking only two values λ_1 and λ_2. Thus based on Theorems 2.6.20, we have

Theorem 3.6.12 *Given an $(n, \{k_1, \ldots, k_l\}; \lambda_1, \lambda_2)$-$\widetilde{\text{R}}$PPBD of class type $k_1^{r_1} \ldots k_l^{r_l}$ satisfying the condition $|\lambda_1 - \lambda_2| \leqslant 1$, then a uniform design $U_n((\frac{n}{k_1})^{r_1} \cdots (\frac{n}{k_l})^{r_l})$ can be derived through Algorithm 3.6.1, and there are no fully aliased columns in this uniform design.*

Example 3.6.9 Applying Algorithm 3.6.1 to the $\widetilde{\text{R}}$PPBD shown in Example 3.6.8, we obtain a uniform design $U_8(2^4 4^1)$ shown below. It is interesting to note that this design is indeed an asymmetrical orthogonal array of strength 2.

Row	1 2 3 4 5
1	1 1 1 1 1
2	2 2 2 2 1
3	1 1 2 2 2
4	2 2 1 1 2
5	1 2 1 2 3
6	2 1 2 1 3
7	1 2 2 1 4
8	2 1 1 2 4

3.6.3 Construction Approach via Saturated Orthogonal Arrays

Saturated orthogonal arrays have some interesting properties that can be used for the construction of supersaturated and uniform designs. Lin (1993) provided a method for constructing two-level supersaturated designs of size $(n, s) = (2t, 4t - 2)$ using a half fraction of a Hadamard matrix (HFHM, for short), where a *Hadamard matrix*, $\boldsymbol{H}$, of order n is an $n \times n$ matrix with elements 1 and -1, which satisfies $\boldsymbol{H}'\boldsymbol{H} = n\boldsymbol{I}_n$. Later, Cheng (1997) gave a theoretical justification of HFHM. The following theorem plays an important role in this construction method, and its extension for more general cases can be found in Fang et al. (2003b).

Theorem 3.6.13 *Suppose $\boldsymbol{U}$ is a saturated orthogonal array $L_n(q^s)$, where*
Case (i) q is a prime power, $n = q^t$, $s = (n - 1)/(q - 1)$ and $t \geqslant 2$, or
Case (ii) $q = 2$, $n = 4t$, $s = 4t - 1$ and $t \geqslant 1$,
then the Hamming distances between different rows are equal to q^{t-1} in Case (i) or $2t$ in Case (ii). That is, the design is a uniform design under the discrete discrepancy.

The Hadamard matrices are saturated orthogonal arrays with parameters satisfying Case (ii) of Theorem 3.6.13. Fang et al. (2003b) generalized the HFHM method to the *fractions of saturated orthogonal arrays* (FSOA, for short) method, for constructing asymmetrical uniform designs from saturated orthogonal arrays. The FSOA method is presented as follows:

Algorithm 3.6.3 (*FSOA method*)

Step 1. Let $\boldsymbol{U}$ be a saturated orthogonal array of strength 2. Choose a column of $\boldsymbol{U}$, say the kth column (**k**), and split the total n rows of $\boldsymbol{U}$ into q groups, such that group i has all the $n/q = q^{t-1}$ level i's in column (**k**). We call this column (**k**) the *branching column*.

Step 2. Given p $(2 \leqslant p < q)$, taking any p groups results in an asymmetrical supersaturated design $U(pq^{t-1}; p^1q^{s-1})$ to examine one p-level factor on the branching column and $(s - 1)$ q-level factors on other columns.

Table 3.18 Uniform designs derived from $L_{16}(4^5)$ (using 1 as the branching column)

$U_{12}(3^1 4^4)$	$U_8(2^1 4^4)$	Row	1	2	3	4	5
1		1	1	1	1	1	1
2		2	1	2	2	2	2
3		3	1	3	3	3	3
4		4	1	4	4	4	4
5	1	5	2	1	2	3	4
6	2	6	2	2	1	4	3
7	3	7	2	3	4	1	2
8	4	8	2	4	3	2	1
	5	9	3	1	3	4	2
	6	10	3	2	4	3	1
	7	11	3	3	1	2	4
	8	12	3	4	2	1	3
9		13	4	1	4	2	3
10		14	4	2	3	1	4
11		15	4	3	2	4	1
12		16	4	4	1	3	2

The resulting supersaturated designs are uniform designs, as shown in Theorem 3.6.13. Now, let us take the saturated orthogonal array $L_{16}(4^5)$ shown in Table 3.18 for an illustration. If we take column (**1**) as the *branching column*, then the total $n = 16$ rows can be split into $q = 4$ groups, i.e., rows 1–4, 5–8, 9–12, and 13–16. Any three groups of rows can be used to form an asymmetrical supersaturated design $U_{12}(3^1 4^4)$ to examine one 3-level factor and four 4-level factors, e.g., rows 1–8 and 13–16. Similarly, any two groups of rows can be used to form an asymmetrical supersaturated design $U_8(2^1 4^4)$, to examine one 2-level factor and four 4-level factors, e.g., rows 5–12. In Table 3.18, we give two such designs whose designs are entitled $U_{12}(3^1 4^4)$ and $U_8(2^1 4^4)$, respectively.

As for the properties of the constructed designs, Fang et al. (2003b) obtained the following result.

Theorem 3.6.14 *The asymmetrical supersaturated designs obtained by the FSOA method described above are uniform designs, i.e.,* $U_{pq^{t-1}}(p^1 q^{s-1})$*'s, and the* p $(2 \leqslant p < q)$*-level factor is orthogonal to those* q*-level factors, no matter which column is selected as the branching column and which groups are selected. Moreover, there are no fully aliased factors in these uniform designs.*

Saturated orthogonal arrays of strength 2 are available in many design books, such as Hedayat et al. (1999). The examples cited in this book can be found at http://neilsloane.com/oadir/. So based on the FSOA method, many uniform designs can be obtained from the saturated orthogonal arrays with parameters satisfying Case (i) of Theorem 3.6.13.

3.6.4 Further Results

In the above subsections, we mainly discussed the construction of uniform designs in terms of the discrete discrepancy (2.5.6), and most of the generated designs achieve the lower bound given in Theorem 2.6.19. Based on Theorem 2.6.20, removing any column from or adding a level-balanced column to any of these uniform designs still results a uniform design, as the numbers of coincidences between any two rows of the design thus obtained differ by at most one. From (2.5.6), we see that this discrepancy regards factors with different levels of the same importance. While the discrete discrepancy (2.5.6) and projection discrepancy (6.5.3) consider different weights for factors with different levels, here we will provide some results on two-dimensional projection uniform designs under the projection discrepancy (6.5.3) by taking $j = 2$ and $\gamma = 1$.

Note that all the uniform designs $U_n(q^s)$ discussed above are constructed based on U-type designs and thus require the number of experimental runs n to be a multiple of the number of factor levels q. In this subsection, we will also provide some results on nearly U-type designs and uniform designs when n is not divisible by q.

A. Two-Dimensional Projection Uniform Designs

For a factorial design $\mathcal{P} = \{\boldsymbol{x}_1, \ldots, \boldsymbol{x}_n\}$ with n runs and s factors, and some weights $w_k > 0$, let

$$\delta_{ij}^{\mathrm{w}}(\mathcal{P}) = \sum_{k=1}^{s} w_k \delta_{x_{ik} x_{jk}}. \tag{3.6.3}$$

$\delta_{ij}^{\mathrm{w}}(\mathcal{P})$ is a generalization of (2.5.2) and is called the *weighted coincidence number* between two rows $\boldsymbol{x}_i$ and $\boldsymbol{x}_j$. Let

$$\delta = \sum_{k=1}^{s} w_k (n/q_k - 1)/(n-1), \text{ and}$$

$$\Delta = \left\{ \sum_{k=1}^{s} w_k \delta^{(k)} : \delta^{(k)} = 0, 1, \text{ for } k = 1, \ldots, m \right\}.$$

Among the values in Δ, let δ_L and δ_U be the two nearest ones to δ, satisfying $\delta_L \leqslant \delta < \delta_U$. Then, based on the majorization theory (Marshall and Olkin 1979) and the connections between the projection uniformity in terms of (6.5.3) and some other criteria (i.e., the generalized minimum aberration, minimum moment aberration, and minimum χ^2 criterion), Liu et al. (2006) obtained the following lower bound for the two-dimensional projection discrepancy.

Theorem 3.6.15 *Suppose the weights* $w_k = \lambda q_k$ *for* $1 \leqslant k \leqslant m$, *then for any* $U(n; q_1 \ldots q_s)$ *design* $\boldsymbol{U}$, *the* $[D_{(2)}(\boldsymbol{U}; \mathcal{K})]^2$ *in (6.5.3) with* $\gamma = 1$ *satisfies*

$$[D_{(2)}(\boldsymbol{U};\mathcal{K})]^2 \geqslant \frac{n-1}{2n\lambda^2}\Big\{(\delta_U+\delta_L)\delta-\delta_U\delta_L\Big\}$$
$$-\frac{1}{2}\left\{s(s-1)+\sum_{k=1}^{s}q_k-\frac{1}{n}\left(\sum_{k=1}^{s}q_k\right)^2\right\}.$$

The equality holds if and only if for any i, among the $(n-1)$ *values of* $\delta^{\rm w}_{1i}(\boldsymbol{U}),\ldots,$ $\delta^{\rm w}_{(i-1)i}(\boldsymbol{U})$, $\delta^{\rm w}_{i(i+1)}(\boldsymbol{U}),\ldots,\delta^{\rm w}_{in}(\boldsymbol{U})$, *there are* $(n-1)\frac{\delta_U-\delta}{\delta_U-\delta_L}$ *with the value* δ_L *and* $(n-1)\frac{\delta-\delta_L}{\delta_U-\delta_L}$ *with the value* δ_U.

We should notice that most of the designs achieving the lower bound provided above are supersaturated designs, and all those symmetrical designs obtained in Sect. 3.6.2 with constant number of coincidences between any two rows are also optimal according to $D_{(2)}(\boldsymbol{U};\mathcal{K})$.

Theorem 3.6.15 provides a condition when the lower bounds can be achieved. For some values of $(n, s, q_1, \ldots, q_s)$, these lower bounds are attainable. For example, when $\boldsymbol{U}$ is a saturated $OA(n; q_1, \ldots, q_s; 2)$, they are attained, as $\boldsymbol{\delta}(\boldsymbol{U}) = (\lambda(s-1), \ldots, \lambda(s-1))$ (Mukerjee and Wu 1995). Based on this fact, Li et al. (2004) extended the FSOA method introduced in Algorithm 3.6.3 to the construction of $\chi^2(\boldsymbol{U})$-optimal asymmetrical supersaturated designs and studied the properties of the resulting designs. Based on the connection between $D_{(2)}(\boldsymbol{U};\mathcal{K})$ and $\chi^2(\boldsymbol{U})$ (see, Liu et al. 2006), the designs constructed from their methods are also uniform according to $D_{(2)}(\boldsymbol{U};\mathcal{K})$.

Another paper concerning the construction of asymmetrical supersaturated designs is due to Fang et al. (2004a), as just introduced in Sect. 3.6.2. From Fang et al. (2004a)'s concluding remarks, we know that all their designs are of *one coincidence position* between any two distinct rows. Also, we can see that most of their designs are of the form $U(n; p^1q^{s-1})$. For $U(n; p^1q^{s-1})$ designs, Liu et al. (2006) had the following result.

Theorem 3.6.16 *Let* $\boldsymbol{U}$ *be a* $U(n; p^1q^{s-1})$ *design, where* $p \leqslant q$ *and* $n/p + (s-1)n/q - s = n-1$. *If there exists exactly* one coincidence position *between any two distinct rows of* $\boldsymbol{U}$, *then* $\boldsymbol{U}$ *is optimal according to* $D_{(2)}(\boldsymbol{U};\mathcal{K})$.

Hence from this theorem, we can easily have the conclusion that the uniform designs $U_n(p^1q^{s-1})$ with $p \leqslant q$ obtained in Sect. 3.6.2 are still uniform according to $D_{(2)}(\boldsymbol{U};\mathcal{K})$.

The *column juxtaposition* method can also be used to construct asymmetrical uniform designs. From Li et al. (2004) and Liu et al. (2006), we have

Corollary 3.6.1 *Let* $\boldsymbol{U}_k$ *for* $1 \leqslant k \leqslant l$ *be balanced designs with the same number of runs. If the natural weights* $w_k = \lambda q_k$ *for* $k = 1, \ldots, m$ *are assumed, and the weighted coincidence numbers between any two distinct rows are constant for each design* $\boldsymbol{U}_k$, *then* $\boldsymbol{U} = (\boldsymbol{U}_1, \ldots, \boldsymbol{U}_l)$ *is optimal according to* $D_{(2)}(\boldsymbol{U};\mathcal{K})$.

Based on this corollary, many uniform designs according to $D_{(2)}(\boldsymbol{U};\mathcal{K})$ can be constructed, not only from saturated orthogonal arrays of strength 2, but also from designs with the given property as shown in the corollary, such as the designs due to Liu and Zhang (2000), Fang et al. (2000), and those introduced in Sects. 3.6.2 and 3.6.3.

B. Nearly U-Type Designs and Uniform Designs

When n is not divisible by q, let $n = qt + r$ $(0 < r < q)$. In this case, we arrange the design $\boldsymbol{U}$ so that $q - r$ levels occur t times, while the remaining r values occur $t + 1$ times in each column of $\boldsymbol{U}$. This guarantees every level appears in each column of $\boldsymbol{U}$ as equally as possible. Such a design is called a *nearly U-type design* and denoted by $\mathrm{NU}(n, q^s)$ (Fang et al. 2004d). For completeness, we admit $r = 0$. An $\mathrm{NU}(n, q^s)$ with $n = qt$ is just a U-type design.

For the discrete discrepancy (2.5.7), Fang et al. (2004d) obtained the following lower bound which generalized the one given in Theorem 2.6.19.

Theorem 3.6.17 *Let n, s, and q be positive integers and $n = qt + r, 0 \leqslant r \leqslant q - 1$. Let $\boldsymbol{U}$ be a fractional factorial design of n runs and s q-level factors. Suppose $\boldsymbol{u} = \frac{st(n-q+r)}{n(n-1)}$ and $\lambda = \lfloor \boldsymbol{u} \rfloor$. Then for the discrete discrepancy (2.5.7)*

$$[DD(\boldsymbol{U})]^2 \geqslant -\left(\frac{a + (q-1)b}{q}\right)^s + \frac{a^s}{n} + \frac{b^s(n-1)}{n}\left[(\lambda + 1 - \mu)\left(\frac{a}{b}\right)^{\lambda} + (\mu - \lambda)\left(\frac{a}{b}\right)^{\lambda+1}\right] \quad (3.6.4)$$

and the lower bound of $DD(\boldsymbol{U})$ on the right-hand side of (3.6.4) *can be achieved if and only if all the $\delta_{ij}(\boldsymbol{U})$'s defined in (2.5.2) for $i \neq j$ take the same value λ, or take only two values λ and $\lambda + 1$.*

Now we call an $\mathrm{NU}(n; q^s)$ a *uniform design* under $\mathrm{DD}(\boldsymbol{U})$, also denoted by $U_n(q^s)$, if it's discrepancy $\mathrm{DD}(\boldsymbol{U})$ achieves the minimum value among all such $\mathrm{NU}(n; q^s)$'s. Obviously, a design $\boldsymbol{U}$ whose $[\mathrm{DD}(\boldsymbol{U})]^2$ value equals to the lower bound in (3.6.4) is a uniform design.

Fang et al. (2004d) further developed a link between such uniform designs and resolvable packings and coverings in combinatorial design theory. Through resolvable packings and coverings without identical parallel classes, many infinite classes of new uniform designs were then produced. Readers can refer to Fang et al. (2004d) for the detailed results.

Exercises

3.1

For the case of $n = 8, s = 2$, give the design space $\mathcal{D}(8; C^2), \mathcal{D}(8; , 8^2), \mathcal{U}(8; 8^2)$ and $\mathcal{U}(8; 2 \times 4)$.

3.2

Transfer the design $\boldsymbol{U}$ in Example 3.1.2 into $\boldsymbol{X}_{lft}$, $\boldsymbol{X}_{ctr}$, $\boldsymbol{X}_{ext}$ and $\boldsymbol{X}_{mis}$ in C^4, respectively. Calculate WD, CD, and MD for these four designs in C^4. Are these four designs equivalent?

3.3

Give necessary conditions for a uniform design table $U_n(n^s)$.

3.4

By the use of the Fibonacci sequence introduced in Sect. 3.3.1, construct $U(n; n^2)$ for $n = 5, 8, 13$ and compare their CD-values with designs in Tables 3.1 and 3.2.

3.5

Suppose an experiment has three factors, temperature, pressure, and reaction time, and the ranges are [50, 100] °C, [3, 5] atm and [10, 25] min, respectively. Then, the experimental domain $\mathcal{X} = [50, 80] \times [3, 6] \times [10, 25]$. Suppose each factor has four levels. Two possible choices of design matrices $\boldsymbol{Z}_1$ and $\boldsymbol{Z}_2$ and their mapping to $[0, 1]^2$ are as follows:

$$\boldsymbol{Z}_1 = \begin{pmatrix} 60 & 6 & 25 \\ 70 & 6 & 15 \\ 60 & 3 & 15 \\ 80 & 5 & 20 \\ 50 & 5 & 10 \\ 70 & 3 & 25 \\ 80 & 4 & 10 \\ 50 & 4 & 20 \end{pmatrix} \Rightarrow \boldsymbol{X}_1 = \begin{pmatrix} 0.3333 & 1 & 1 \\ 0.6667 & 1 & 0.3333 \\ 0.3333 & 0 & 0.3333 \\ 1 & 0.6667 & 0.6667 \\ 0 & 0.6667 & 0 \\ 0.6667 & 0 & 1 \\ 1 & 0.3333 & 0 \\ 0 & 0.3333 & 0.6667 \end{pmatrix},$$

$$\boldsymbol{Z}_2 = \begin{pmatrix} 61.25 & 5.625 & 23.125 \\ 68.75 & 5.625 & 15.625 \\ 61.25 & 3.375 & 15.625 \\ 76.25 & 4.875 & 19.375 \\ 53.75 & 4.875 & 11.875 \\ 68.75 & 3.375 & 23.125 \\ 76.25 & 4.125 & 11.875 \\ 53.75 & 4.125 & 19.375 \end{pmatrix} \Rightarrow \boldsymbol{X}_2 = \begin{pmatrix} 0.375 & 0.875 & 0.875 \\ 0.625 & 0.875 & 0.375 \\ 0.375 & 0.125 & 0.375 \\ 0.875 & 0.625 & 0.625 \\ 0.125 & 0.625 & 0.125 \\ 0.625 & 0.125 & 0.875 \\ 0.875 & 0.375 & 0.125 \\ 0.125 & 0.375 & 0.625 \end{pmatrix}.$$

Answer the following questions:

(1) Compare designs $\boldsymbol{Z}_1$ and $\boldsymbol{Z}_2$ and give your comments.
(2) For obtaining $\boldsymbol{X}_1$ and $\boldsymbol{X}_2$, one mapping includes 0 and 1, but another does not involve 0 and 1. Find the formulae that use for the two mappings.

3.6

Find the Euler function $\phi(n)$ for $n = 8, 9, 10, 14, 15$ and the corresponding generating vectors $\mathcal{H}_n$.

3.7

Give three cases satisfying $\phi(n) > \phi(n+1)$. In these cases, the leave-one-out *glpm* is not worth to be recommended.

3.8

Give the cardinality of $\mathcal{G}_{8,2}$. Under the discrepancy MD, put designs of $\mathcal{G}_{8,2}$ into groups such that designs in the same group are equivalent and designs in different group have different MD-values.

3.9

For given (n, s), find the cardinality of $\mathcal{A}_{n,s}$ for $n = 11, s = 2, 3, 4, 5$; and $n = 30$, $s = 2, 3, 4, 5$, where

$$\mathcal{A}_{n,s} = \{a : a < n,\ \gcd(a^j, n) = 1,\ j = 1, \ldots, s-1; 1, a, a^2, \ldots, a^{s-1} \widetilde{(\text{mod } n)} \\ \text{are distinct each other}\}.$$

3.10

By use of the design $U_{47}(47^3)$, construct a nearly uniform design $U_8(8^3)$ under MD by the cutting method.

3.11

Let $n = 47$ and $s = 46$. Find the L_1-distance and MD of the corresponding glp set U. Moreover, consider the simple linear level permutations $U + i \widetilde{(\text{mod } n)}$ and give your conclusion.

3.12

A Latin square of order n is an $n \times n$ matrix filled with n different symbols, each symbol in each row/column appears once and only once.

(1) Give a Latin square for order 3 and order 4.

(2) Find the definition for the concept of orthogonal Latin squares.

(3) Find the way to use orthogonal Latin squares to the construction of an orthogonal designs $L_{n^2}(n^{n-1})$.

(4) Sudoku puzzle such that the final 9×9 matrix to be a Latin square. Fill the following Sudoku puzzle:

9		4	2			8	3	
	5	7	4		9			
6	2			7	1		9	5
8	9	2	7	3			1	4
1			6		4	9		
4	7				8		5	
5				4		3	8	2
	8		5					9
			8	9	2		6	1

References

Beth, T., Jungnickel, D., Lenz, H.: Design theory. Volume II. Encyclopedia of Mathematics and Its Applications, vol. 78. Cambridge University Press, Cambridge (1999)

Caliński, T., Kageyama, S.: Block Designs: A Randomization Approach. Volume I. Lecture Notes in Statistics, vol. 150. Springer, New York (2000)

Cheng, C.S.: E(s^2)-optimal supersaturated designs. Stat. Sin. **7**, 929–939 (1997)

Colbourn, C.J., Dinita, J.H.: CRC Handbook of Combinatorial Designs. CRC Press, New York (1996)

Colbourn, C.J., Dinitz, J.H., Stinson, D.R.: Applications of combinatorial designs to communications, cryptography, and networking. Surveys in Combinatorics (Canterbury). London Mathematical Society Lecture Note Series, vol. 267, pp. 37–100. Cambridge University Press, Cambridge (1999)

Fang, K.T.: The uniform design: application of number-theoretic methods in experimental design. Acta Math. Appl. Sin. **3**, 363–372 (1980)

Fang, K.T., Hickernell, F.J.: The uniform design and its applications. Bulletin of the International Statistical Institute, vol. 1 (50th Session), pp. 339–349. Beijing (1995)

Fang, K.T., Li, J.K.: Some new results on uniform design. Chin. Sci. Bull. **40**, 268–272 (1995)

Fang, K.T., Ma, C.X.: Wrap-around L_2-discrepancy of random sampling, Latin hypercube and uniform designs. J. Complex. **17**, 608–624 (2001a)

Fang, K.T., Ma, C.X.: Orthogonal and Uniform Experimental Designs. Science Press, Beijing (2001b)

Fang, K.T., Wang, Y.: A note on uniform distribution and experiment design. Chin. Sci. Bull. **26**, 485–489 (1981)

Fang, K.T., Wang, Y.: Number-Theoretic Methods in Statistics. Chapman and Hall, London (1994)

Fang, K.T., Lin, D.K.J., Ma, C.X.: On the construction of multi-level supersaturated designs. J. Stat. Plan. Inference **86**, 239–252 (2000)

Fang, K.T., Ge, G.N., Liu, M.Q., Qin, H.: Optimal supersaturated designs and their constructions, Technical report MATH-309, Hong Kong Baptist University (2001)

Fang, K.T., Ge, G.N., Liu, M.Q.: Construction of $E(f_{NOD})$-optimal supersaturated designs via room squares. In: Chaudhuri, A., Ghosh, M. (eds.) Calcutta Statistical Association Bulletin, vol. 52, pp. 71–84 (2002a)

Fang, K.T., Ge, G.N., Liu, M.Q.: Uniform supersaturated design and its construction. Sci. China Ser. A **45**, 1080–1088 (2002b)

Fang, K.T., Ma, C.X., Winker, P.: Centered L_2-discrepancy of random sampling and Latin hypercube design, and construction of uniform designs. Math. Comput. **71**, 275–296 (2002c)

Fang, K.T., Ge, G.N., Liu, M.Q., Qin, H.: Construction on minimum generalized aberration designs. Metrika **57**, 37–50 (2003a)

Fang, K.T., Lin, D.K.J., Liu, M.Q.: Optimal mixed-level supersaturated design. Metrika **58**, 279–291 (2003b)

Fang, K.T., Ge, G.N., Liu, M.Q., Qin, H.: Combinatorial constructions for optimal supersaturated designs. Discret. Math. **279**, 191–202 (2004a)
Fang, K.T., Ge, G.N., Liu, M.Q., Qin, H.: Construction of uniform designs via super-simple resolvable t-designs. Util. Math. **66**, 15–32 (2004b)
Fang, K.T., Ge, G.N., Liu, M.Q.: Construction of optimal supersaturated designs by the packing method. Sci. China Ser. A **47**, 128–143 (2004c)
Fang, K.T., Lu, X., Tang, Y., Yin, J.: Constructions of uniform designs by using resolvable packings and coverings. Discret. Math. **274**, 25–40 (2004d)
Fang, K.T., Li, R., Sudjianto, A.: Design and Modeling for Computer Experiments. Chapman and Hall/CRC, New York (2006a)
Fang, K.T., Tang, Y., Yin, J.X.: Resolvable partially pairwise balanced designs and their applications in computer experiments. Util. Math. **70**, 141–157 (2006b)
Hedayat, A.S., Sloane, N.J.A., Stufken, J.: Orthogonal Arrays: Theory and Applications. Springer, New York (1999)
Hua, L.K., Wang, Y.: Applications of Number Theory to Numerical Analysis. Springer and Science Press, Berlin and Beijing (1981)
Johnson, M.E., Moore, L.M., Ylvisaker, D.: Minimax and maxmin distance desis. J. Stat. Plan. Inference **26**, 131–148 (1990)
Korobov, N.M.: The approximate computation of multiple integrals. Dokl. Akad. Nauk. SSSR **124**, 1207–1210 (1959)
Li, P.F., Liu, M.Q., Zhang, R.C.: Some theory and the construction of mixed-level supersaturated designs. Stat. Probab. Lett. **69**, 105–116 (2004)
Lin, D.K.J.: A new class of supersaturated designs. Technometrics **35**, 28–31 (1993)
Liu, M.Q., Fang, K.T.: Some results on resolvable incomplete block designs. Sci. China Ser. A **48**, 503–512 (2005)
Liu, M.Q., Zhang, R.C.: Construction of $E(s^2)$ optimal supersaturated designs. J. Stat. Plan. Inference **86**, 229–238 (2000)
Liu, M.Q., Fang, K.T., Hickernell, F.J.: Connections among different criteria for asymmetrical fractional factorial designs. Stat. Sin. **16**, 1285–1297 (2006)
Ma, C.X., Fang, K.T.: A new approach to construction of nearly uniform designs. Int. J. Mater. Prod. Technol. **20**, 115–126 (2004)
Marshall, A.W., Olkin, I.: Inequalities: Theory of Majorization and Its Applications. Academic, New York (1979)
Morris, M.D.: Factorial sampling plans for preliminary computational experiments. Technometrics **33**, 243–255 (1991)
Mukerjee, R., Wu, C.F.J.: On the existence of saturated and nearly saturated asymmetrical orthogonal arrays. Ann. Stat. **23**, 2102–2115 (1995)
Nguyen, N.K.: An algorithmic approach to constructing supersaturated designs. Technometrics **38**, 69–73 (1996)
Sacks, J., Schiller, S.B., Welch, W.J.: Designs for computer experiments. Technometrics **31**, 41–47 (1989)
Santner, T.J., Williams, B.J., Notz, W.I.: The Design and Analysis of Computer Experiments. Springer, New York (2003)
Shaw, J.E.H.: A quasirandom approach to integration in Bayesian statistics. Ann. Stat. **16**, 859–914 (1988)
Stinson, D.R.: A survey of Kirkman triple systems and related designs. Discret. Math. **92**(1–3), 371–393 (1991)
Stinson, D.R.: Packings. In: Colbourn, C.J., Dinitz, J.H. (eds.) CRC Press Series on Discrete Mathematics and Its Applications, pp. 409–413. CRC Press, Boca Raton (1996)
Tang, Y., Xu, H., Lin, D.K.J.: Uniform fractional factorial designs. Ann. Stat. **40**, 891–907 (2012)
Wang, Y., Fang, K.T.: A note on uniform distribution and experimental design. Chin. Sci. Bull. **26**, 485–489 (1981)

Yamada, S., Ikebe, Y.T., Hashiguchi, H., Niki, N.: Construction of three-level supersaturated design. J. Stat. Plan. Inference **81**, 183–193 (1999)

Yuan, R., Lin, D.K.J., Liu, M.Q.: Nearly column-orthogonal designs based on leave-one-out good lattice point sets. J. Stat. Plan. Inference **185**, 29–40 (2017)

Zhou, Y.D., Xu, H.: Space-filling fractional factorial designs. J. Am. Stat. Assoc. **109**, 1134–1144 (2014)

Zhou, Y.D., Xu, H.: Space-filling properties of good lattice point sets. Biometrika **102**, 959–966 (2015)

Zhou, Y.D., Fang, K.T., Ning, J.H.: Mixture discrepancy for quasi-random point sets. J. Complex. **29**, 283–301 (2013)

Chapter 4
Construction of Uniform Designs—Algorithmic Optimization Methods

In the previous chapter, the deterministic methods such as good lattice point method and its modifications were introduced. Generally, most existing uniform designs are obtained by numerical search on the U-type design space $\mathcal{U}(n; n^s)$ or $\mathcal{U}(n; q_1, \ldots, q_s)$. This chapter introduces some powerful optimization methods and related algorithms for generating uniform designs. Especially, two stochastic optimization methods, threshold-accepting method and integer programming problem method, will be given in details in Sects. 4.2 and 4.3, respectively.

4.1 Numerical Search for Uniform Designs

From the definition of uniform design in Definition 3.1.2, the optimization problem in construction of n-run uniform designs for given uniformity criterion D is to find a design $\mathcal{P}^* \in \mathcal{D}(n; C^s)$ such that

$$D(\mathcal{P}^*) = \min_{\mathcal{P} \in \mathcal{D}(n; C^s)} D(\mathcal{P}), \tag{4.1.1}$$

where $\mathcal{D}(n; C^s) = \{\{\boldsymbol{x}_1, \ldots, \boldsymbol{x}_n\}, \boldsymbol{x}_i \in C^s, i = 1, \ldots, n\}$. It seems that the traditional optimization such as the Gauss–Newton method can be applied to this problem.

Recall the mathematical optimization problem. Let $f(\boldsymbol{x}) = f(x_1, \ldots, x_m)$ be a continuous function on a set $\mathcal{X}$. One wants to find $\boldsymbol{x}^* \in \mathcal{X}$ such that

$$f(\boldsymbol{x}^*) = \min_{\boldsymbol{x} \in \mathcal{X}} f(\boldsymbol{x}), \tag{4.1.2}$$

where $\boldsymbol{x}^*$ is called the *minimum point* and $f(\boldsymbol{x}^*)$ the minimal f-value on $\mathcal{X}$. The function f is called, variously, an *objective function* or a *loss function*. Typically, $\mathcal{X}$

K.-T. Fang et al., *Theory and Application of Uniform Experimental Designs*, Lecture Notes in Statistics 221,
https://doi.org/10.1007/978-981-13-2041-5_4

is some subset of the Euclidean space R^m and called the *search domain* or *search space*; the points of $\mathcal{X}$ are called *candidate solutions* or *feasible solutions*. Often there are some constraints that the elements of $\mathcal{X}$ have to be satisfied. A *local minimum* $\boldsymbol{x}_{local} \in \mathcal{X}$ satisfies

$$f(\boldsymbol{x}_{local}) \leqslant f(\boldsymbol{x}), \text{ for any } \boldsymbol{x} \text{ satisfying } ||\boldsymbol{x} - \boldsymbol{x}_{local}|| \leqslant \delta, \text{ and some } \delta > 0.$$

When there is only one local minimum in the search domain $\mathcal{X}$, there are many powerful algorithms. Among those algorithms, the *Gauss–Newton method* has been widely used. It starts from an *initial point* $\boldsymbol{x}_0 \in \mathcal{X}$ and iteratively finds the value of the variables which minimize the objective function. Denote $\boldsymbol{x}_c = \boldsymbol{x}_0$. For each iteration, one needs to calculate derivatives of the function f and then choose a new point $\boldsymbol{x}_{new} \in \mathcal{X}$. If $f(\boldsymbol{x}_{new}) < f(\boldsymbol{x}_c)$, let $\boldsymbol{x}_{new}$ replace $\boldsymbol{x}_c$ and go to next iteration. We end the iteration process when $\boldsymbol{x}_{new} = \boldsymbol{x}_c$. However, the Gauss–Newton method only provides a local minimal solution if the objective function f has several local minima. The optimization problem (3.2.1) for searching uniform designs always has many local minima. For any optimization problem, we have to estimate computational complexity of the problem. From Wikipedia, the free encyclopedia, "A computational problem is understood to be a task that is in principle amenable to being solved by a computer, which is equivalent to stating that the problem may be solved by mechanical application of mathematical steps, such as an algorithm."

For finding a uniform design presented in (4.1.1), there are some serious difficulties:

(1) *The high-dimensional problem.* It is a ns-optimization problem that is too high for minimization when n and s are moderate (see Sect. 3.2.1).
(2) *Domain candidate.* The number of candidates of the domain is infinite.
(3) *Multiple local minima.* There are many local minima.

Due to these difficulties to find a uniform design $U_n(n^s)$ is an *NP-hard problem*, where NP stands for "non-deterministic polynomial time." It is widely suspected that there are no polynomial-time algorithms for NP-hard problems. Therefore, we have to reduce the design space, for example, to U-type design set $\mathcal{U}(n; n^s)$, or more generally, to $\mathcal{U}(n; q_1, \ldots, q_s)$, where each design has n runs and s factors each having q_j levels. For simplicity, denote $\mathcal{U}$ as the U-type design space $\mathcal{U}(n; q_1, \ldots, q_s)$. Then, the optimization problem of construction of uniform designs in (4.1.1) is reduced to find a design $\mathcal{P}^* \in \mathcal{U}$ such that

$$D(\mathcal{P}^*) = \min_{\mathcal{P} \in \mathcal{U}} D(\mathcal{P}). \tag{4.1.3}$$

Here, the discrepancy can be the centered L_2-discrepancy (CD), wrap-around L_2-discrepancy (WD), mixture discrepancy (MD), or others. The minimization problem in (4.1.3) is still a challenging problem as there are several difficulties stated below:

- The design space is not a compact area, and the traditional optimization methods are useless as there is no continuous concept as well as derivatives. We need modern optimization techniques, among which many *stochastic optimization algorithms* are recommended.

- The number of candidates in this design space increases exponentially in both n and s. Searching a uniform design on $\mathcal{U}$ is an NP-hard problem.
- Arbitrarily permuting the order of the runs or the order of the factors does not change the discrepancy. Thus, the uniform design is not unique.
- The discrepancy as the objective function is a multi-modal function with many local minima that are not global minima. It leads to difficulty how to make a termination rule in the algorithm.
- When a $\boldsymbol{x}_{new}$ attains the lower bound of the objective function (refer to Sect. 2.6), the process will be terminated. However, to find a *tight lower bound* of the discrepancy is an open problem in most cases. A lower bound is called tight if it is attainable; otherwise, it is called conservative.

A good search algorithm for uniform designs must face the challenges listed above. In the literature, many stochastic optimization algorithms, such as the *simulated annealing algorithm* (Kirkpatrick et al. 1983; Morris and Mitchell 1995), *stochastic evolutionary* (Jin et al. 2005), *threshold-accepting heuristic* method (Dueck and Scheuer 1990; Winker and Fang 1998; Fang et al. 2002, 2003, 2005, 2006b), and *integer programming* method (Fang and Ma 2001b; Zhou et al. 2013) had been proposed to construct uniform designs.

The general procedure of the stochastic optimization algorithms for construction of uniform designs is as follows. Choose an initial design $\boldsymbol{U}_0$ as the current design $\boldsymbol{U}_c$; find another design $\boldsymbol{U}_{new}$ in the neighborhood of the current design $\boldsymbol{U}_c$, if the acceptance criterion is satisfied, let the current design $\boldsymbol{U}_c$ replace by $\boldsymbol{U}_{new}$; continue this procedure until some stopping rule is satisfied. Then, a stochastic optimization algorithm may be described in terms of the following general framework.

Algorithm 4.1.1 (*Framework for Constructing Uniform Designs*)

Input. Give n, the number of runs, s, the number of factors, $\mathcal{U}$, the design space (a set of candidate pool of designs) and a discrepancy $D(\cdot)$ as the objective function in optimization.

Step 1. *Initialization*. Choose an initial design $\boldsymbol{U}_0 \in \mathcal{U}$, and let $\boldsymbol{U}_c = \boldsymbol{U}_0$.

Step 2. *Evaluation and Iteration*. Choose a new design, denoted by $\boldsymbol{U}_{new}$, in the neighborhood of $\boldsymbol{U}_c \subset \mathcal{U}$. Compare discrepancy values between $\boldsymbol{U}_{new}$ and $\boldsymbol{U}_c$ and decide whether to replace the current design $\boldsymbol{U}_c$ with $\boldsymbol{U}_{new}$. If yes, repeat this step, otherwise jump to some new design as $\boldsymbol{U}_c$ by a given jumping rule.

Step 3. *Termination*. If the jumping rule is terminated, the current design $\boldsymbol{U}_c$ in the set $\mathcal{U}$ has sufficiently good uniformity, then output the design and stop.

Each step in this framework can be implemented in different ways, depending on the complexity of the design space and structure of the chosen discrepancy, the qualities desired for a design, and the computational time. For example, the design space can be the set of design with n runs and s factors or its subset. The discrepancy can choose WD, CD, MD, or others. The initial design can be randomly chosen from the design space or a design with a good quality. There are many ways for defining a neighborhood of a design in $\mathcal{U}$ and switching from the current design to some new design. There are different jumping rules to jump out from a local minimum. The various ways of implementing each step are discussed below. For the sake of efficiency, there may be a reordering or relabeling of the steps in some cases that differs from that of Algorithm 4.1.1.

Moreover, the lower bounds of the chosen discrepancy are useful in Algorithm 4.1.1, since we stop the iteration process as soon as the discrepancy of the new design in the iteration process attains the lower bound. Some lower bounds of WD, CD, and MD are given in Sect. 2.6.

4.2 Threshold-Accepting Method

The *threshold-accepting* (TA) method is a powerful optimization algorithm for constructing uniform designs. It was proposed by Dueck and Scheuer (1990). Winker and Fang (1997) were the first to use the TA algorithm to quasi-Monte Carlo methods, especially to evaluate the star discrepancy. Late they obtained U-type designs with lower star discrepancy (Winker and Fang 1998). Fang et al. (2000) applied the TA algorithm for several types of L_2-discrepancies and pointed out that many existing orthogonal designs can be obtained by TA under CD or WD. Fang et al. (2002) gave a comprehensive study on the Latin hypercube sampling and provided some useful information in numerical search for uniform designs. Based on this information, they modified the TA algorithm and applied it for finding uniform designs under CD. Fang et al. (2003) further used TA under both the CD and WD. Recent studies can refer to Fang et al. (2005), Fang et al. (2006b), and Zhou and Fang (2013), where the TA algorithm was used to construct uniform designs with large size. In this section, we introduce the methodology of the TA algorithm.

TA algorithm can be used for the discrete optimization problem. Several key issues of the TA algorithm for searching uniform designs are as follow:

Objective. The TA tackles the optimization problem

$$\min_{U \in \mathcal{U}} D(\boldsymbol{U}), \tag{4.2.1}$$

where $\mathcal{U}$ is a subset of the set of all U-type designs $\mathcal{U}(n; q_1, \ldots, q_s)$, and D is a specified measure of uniformity, such as CD or MD.

Figure 4.1 presents a flowchart of the TA implementation for minimizing the discrepancy of uniform designs on the design space $\mathcal{U}$. Various modifications can refer to the above literature. A design $\mathcal{P} \in \mathcal{U}$ can be expressed as a $n \times s$ matrix $\boldsymbol{U} = (u_{ij})$, $u_{ij} \in [0, 1]$. In the text of this book, both $\boldsymbol{U}$ and $\mathcal{P}$ denote a design depending on expression convenience. Some concepts in the TA algorithm are explained as follows.

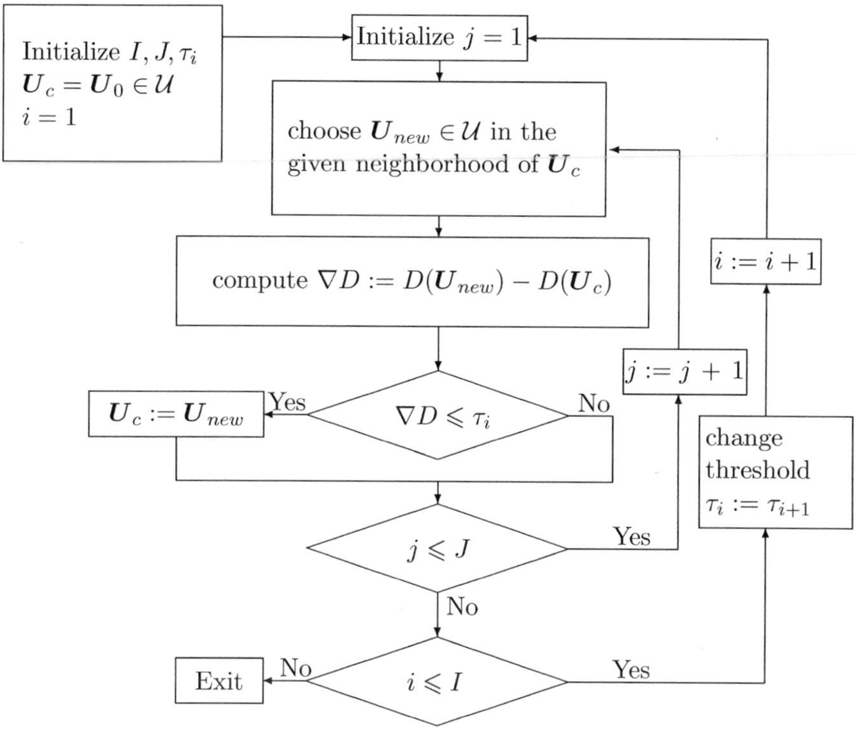

Fig. 4.1 Threshold-accepting algorithm for finding uniform designs

Initial designs. The initial design $\boldsymbol{U}_0$ can be randomly chosen from the design space $\mathcal{U}$, or be chosen by the good lattice point method, or some known design in $\mathcal{U}$ with a lower discrepancy.

For the design space $\mathcal{U}(n, q^s)$ or $\mathcal{U}(n, q_1, \ldots, q_s)$, $s < n$, one can choose a design $U(n; n^s)$ and then this design can be transferred into $U(n; q^s)$ by the *pseudo-level transformation* (Fang et al. 2006a):

$$\{(i-1)n/q + 1, \ldots, in/q\} \to i, \quad i = 1, \ldots, q, \tag{4.2.2a}$$

or

$$\{(i-1)r_j + 1, \ldots, ir_j\} \to i, \quad i = 1, \ldots, q_j, \tag{4.2.2b}$$

for symmetrical design or asymmetrical designs, respectively, where q_j is the number of level of the jth factor, $r_j = n/q_j$.

Threshold and Replacement rule. The threshold value τ is a nonnegative number and is getting smaller during the iterations. The traditional *local search* (LS) algorithm leads the objective function to get smaller and decides to replace the current design $\boldsymbol{U}_c$ by $\boldsymbol{U}_{new}$ if $\nabla D = D(\boldsymbol{U}_{new}) - D(\boldsymbol{U}_c) \leqslant 0$. The LS algorithm is easy to stick at a local minimum discrepancy design that might be far from the global minimum discrepancy designs. TA algorithm uses alternative rule: The current design $\boldsymbol{U}_c$ is replaced by $\boldsymbol{U}_{new}$ if $\nabla D = D(\boldsymbol{U}_{new}) - D(\boldsymbol{U}_c) \leqslant \tau$, where $\tau \geqslant 0$. When $\tau > 0$, it gives a chance to let the iteration process jumping out from a local minimum. When τ is not small, the iteration process likes a random walk; when τ is smaller, the iteration process can move only on a small neighborhood of $\boldsymbol{U}_c$; and when $\tau = 0$, the iteration process will end at the local minimum that is in the neighborhood of $\boldsymbol{U}_c$. The definition of the neighborhood of $\boldsymbol{U} \in \mathcal{U}$ is pre-decided and is discussed late.

The threshold can affect the result significantly. In the literature, one might choose a suitable positive integer I and determines *a series of thresholds* $\tau_1 > \tau_2 > \cdots > \tau_{I-1} > \tau_I = 0$. In the iteration $I = i$ circle if $\nabla D > \tau_i$ and $j < J$, a new design is randomly chosen from the neighborhood of $\boldsymbol{U}_c$ and let $j := j + 1$. Usually, the number of out circles I should be much less than the number of inner circles J. Please refer to Fig. 4.1 for the role of I and J. The choices of I and J depend on size of the design. In the literature, it was suggested that $I \in [10, 100]$ and $J \in [10^4, 10^5]$ when the number of runs $n \leqslant 1000$, and I and J may increase as n increases (Winker and Fang 1997; Zhou and Fang 2013). For example, one can choose $I = 10$, $J = 10^5$ when $n \leqslant 100$ and $I = 100$, $J = 10^5$ when $n \geqslant 300$. For obtaining suitable I and J, one can do some empirical study in advance.

For designing a *threshold sequence* $[\tau_1, \ldots, \tau_I]$, a set of M designs are randomly generated from $\mathcal{U}$. Then, calculate their objective function values and the range (denoted by R) of these M-values. Denote the largest and the smallest objective function values by $D_{\max}$ and $D_{\min}$, respectively, and $R = D_{\max} - D_{\min}$. The first threshold (τ_1) is set as a fraction α ($0 < \alpha < 1$) of R. The remaining $I - 1$ thresholds are usually determined by iteration formula $\tau_i = f(\tau_{i-1})$, and $f(\cdot)$ is a linear function or other functions, for example,

$$\tau_i = \frac{I - i}{I}\tau_{i-1}, i = 2, \ldots, I. \tag{4.2.3}$$

One can choose several α-values in advance and compare their performance, and then, choose a better one.

Another way to determine τ_1 is to choose a design $\boldsymbol{U}$ in $\mathcal{U}$ randomly and also generate N designs $\boldsymbol{U}_j$ in the neighborhood of $\boldsymbol{U}$ randomly. Calculate N differences $\nabla D_j = D(\boldsymbol{U}_j) - D(\boldsymbol{U})$, and draw an empirical distribution of these ∇D_j's, denote by F_1. Then, τ_1 can be chosen as the t percentile of F_1, where t is less than 5 for example.

Neighborhood. The design space is a finite set and there is no continuity concept for the objective function. Therefore, the *neighborhood* concept was proposed for replacement of the continuity. Two designs in the same neighborhood should be close to each other in a certain sense of the design structure. Let $\boldsymbol{U}_c$ be a design matrix in $\mathcal{U}$.

The definition of neighborhoods $\mathcal{N}(\boldsymbol{U}_c)$ has to take into account several conditions. First, as already pointed out above, $\mathcal{N}(\boldsymbol{U}_c) \subset \mathcal{U}$ and each $\boldsymbol{U}_{new} \in \mathcal{N}(\boldsymbol{U}_c)$ should also be a U-type design. Second, in order to impose a real "local" structure, the designs $\boldsymbol{U}_c$ and those in $\mathcal{N}(\boldsymbol{U}_c)$ should not differ too much. Third, the computational complexity of the algorithm depends to a large extent on calculating ∇D, i.e., the difference in the objective function when moving from $\boldsymbol{U}_c$ to $\boldsymbol{U}_{new}$. Thus, if ∇D can be obtained without calculating $\boldsymbol{U}_{new}$ from scratch, a significant speedup might result. All three requirements can be easily fulfilled by selecting one or more columns of $\boldsymbol{U}_c$ and exchanging two elements within each selected column. For example, one can randomly choose one column of $\boldsymbol{U}_c$, randomly choose two elements in this column, and exchange these two elements to form a new design. All these new designs form a neighborhood of $\boldsymbol{U}_c$

$$\mathcal{N}_1(\boldsymbol{U}_c) = \{\boldsymbol{U} : \text{by exchanging two elements in any one column of } \boldsymbol{U}_c\}.$$

Of course, one can define a larger neighborhood. For example,

$$\mathcal{N}_2(\boldsymbol{U}_c) = \{\boldsymbol{U} : \text{by exchanging two elements in any two columns of } \boldsymbol{U}_c\}.$$

Winker and Fang (1997) gave a discussion on the choice of the neighborhood size. The larger size of the neighborhood we define, the more freedom we choose $\boldsymbol{U}_{new}$.

In fact, a neighborhood is a small perturbation of $\boldsymbol{U}_c$ in most TA algorithm versions. Unlike the pre-decided neighborhood $\mathcal{N}_1(\boldsymbol{U}_c)$ or $\mathcal{N}_2(\boldsymbol{U}_c)$ where the column is randomly chosen, Fang et al. (2005) proposed two other ways of choosing a new design.

1. *"maximal and minimal distances of row pairs" method.* Denote by $(\boldsymbol{x}_{i1}, \boldsymbol{x}_{i2})$ and $(\boldsymbol{x}_{j1}, \boldsymbol{x}_{j2})$ the respective row pairs with maximal and minimal distances for the current design $\boldsymbol{U}_c$; here, the distance can be chosen as the L_p-distance or other distances. We randomly select a row $\boldsymbol{x}_i$ from $\boldsymbol{x}_{i_1}$ or $\boldsymbol{x}_{i_2}$ and a row $\boldsymbol{x}_j$ from $\boldsymbol{x}_{j_1}$ or $\boldsymbol{x}_{j_2}$. Then, randomly select a column k. If the kth element, x_{ik}, in the row $\boldsymbol{x}_i$ is not equal to x_{jk}, the kth element in the row $\boldsymbol{x}_j$, then exchange x_{ik} and x_{jk} to obtain a new design $\boldsymbol{U}_{new}$.
2. *"single row with maximal and minimal sum of distances" method.* Denote $d_{ij} = d(\boldsymbol{x}_i, \boldsymbol{x}_j)$, where $d(\cdot, \cdot)$ is a pre-decided distance. Denote by the row $\boldsymbol{x}_i$ and row $\boldsymbol{x}_j$ the respective rows with maximal and minimal distances for the current design $\boldsymbol{U}_c$. This means $\sum_{t \neq i} d_{ti}$ is maximal and $\sum_{t \neq j} d_{tj}$ is minimal among $\sum_{t \neq k} d_{tk}, k = 1, \ldots, s$. Now, randomly select a column k. If x_{ik}, the kth element in the row $\boldsymbol{x}_i$, is not equal to x_{jk}, the kth element in the row $\boldsymbol{x}_j$, then exchange x_{ik} and x_{jk} to obtain a new design $\boldsymbol{U}_{new}$.

Each method has its own advantages. Compared with "*maximal and minimal distances of row pairs*" method, "*single row with maximal and minimal sum of distances*" method is expected to accelerate the searching more, while the former method can provide more chances of jumping out from a local minimal status. The main idea of

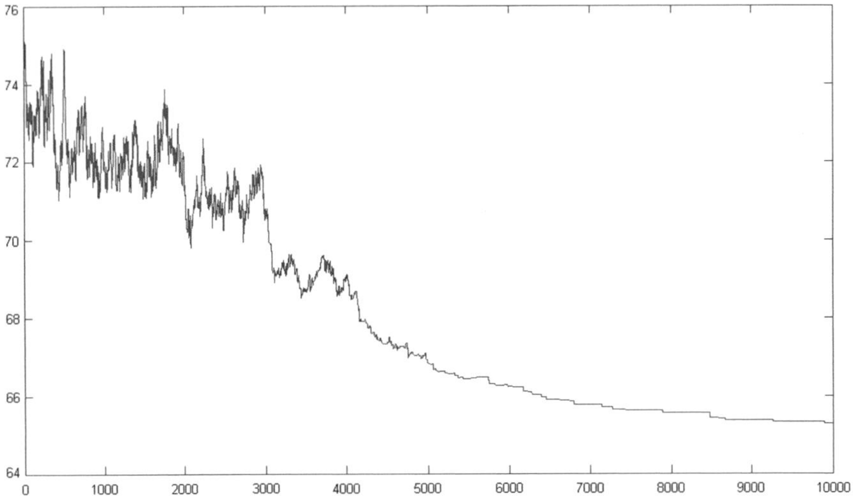

Fig. 4.2 The trace of the TA process

randomly using these two preselection methods to determine the neighborhood for each iteration is both to accelerate the speed and to jump out from a local minimal status. Moreover, experiments also show that when a single preselection method is used, the result will always be worse.

Trace plot. The trace plot is useful to find the iteration process behavior. It shows the objective function value in each iteration step. An example is provided in Fig. 4.2, where the x-axis stands for the number of iterations and the corresponding y-axis indicates the MD-value of $\boldsymbol{U}_c$ at the current stage. From Fig. 4.2, we can observe that the MD-value starts with a period of random walk and then shrink suddenly toward to the optimum after a number of iterations. Figure 4.3 is the tract plot for the same problem with different parameters I, J, and τ_i's. The reader can find their different iteration process behavior.

Historical optimum reversion. When $\boldsymbol{U}_0$ is a good design with lower discrepancy, TA may deliver a $\boldsymbol{U}_{opt}$ worse than $\boldsymbol{U}_0$. This problem happens because TA always encourages the current design $\boldsymbol{U}_c$ to jump out of a local optimum if the current threshold τ value is not small. To overcome this difficulty, Fang et al. (2016) proposed a new mechanism called the *historical optimum reversion*. This mechanism allows the current design $\boldsymbol{U}_c$ to return to the historical optimal design of TA at certain moments. For example, we embed a "judgement" before the iterations under each threshold value change. That is, if we choose MD as the objective function, once the threshold value changes in TA, a comparison will be made immediately on MD($\boldsymbol{U}_c$) and MD($\boldsymbol{U}_h$), where $\mathbf{U}_h$ is the historical optimal design of TA (when TA has multiple historical optimal designs, $\mathbf{U}_h$ is assigned to one of them randomly). If MD($\mathbf{U}_c$) $>$ MD($\boldsymbol{U}_h$), we will let $\boldsymbol{U}_c = \boldsymbol{U}_h$ and proceed the ensuing iterations.

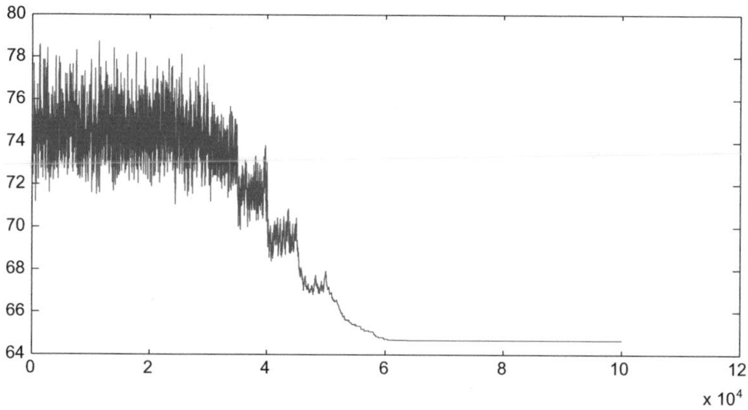

Fig. 4.3 The traces of the TA process by different parameters

Phases in TA algorithm. The initial design $\boldsymbol{U}_0$ affects the quality of output design. One way is randomly choose m initial designs and run TA m times. The best output among the m output designs is recommended as the output UD. Alternative suggestion for an initial design is to choose a design with lower discrepancy as initial design. Based on this idea, we utilize TA with several phases. In the first phase, we use the traditional TA to find an output design that is recorded as $\boldsymbol{U}_1$. In the second phase, we set this $\boldsymbol{U}_1$ as the new initial design and utilize TA again with modified settings to phase 1. A new output, $\boldsymbol{U}_2$, can be obtained. Then, use $\boldsymbol{U}_2$ as the initial design for the third phase with a similar procedure. So on and so forth, we stop the process and deliver the final optimal design $\boldsymbol{U}_k$ at the end of the kth phase if $\text{MD}(\boldsymbol{U}_k) - \text{MD}(\boldsymbol{U}_{k-1}) = 0$. Usually, to ensure the convergency of this strategy, the α values for determining the thresholds in each phases, denoted as $[\alpha_1, \ldots, \alpha_k]$, should be set in a descending manner.

Example 4.2.1 Suppose that one wants to find a uniform design table $U_{27}(3^{13})$ on the design space $\mathcal{U}(27; 3^{13})$. An initial design $\boldsymbol{U}_0$ is randomly chosen from the design space and its $\text{MD}(\boldsymbol{U}_0) = 75.61$. Consider to use TA by phases to optimize $\boldsymbol{U}_0$ again. In this case, we embed the two adjustments to TA, i.e., historical optimum reversion and multiple phases. The setting in this example is chosen by $I = 20$, $J = 5000$, $[\alpha_1, \ldots, \alpha_5] = [0.15\ 0.016\ 0.01\ 0.002\ 0.0005]$. This setting has the same I and J inputs as the TA in phase 1. The threshold sequence is adapted from (4.2.3). During the TA process, one obtained the optimal design from each phase, denoted as $[\mathbf{U}_1^a \cdots \mathbf{U}_5^a]$. The output design $\boldsymbol{U}_{opt}^a = \boldsymbol{U}_5^a$ with $\text{MD}(\boldsymbol{U}_{opt}^a) = 64.1888$. This result significantly out performs the best outcome ($\text{MD}(\boldsymbol{U}_2) = 64.3689$) in phase 2. Figure 4.4 presents the trace plot of the TA process. The effectiveness of two adjustments (historical optimum reversion and multiple phases in TA algorithm) is very intuitive from Fig. 4.4. On the one hand, in phases 2 and 3, the historical optimum reversion mechanism eliminates the variation of the process. Although $\boldsymbol{U}_c$ may evacuate from their good initial designs, adjusted TA allows the current design to retrieve its his-

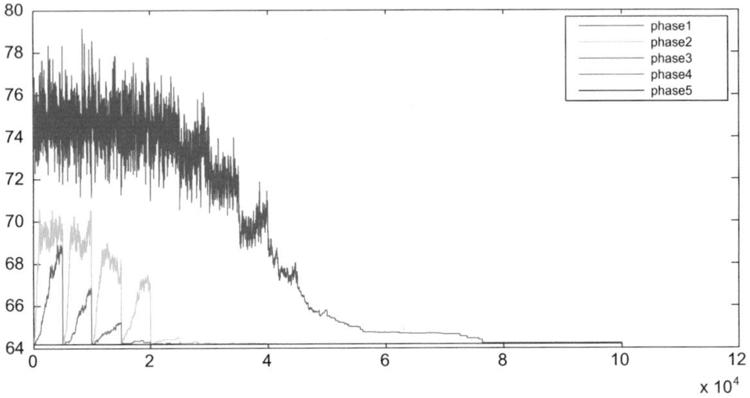

Fig. 4.4 MD trace plot of Example 4.2.1 by the adjusted TA

torical optimum from time to time. This mechanism prevents TA from abandoning the best local structure it once obtained. Because every threshold sequences end up with value 0, this algorithm also guarantees the output design has a good uniformity in each phase. For phase 1, since the MD of $\boldsymbol{U}_0$ is relatively large, it behoves TA to involve more effective iterations for optimization. We observe that $\boldsymbol{U}_c$ in phases 4 and 5 stays in its optimal design and does not make any change.

Iteration formulae. The discrepancy commonly used to differentiate designs is as the objective function in the TA process. In each iteration, one needs to calculate $D(\boldsymbol{U}_{new})$. Because $D(\boldsymbol{U}_{new})$ is in the neighborhood $\mathcal{N}(\boldsymbol{U}_c)$, it should have a close relationship of the computational formula between $D(\boldsymbol{U}_{new})$ and $D(\boldsymbol{U}_c)$. If we can find some iteration formula from $D(\boldsymbol{U}_c)$ to $D(\boldsymbol{U}_{new})$, we can save computing time for $D(\boldsymbol{U}_{new})$. For different discrepancies, there are different formulae.

(A) *Iteration Formula for the Centered L_2-Discrepancy*

Jin et al. (2005) proposed a simple way to evaluating the centered CD. Let $\boldsymbol{U}_c = (u_{ik})$ be the current design, $\boldsymbol{X} = (x_{ik})$ its reduced design matrix, and $\boldsymbol{Z} = (z_{ik})$ the corresponding centered design matrix, where $z_{ik} = x_{ik} - 0.5$. Let $\boldsymbol{C} = (c_{ij})$ be a symmetrical matrix, whose elements are:

$$c_{ij} = \begin{cases} \frac{1}{n^2}\prod_{k=1}^{s}\frac{1}{2}(2+|z_{ik}|+|z_{jk}|-|z_{ik}-z_{jk}|), & \text{if } i \neq j; \\ \frac{1}{n^2}\prod_{k=1}^{s}(1+|z_{ik}|) - \frac{2}{n}\prod_{k=1}^{s}\left(1+\frac{1}{2}|z_{ik}|-\frac{1}{2}z_{ik}^2\right), & \text{otherwise.} \end{cases}$$

It can be verified that

$$[\mathrm{CD}(\boldsymbol{U}_c)]^2 = \left(\frac{13}{12}\right)^2 + \sum_{i=1}^{n}\sum_{j=1}^{n} c_{ij}.$$

For any $1 \leqslant t \leqslant n$ and $t \neq i, j$, let

$$\gamma(i, j, k, t) = \frac{2 + |z_{jk}| + |z_{tk}| - |z_{jk} - z_{tk}|}{2 + |z_{ik}| + |z_{tk}| - |z_{ik} - z_{tk}|}.$$

After an exchange of x_{ik} and x_{jk}, the square CD-value of the new design $\boldsymbol{U}_{new}$ is given by

$$[\mathrm{CD}(\boldsymbol{U}_{new})]^2 = [\mathrm{CD}(\boldsymbol{U}_c)]^2 + \nabla(\mathrm{CD}),$$

where

$$\nabla(\mathrm{CD}) = c_{ii}^* - c_{ii} + c_{jj}^* - c_{jj} + 2 \sum_{t=1, t\neq i,j}^{n} (c_{it}^* - c_{it} + c_{jt}^* - c_{jt})$$

and $c_{it}^* = \gamma(i, j, k, t)c_{it}$, $c_{jt}^* = c_{jt}/\gamma(i, j, k, t)$. By the above iteration formula, the computational complexity is about $O(n)$. Fang et al. (2006b) suggested a different iteration formula from the above ones.

(B) *Iteration Formula for the Wrap-Around L_2-Discrepancy*

For WD, it also exists some iteration formula. Denote

$$\alpha_{ij}^k \equiv |x_{ik} - x_{jk}|(1 - |x_{ik} - x_{jk}|),$$

$$\delta_{ij} = \sum_{k=1}^{s} \ln\left(\frac{3}{2} - \alpha_{ij}^k\right), i, j = 1, \ldots, n, i \neq j, k = 1, \ldots, s.$$

Then, the WD can be expressed in terms of the sum of $e^{\delta_{ij}}$'s. And for a single exchange of two elements in the selected column, there are altogether $2(n-2)$ distances (δ_{ij}'s) updated. Suppose the kth elements in rows $\boldsymbol{x}_i$ and $\boldsymbol{x}_j$ are exchanged, then for any row $\boldsymbol{x}_t$ other than $\boldsymbol{x}_i$ or $\boldsymbol{x}_j$, the distances of row pair $(\boldsymbol{x}_i, \boldsymbol{x}_t)$ and row pair $(\boldsymbol{x}_j, \boldsymbol{x}_t)$ will be changed. Denote $\tilde{\delta}_{ti}$ and $\tilde{\delta}_{tj}$ as the new distances between row pair $(\boldsymbol{x}_i, \boldsymbol{x}_t)$ and row pair $(\boldsymbol{x}_j, \boldsymbol{x}_t)$; then,

$$\tilde{\delta}_{it} = \delta_{it} + \ln(3/2 - \alpha_{jt}^k) - \ln(3/2 - \alpha_{it}^k);$$

$$\tilde{\delta}_{jt} = \delta_{jt} + \ln(3/2 - \alpha_{it}^k) - \ln(3/2 - \alpha_{jt}^k).$$

The objective function change is given as

$$\nabla(\mathrm{WD}) = \frac{2}{n^2} \sum_{t\neq i,j} \left(e^{\tilde{\delta}_{it}} - e^{\delta_{it}} + e^{\tilde{\delta}_{jt}} - e^{\delta_{jt}}\right).$$

Moreover, the iteration formula for MD is similar to that of CD. We leave it as an exercise for reader.

Jumping rule. The search process may stick at a local minimum design. Many existing stochastic optimization algorithms, such as the simulated annealing algorithm (Kirkpatrick et al. (1983)) and the stochastic evolutionary algorithm (Jin et al. (2005)), use some rules for jumping out from a current local minimum. There are different considerations for jumping out. The TA algorithm in Fig. 4.1 employs a series of thresholds τ_r for jumping out. When τ_r is large, it is easy to jump out from a local minimum design. Fang et al. (2006b) suggested some jumping rules into the TA algorithm and proposed so-called balance-pursuit heuristic algorithm.

Moreover, the lower bounds of a pre-decided discrepancy, given in Sect. 2.6, are useful in the TA algorithm, i.e., the algorithm may stop when the discrepancy of the design U_{new} arrives the lower bound. Algorithm 4.2.1 provides the pseudo-code for the TA implementation sketched by Fang et al. (2003).

Algorithm 4.2.1 (*Pseudo-code for Threshold Accepting*)

Initialize I, J and the sequence of thresholds $\tau_i, i = 1, 2, \ldots, I$
Generate starting design $\boldsymbol{U}_0 \in \mathcal{U}(n; q^s)$ and let
for $i = 1$ to I **do**
    for $j = 1$ to J **do**
        Generate $\boldsymbol{U}_1 \in \mathcal{N}(\boldsymbol{U}_0)$ (neighbor of $\boldsymbol{U}_0$)
        if $D(\boldsymbol{U}_1)$ arrives the lower bound, stop the iteration
        elseif $D(\boldsymbol{U}_1) < D(\boldsymbol{U}_0) + \tau_i$ **then**
            $\boldsymbol{U}_0 = \boldsymbol{U}_1$
        end if
    end for
end for

Essentially, the TA heuristic is a refined local search algorithm operating on the set of U-type designs $\mathcal{U}(n; q^s)$ for given n, q, and s. The algorithm starts with a randomly generated U-type design $\boldsymbol{U}_0$ and proceeds by iterating local search steps. Each search step consists in choosing a U-type design $\boldsymbol{U}_1$ in the neighborhood of the current design and to compare the discrepancy of both designs. In contrast to a simple local search algorithm, the new design is not only accepted if its discrepancy is lower than the one of the current design, but also if it is not too much higher. The extent to which such a worsening is accepted is defined by the threshold sequence (τ_r), which typically decreases to zero with the number of iterations reaching a preset maximum, i.e., at the beginning of a run of the algorithm, comparatively large increases of the discrepancy are accepted when moving from $\boldsymbol{U}_0$ to $\boldsymbol{U}_1$, while toward the end of a run, only local improvements are accepted.

4.3 Construction Method Based on Quadratic Form

For searching a uniform design under a given discrepancy, one can consider the property of the expression of the discrepancy. In this section, it is shown that the discrepancy defined by the reproducing kernel in (2.4.6) can be expressed by the quadratic form. Searching uniform design then becomes a *quadratic zero-one integer programming* problem, and the *integer programming* problem method is used for solving such problem.

4.3.1 Quadratic Forms of Discrepancies

The generalized L_2-discrepancies such as CD, WD, MD, and LD in Sects. 2.3 and 2.5 are defined by the reproducing kernel in (2.4.6). Their computational formulae can be expressed as the following unified form

$$\text{constant} + \frac{1}{n^2}\sum_{i=1}^{N}\sum_{j=1}^{N}\prod_{k=1}^{m} f(d_{ik}, d_{jk}, q_k) - \frac{2}{n}\sum_{i=1}^{N}\prod_{k=1}^{m} g(d_{ik}, q_k),$$

where $f(\cdot,\cdot,\cdot)$ and $g(\cdot,\cdot)$ are different types of functions according to different discrepancies.

Let $\mathcal{V}_{q_i} = \{1, 2, \ldots, q_i\}$ and $N = q_1 \ldots q_s$. Their cartesian product $\mathcal{V}^s = \mathcal{V}_{q_1} \times \cdots \times \mathcal{V}_{q_s}$. For a U-type design $\mathcal{P} \in \mathcal{U}(n; q_1, \ldots, q_s)$, let $n(i_1, \ldots, i_s)$ denote the number of times that the point $(i_1, \ldots, i_s)$ occurs in $\mathcal{P}$. Then, the design $\mathcal{P}$ can be uniquely determined by the column vector of length N given by

$$\boldsymbol{y}_{\mathcal{P}} = (n(i_1, \ldots, i_s))_{(i_1,\ldots,i_s)\in\mathcal{V}^s}, \tag{4.3.1}$$

where all $(i_1, \ldots, i_s)$ points in $\mathcal{V}^s$ are arranged in the *lexicographical order*. $\boldsymbol{y}_{\mathcal{P}}$ is also called as the *frequency vector* (Fang and Ma 2001b; Zhou et al. 2012). For example, when $q = 2$ and $s = 3$, the elements of the frequency vector will be $n(1, 1, 1)$, $n(1, 1, 2)$, $n(1, 2, 1)$, $n(1, 2, 2)$, $n(2, 1, 1)$, $n(2, 1, 2)$, $n(2, 2, 1)$, and $n(2, 2, 2)$. In particular, a full factorial $U(N; q_1, \ldots, q_s)$ design corresponds to $\boldsymbol{y}_{\mathcal{P}} = \mathbf{1}_N$, where $\mathbf{1}_N$ is the N-vector of ones. Now, we can write

$$\begin{aligned}&\sum_{i=1}^{N}\sum_{j=1}^{N}\prod_{k=1}^{m} f(d_{ik}, d_{jk}, q_k)\\ &= \sum_{(i_1,\ldots,i_s)\in\mathcal{V}^s}\ \sum_{(j_1,\ldots,j_s)\in\mathcal{V}^s} n(i_1, \ldots, i_s)\, n(j_1, \ldots, j_s)\prod_{k=1}^{m} f(i_k, j_k, q_k)\end{aligned}$$

and

$$\sum_{i=1}^{N}\prod_{k=1}^{m} g(d_{ik}, q_k) = \sum_{(i_1,\ldots,i_s)\in\mathcal{V}^s} n(i_1, \ldots, i_s)\prod_{k=1}^{m} g(i_k, q_k).$$

The squared discrepancies $\text{CD}^2(\mathcal{P})$, $\text{WD}^2(\mathcal{P})$, and $\text{MD}^2(\mathcal{P})$ can then be written as quadratic forms of $\boldsymbol{y}_{\mathcal{P}}$.

Lemma 4.3.1 *For a design $\mathcal{P} \in \mathcal{D}(n; q_1, \ldots, q_s)$, we have*

$$\text{WD}^2(\mathcal{P}) = -\left(\frac{4}{3}\right)^s + \frac{1}{n^2}\boldsymbol{y}_{\mathcal{P}}^T \boldsymbol{W} \boldsymbol{y}_{\mathcal{P}}, \tag{4.3.2}$$

$$\mathrm{CD}^2(\mathcal{P}) = \left(\frac{13}{12}\right)^s - \frac{2}{n}\boldsymbol{c}^T\boldsymbol{y}_{\mathcal{P}} + \frac{1}{n^2}\boldsymbol{y}_{\mathcal{P}}^T\boldsymbol{C}\boldsymbol{y}_{\mathcal{P}}, \tag{4.3.3}$$

and

$$\mathrm{MD}^2(\mathcal{P}) = \left(\frac{19}{12}\right)^s - \frac{2}{n}\boldsymbol{m}^T\boldsymbol{y} + \frac{1}{n^2}\boldsymbol{y}_{\mathcal{P}}^T\boldsymbol{M}\boldsymbol{y}_{\mathcal{P}}, \tag{4.3.4}$$

where $\boldsymbol{W} = \boldsymbol{W}_1 \otimes \cdots \otimes \boldsymbol{W}_s$, $\boldsymbol{W}_k = (w_{ij}^k)$,

$$w_{ij}^k = 1.5 - |i-j|(q_k - |i-j|)/q_k^2,$$

$$\boldsymbol{c} = \boldsymbol{c}_1 \otimes \cdots \otimes \boldsymbol{c}_s,\ \boldsymbol{c}_k = (c_1^k, \ldots, c_{q_k}^k)^T,\ \boldsymbol{C} = \boldsymbol{C}_1 \otimes \cdots \otimes \boldsymbol{C}_s,\ \boldsymbol{C}_k = (c_{ij}^k),$$

$$\begin{aligned} c_i^k &= 1 + |2i-1-q_k|/(4q_k) - |2i-1-q_k|^2/(8q_k^2), \\ c_{ij}^k &= 1 + |2i-1-q_k|/(4q_k) + |2j-1-q_k|/(4q_k) - |i-j|/(2q_k), \end{aligned}$$

$\boldsymbol{m} = \boldsymbol{m}_1 \otimes \boldsymbol{m}_2 \otimes \cdots \otimes \boldsymbol{m}_s$, $\boldsymbol{m}_k = (m_1^k, \ldots, m_{q_k}^k)$, $\quad \boldsymbol{M} = \boldsymbol{M}_1 \otimes \boldsymbol{M}_2 \otimes \cdots \otimes \boldsymbol{M}_s$, $\boldsymbol{M}_k = (m_{ij}^k)$,

$$\begin{aligned} m_i^k &= \frac{5}{3} - \frac{1}{4}\left|\frac{2i-1-q_k}{2q_k}\right| - \frac{1}{4}\left|\frac{2i-1-q_k}{2q_k}\right|^2, \\ m_{ij}^k &= \frac{15}{8} - \frac{1}{4}\left|\frac{2i-1-q_k}{2q_k}\right| - \frac{1}{4}\left|\frac{2j-1-q_k}{2q_k}\right| - \frac{3}{4}\left|\frac{i-j}{q_k}\right| + \frac{1}{2}\left|\frac{i-j}{q_k}\right|^2, \end{aligned}$$

$i, j = 1, \ldots, q_k$, $k = 1, \ldots, s$, *and* $\otimes$ *is the Kronecker product.*

It should be mentioned that the expressions of $\mathrm{WD}^2(\mathcal{P})$ were introduced by Fang and Ma (2001b), the expressions of $\mathrm{CD}^2(\mathcal{P})$ were presented in Fang and Qin (2003), and the expression of $\mathrm{MD}^2(\mathcal{P})$ were given in Chen et al. (2015).

4.3.2 Complementary Design Theory

A full factorial design $U(N; q_1, \ldots, q_s)$ with $N = q_1 \ldots q_s$ can be split into two subdesigns according to the design points, $\mathcal{P}$ and $\overline{\mathcal{P}}$, with n runs and $N-n$ runs, respectively. One is called the *complementary design* of the other. Denote $Z_t = \{0, 1, 2, \ldots, t\}$ and $Z_t^N = Z_t \times \cdots \times Z_t$. Then, $\boldsymbol{y}_{\mathcal{P}} \in Z_1^N$ means every element of $\boldsymbol{y}_{\mathcal{P}}$ is 0 or 1, and $\boldsymbol{y}_{\mathcal{P}} = \boldsymbol{1}_N - \boldsymbol{y}_{\overline{\mathcal{P}}}$. It is worth to investigate the relationship between the discrepancy of $\mathcal{P}$ and that of its complementary design $\overline{\mathcal{P}}$. In this subsection, these relationships under the uniformity criteria CD, WD, and MD are given. Such theoretical results can be used for the construction of uniform designs with a large number of runs through the corresponding complementary designs, which have a small number of runs.

Based on Lemma 4.3.1, the relationships between the discrepancies of design $\mathcal{P}$ and its complementary design $\overline{\mathcal{P}}$ were obtained by Jiang and Ai (2014) and Chen et al. (2015). The results are summarized in the following theorem.

Theorem 4.3.1 *For any design $\mathcal{P} \in \mathcal{D}(n; q_1 \ldots q_s)$, the discrepancies of design $\mathcal{P}$ and its complementary design $\overline{\mathcal{P}}$ have the following relationships*

$$\begin{aligned} \mathrm{WD}^2(\mathcal{P}) = & -\left(\frac{4}{3}\right)^s + \frac{2n-N}{n^2} \prod_{k=1}^{s} \left(\frac{4q_k}{3} + \frac{1}{6q_k}\right) \\ & + \frac{(N-n)^2}{n^2}\left[\mathrm{WD}^2(\overline{\mathcal{P}}) + \left(\frac{4}{3}\right)^s\right], \end{aligned} \tag{4.3.5}$$

$$\begin{aligned} \mathrm{CD}^2(\mathcal{P}) = & \left(\frac{13}{12}\right)^s + \frac{1}{n^2}\left(\mathbf{1}_N^T \boldsymbol{C} \mathbf{1}_N - 2n\boldsymbol{c}^T \mathbf{1}_N\right) \\ & + \frac{1}{n^2}\left[-2\boldsymbol{y}_{\overline{\mathcal{P}}}^T(\boldsymbol{C}\mathbf{1}_N - N\boldsymbol{c}) + (N-n)^2\left(\mathrm{CD}^2(\overline{\mathcal{P}}) - \left(\frac{13}{12}\right)^s\right)\right], \end{aligned} \tag{4.3.6}$$

$$\begin{aligned} \mathrm{MD}^2(\mathcal{P}) = & \left(\frac{19}{12}\right)^s + \frac{1}{n^2}\left(\mathbf{1}_N^T \boldsymbol{M} \mathbf{1}_N - 2n\boldsymbol{m}^T \mathbf{1}_N\right) \\ & + \frac{1}{n^2}\left(-2\boldsymbol{y}_{\overline{\mathcal{P}}}^T(\boldsymbol{M}\mathbf{1}_N - N\boldsymbol{m}) + (N-n)^2\left(\mathrm{MD}^2(\overline{\mathcal{P}}) - \left(\frac{19}{12}\right)^s\right)\right), \end{aligned} \tag{4.3.7}$$

where $\boldsymbol{C}$, $\boldsymbol{c}$, $\boldsymbol{M}$, and $\boldsymbol{m}$ are defined in Lemma 4.3.1.

Proof We only prove the result for MD and the proof for other two discrepancies are similar. Since $\boldsymbol{y}_{\mathcal{P}} = \mathbf{1}_N - \boldsymbol{y}_{\overline{\mathcal{P}}}$, from (4.3.4) we have

$$\begin{aligned} \mathrm{MD}^2(\mathcal{P}) &= \left(\frac{19}{12}\right)^s - \frac{2}{n}\boldsymbol{m}^T(\mathbf{1}_N - \boldsymbol{y}_{\overline{\mathcal{P}}}) + \frac{1}{n^2}\left(\mathbf{1}_N - \boldsymbol{y}_{\overline{\mathcal{P}}}\right)^T \boldsymbol{M}\left(\mathbf{1}_N - \boldsymbol{y}_{\overline{\mathcal{P}}}\right) \\ &= \left(\frac{19}{12}\right)^s - \frac{2}{n}\boldsymbol{m}^T \mathbf{1}_N + \frac{2}{n}\boldsymbol{m}^T \boldsymbol{y}_{\overline{\mathcal{P}}} \\ &\quad + \frac{1}{n^2}\left(\mathbf{1}_N^T \boldsymbol{M}\mathbf{1}_N - 2\boldsymbol{y}_{\overline{\mathcal{P}}}^T \boldsymbol{M}\mathbf{1}_N + \boldsymbol{y}_{\overline{\mathcal{P}}}^T \boldsymbol{M}\boldsymbol{y}_{\overline{\mathcal{P}}}\right), \end{aligned} \tag{4.3.8}$$

$$\mathrm{MD}^2(\overline{\mathcal{P}}) = \left(\frac{19}{12}\right)^s - \frac{2}{N-n}\boldsymbol{m}^T \boldsymbol{y}_{\overline{\mathcal{P}}} + \frac{1}{(N-n)^2}\boldsymbol{y}_{\overline{\mathcal{P}}}^T \boldsymbol{M}\boldsymbol{y}_{\overline{\mathcal{P}}}. \tag{4.3.9}$$

Combine (4.3.8) and (4.3.9), the result in (4.3.7) can be easily obtained. The proof is finished.

Note that the design $\mathcal{P}$ in Theorem 4.3.1 is not necessary a U-type design. According to the relationships (4.3.5)–(4.3.7) in Theorem 4.3.1, the rules for determining a uniform design $\mathcal{P}$ under the discrepancy measures through its complementary design can be easily obtained.

Theorem 4.3.2 *For a design $\mathcal{P}$ in $\mathcal{D}(n; q_1, \ldots, q_s)$,*

(1) the design $\mathcal{P}$ is a uniform design under WD if and only if its complementary design $\overline{\mathcal{P}}$ is a uniform design under WD on $\mathcal{D}(N-n, q_1, \ldots, q_s)$.
(2) if the term $\boldsymbol{y}_{\mathcal{P}}^T(\boldsymbol{C}\mathbf{1}_N - N\boldsymbol{c})$ is a constant for all designs on $\mathcal{D}(n; q_1, \ldots, q_s)$, then a design $\mathcal{P}$ is a uniform design under CD if and only if its complementary design $\overline{\mathcal{P}}$ is a uniform design under CD on $\mathcal{D}(N-n, q_1, \ldots, q_s)$.
(3) if the term $\boldsymbol{y}_{\mathcal{P}}^T(\boldsymbol{M}\mathbf{1}_N - N\boldsymbol{m})$ is a constant, the design $\mathcal{P}$ is a uniform design under MD on $\mathcal{D}(n; q_1 \ldots q_s)$ if and only if its complementary design $\overline{\mathcal{P}}$ is a uniform design under MD on $\mathcal{D}(N-n; q_1, \ldots, q_s)$.

The proof of Theorem 4.3.2 is obvious, and we omit it. Comparing with Theorem 4.3.2(1) for WD, the additional condition that $\boldsymbol{y}_{\mathcal{P}}^T(\boldsymbol{C}\mathbf{1}_N - N\boldsymbol{c})$ or $\boldsymbol{y}_{\mathcal{P}}^T(\boldsymbol{M}\mathbf{1}_N - N\boldsymbol{m})$ is a constant is needed in Theorem 4.3.2(2) for CD or Theorem 4.3.2(3) for MD. However, we can find some special cases that the additional condition may be satisfied. Then, we obtain the following two corollaries.

Corollary 4.3.1 *Under the discrepancy measure CD, if one of the following conditions satisfies*
(1) the levels $q_1, \ldots, q_s$ are all odd,
(2) $s = 2$,
(3) $q_1 = \cdots = q_{s-1} = 2$,
a design $\mathcal{P}$ is a uniform design in $\mathcal{D}(n; q_1, \ldots, q_s)$ if and only if its complementary design $\overline{\mathcal{P}}$ is a uniform design in $\mathcal{D}(N-n, q_1, \ldots, q_s)$.

Proof We should prove that $\boldsymbol{y}_{\mathcal{P}}^T(\boldsymbol{C}\mathbf{1}_N - N\boldsymbol{c})$ is a constant for these cases. It can be easily verified that

$$\boldsymbol{C}_k\mathbf{1}_{q_k} - q_k\boldsymbol{c}_k = \frac{1}{8q_k}\mathbf{1}_{q_k}\, I_{\{q_k \text{ is even}\}} \text{ for } k = 1, \ldots, s,$$

where $I_{\{\cdot\}}$ is the indicator function.

(1) Note that when $q_1, \ldots, q_m$ are all odd, $\boldsymbol{C}_k\mathbf{1}_{q_k} = q_k\boldsymbol{c}_k$, for $k = 1, \ldots, s$. Then, we have $\boldsymbol{C}\mathbf{1}_N = N\boldsymbol{c}$ and $\boldsymbol{y}_{\mathcal{P}}^T(\boldsymbol{C}\mathbf{1}_N - N\boldsymbol{c}) = 0$ is a constant.

(2) When $s = 2$, we have

$$\begin{aligned}
&\boldsymbol{C}\mathbf{1}_N - N\boldsymbol{c}\\
&= \left(q_1\boldsymbol{c}_1 + \frac{1}{8q_1}\mathbf{1}_{q_1}\, I_{\{q_1 \text{ is even}\}}\right) \otimes \left(q_2\boldsymbol{c}_2 + \frac{1}{8q_2}\mathbf{1}_{q_2}\, I_{\{q_2 \text{ is even}\}}\right) - N\boldsymbol{c}\\
&= \frac{q_2}{8q_1}\mathbf{1}_{q_1} \otimes \boldsymbol{c}_2\, I_{\{q_1 \text{ is even}\}} + \frac{q_1}{8q_2}\boldsymbol{c}_1 \otimes \mathbf{1}_{q_2}\, I_{\{q_2 \text{ is even}\}}\\
&\quad + \frac{1}{64q_1q_2}\mathbf{1}_N\, I_{\{\text{both } q_1 \text{ and } q_2 \text{ are even}\}}.
\end{aligned}$$

Since $\mathbf{y}_D^T\mathbf{1}_N = n$, $\mathbf{y}_D^T(\mathbf{1}_{q_1} \otimes \mathbf{c}_2)$ and $\mathbf{y}_D^T(\mathbf{c}_1 \otimes \mathbf{1}_{q_2})$ are all constants for all designs in $\mathcal{D}(n; q_1, q_2)$, then $\boldsymbol{y}_{\mathcal{P}}^T(\boldsymbol{C}\mathbf{1}_N - N\boldsymbol{c})$ is a constant.

(3) When $q_k = 2$, and $\boldsymbol{c}_k = \frac{35}{32}\mathbf{1}_2$ for $k = 1, \ldots, s-1$, we have $\boldsymbol{c} = \left(\frac{35}{32}\right)^{s-1}\mathbf{1}_{2^{s-1}} \otimes \boldsymbol{c}_s$. Moreover, $\boldsymbol{C}_k\mathbf{1}_2 = \frac{9}{4}\mathbf{1}_2$ for $k = 1, \ldots, s-1$ and $\boldsymbol{C}_s\mathbf{1}_{q_s} - q_s\mathbf{c}_s = \frac{1}{8q_s}\mathbf{1}_{q_s} I_{\{q_s \text{ is even}\}}$. Then,

$$\boldsymbol{C}\mathbf{1}_N - N\boldsymbol{c} = \left(\left(\frac{9}{4}\right)^{s-1}\mathbf{1}_{2^{s-1}}\right) \otimes (\boldsymbol{C}_s\mathbf{1}_{q_s}) - N\boldsymbol{c}$$

$$= \left(q_s\left(\frac{9}{4}\right)^{s-1} - N\left(\frac{35}{32}\right)^{s-1}\right)\mathbf{1}_{2^{s-1}} \otimes \mathbf{c}_s + \left(\frac{9}{4}\right)^{s-1}\frac{1}{8q_s}\mathbf{1}_N I_{\{q_s \text{ is even}\}}.$$

Since $\mathbf{y}_D^T\mathbf{1}_N = n$ and $\mathbf{y}_D^T(\mathbf{1}_{2^{s-1}} \otimes \boldsymbol{c}_s)$ is a constant for all designs in $\mathcal{D}(n, 2^{s-1}q_s)$, then $\boldsymbol{y}_{\mathcal{P}}^T(\boldsymbol{C}\mathbf{1}_N - N\boldsymbol{c})$ is a constant. The proof is complete.

Consider the discrepancy measure MD. When $\mathcal{P} \in \mathcal{D}(n; 2^s)$, from the definitions of $\boldsymbol{M}$ and $\boldsymbol{m}$ in Theorem 4.3.2, it is obvious that

$$\boldsymbol{M}\mathbf{1}_N - N\boldsymbol{m} = \left[\left(\frac{13}{4}\right)^s - \left(\frac{305}{96}\right)^s\right]\mathbf{1}_N,$$

then $\boldsymbol{y}_{\mathcal{P}}^T(\boldsymbol{M}\mathbf{1}_N - N\boldsymbol{m}) = \left[\left(\frac{13}{4}\right)^s - \left(\frac{305}{96}\right)^s\right]\boldsymbol{y}_{\mathcal{P}}^T\mathbf{1}_N = (N-n)\left[\left(\frac{13}{4}\right)^s - \left(\frac{305}{96}\right)^s\right]$ is a constant. As a consequence, we have the following result.

Corollary 4.3.2 *When $n < N = 2^s$, a design $\mathcal{P}$ is a uniform design under MD on $\mathcal{D}(n; 2^s)$ if and only if its complementary design $\overline{\mathcal{P}}$ is a uniform design under MD on $\mathcal{D}(N-n; 2^s)$.*

Moreover, when the number of levels is larger than 2, it also can obtain uniform designs or nearly uniform designs through the complementary design theory. Then, the computation complexity for search uniform design with a large number of runs can be reduced significantly. The following example shows that such method is useful to find the uniform designs especially for designs with large size.

The condition that the term $\boldsymbol{y}_{\mathcal{P}}^T(\boldsymbol{C}\mathbf{1}_N - N\boldsymbol{c})$ or $\boldsymbol{y}_{\mathcal{P}}^T(\boldsymbol{M}\mathbf{1}_N - N\boldsymbol{m})$ is a constant in Theorem 4.3.2 is a sufficient condition such that a design is the uniform design if and only if its complementary design is the uniform design. However, the complementary design theory is also useful even such condition is not satisfied. Here has an example.

Example 4.3.1 Let $n = 6, q = 3$ and $s = 3$. Consider to construct uniform design $U_6(3^3)$ as well as the $U_{21}(3^3)$. Choose six different points among the 3^3-full factorial design to form the designs, i.e., the design space is $\mathcal{D}(6; 3^3)$ with no repeated design point. Among the $\binom{27}{6} = 296{,}010$ possible designs, the following design

$$\mathcal{P} = \begin{pmatrix} 1 & 1 & 2 & 2 & 3 & 3 \\ 1 & 2 & 2 & 3 & 1 & 3 \\ 1 & 3 & 2 & 1 & 2 & 3 \end{pmatrix}^T$$

is a uniform design with the minimum MD-value 0.1219. Then, the corresponding complementary design $\overline{\mathcal{P}}$ can be obtained by deleting the six points of $\mathcal{P}$ in the 3^3-full factorial design, and its MD-value is 0.1097. It can be checked that among all the 296,010 possible designs with 21 different points from the 3^3-full factorial design, the MD-value of $\overline{\mathcal{P}}$ is the minimum, which means $\overline{\mathcal{P}}$ is also the uniform design in the design space $\mathcal{D}(21; 3^3)$ with no repeated point.

The complementary design theory is powerful to construct uniform design $\mathcal{P}$ with large number of runs n when n is close to N, the number of runs of full factorial design. The construction procedure is as follows.

Step 1 Given the parameters $n, q_1, \ldots, q_s$, construct the uniform design $\overline{\mathcal{P}}$ with $N - n$ runs by some algorithm.
Step 2 Obtain $\mathcal{P}$ through the complementary design $\overline{\mathcal{P}}$.

For example, when $N = 5^6$ and $n = N - 10$, it is convenient to construct the (nearly) uniform design $U_{10}(5^6)$ through the TA algorithm, and then, the (nearly) uniform design $U_{N-10}(5^6)$ can also be easily obtained by the complementary design theory.

4.3.3 *Optimal Frequency Vector*

According to the quadratic forms of different discrepancies in Sect. 4.3.1, searching a uniform design can be regarded as a quadratic integer programming problem. Note that the theory of optimization including *convex optimization* and *quadratic integer programming* has been rapidly developed in the past decades (see Boyd and Vandenberghe 2004). In this subsection, applications of some efficient algorithms in the quadratic integer programming for construction of uniform designs are given.

First, we give a brief introduction to quadratic programming, convex quadratic programming, semidefinite programming, and integer programming. An optimization program can be described as follows.

$$\begin{aligned} \min\ & f_0(\boldsymbol{y}) \\ s.t.\ & f_i(\boldsymbol{y}) \leqslant 0,\ i = 1, \ldots, t, \\ & h_i(\boldsymbol{y}) = 0,\ i = 1, \ldots, p. \end{aligned} \tag{4.3.10}$$

We want to find an *optimizing variable* $\boldsymbol{y} \in R^m$ that minimizes the *objective function* $f_0(\boldsymbol{y})$ among all $\boldsymbol{y}$ that satisfy the inequality constraints given in (4.3.10). The function $f_0 : R^m \to R$ is also called *cost function*. The functions $f_i : R^m \to R$ and $h_i : R^m \to R$ are called the *inequality constraint functions* and the *equality constraint functions*, respectively. When $t = p = 0$, we say the problem (4.3.10) to be unconstrained.

When $f_0, \ldots, f_t$ are convex functions, and $h_i(\boldsymbol{y}) = \boldsymbol{a}_i^T \boldsymbol{y} - \boldsymbol{b}_i, i = 1, \ldots, p$, the problem (4.3.10) becomes a *convex optimization problem*. The convex optimization problem is called a *quadratic program* (QP) if the objective function f_0 is (convex) quadratic, and the constraint functions are linear. A quadratic program can be expressed in the form

$$\begin{aligned} \min\ & f_0(\mathbf{y}) = \frac{1}{2}\mathbf{y}^T\mathbf{Q}\mathbf{y} + \mathbf{c}^T\mathbf{y} \\ s.t.\ & \mathbf{G}\mathbf{y} \le \mathbf{h}, \\ & \mathbf{A}\mathbf{y} = \mathbf{b}. \end{aligned} \tag{4.3.11}$$

where $\mathbf{Q} \in R^{m\times m}$ is a symmetric real matrix, $\mathbf{c} \in R^m$, $\mathbf{h} \in R^t$, $\mathbf{b} \in R^p$, $\mathbf{G} \in R^{t\times m}$, $\mathbf{A} \in R^{p\times m}$, and the notation $\mathbf{G}\mathbf{y} \leqslant \mathbf{h}$ means that every entry of the vector $\mathbf{G}\mathbf{y}$ is less than or equal to the corresponding entry of the vector $\mathbf{h}$. If $\mathbf{Q}$ is a symmetric positive-semidefinite matrix, the quadratic programming problem (4.3.11) is called the *semidefinite quadratic programming*. For solving the QP in (4.3.11), a variety of methods are commonly used, including subgradient methods (Shor 1985), bundle methods (Hiriart-Urruty and Lemaréchal 1993), interior point methods (Boyd and Vandenberghe 2004), and the ellipsoid method (Shor 1991). If the matrix $\mathbf{Q}$ is positive definite, the ellipsoid method solves the problem in polynomial time (Kozlov et al. 1979). If $\mathbf{Q}$ is indefinite, then the problem QP is NP-hard, even if $\mathbf{Q}$ has only one negative eigenvalue, the problem is NP-hard (Pardalos and Vavasis 1991).

In many applications, each element of the vector $\mathbf{y}$ should be chosen from the discrete set $\{0, 1\}$, and the constraints in the quadratic programming problem (4.3.11) become $\mathbf{y} = (y_1, \ldots, y_m) \in \{0, 1\}^m$, i.e., $y_i \in \{0, 1\}$. Here, we consider the following special quadratic programming problem

$$\begin{aligned} \min\ & f_0(\mathbf{y}) = \mathbf{y}^T\mathbf{Q}\mathbf{y} \\ s.t.\ & \mathbf{y} \in \{0, 1\}^m. \end{aligned} \tag{4.3.12}$$

Such a formulation (4.3.12) is called an *unconstrained binary quadratic programming problem*. This problem is also known as the *unconstrained quadratic bivalent programming problem* or the *unconstrained quadratic zero-one programming problem*. In many cases, we may add the constraint that the summation of the values of the elements of $\mathbf{y}$ equals some constant n, i.e.,

$$\begin{aligned} \min\ & f_0(\mathbf{y}) = \mathbf{y}^T\mathbf{Q}\mathbf{y} \\ s.t.\ & \mathbf{1}^T\mathbf{y} = n,\ \ \mathbf{y} \in \{0, 1\}^m, \end{aligned} \tag{4.3.13}$$

where $\mathbf{1} = (1, \ldots, 1)^T$. Iasemidis et al. (2001) showed that the quadratic binary programming problem (4.3.13) can be reduced to problem (4.3.12). Since the constraint in problem (4.3.12) is not convex, the quadratic binary programming problem is not a convex optimization problem and the complexity of solving such problem is also NP-hard. One has to use some optimization algorithms to search the solution.

Now, consider construction of uniform designs for given parameters $q_1, \ldots, q_s$ and n. Based on the formula (4.3.2), the problem of constructing a uniform design $U_n(q_1, \cdots, q_s)$ under WD can be formulated as the following optimization problem:

$$\begin{cases} \min \ f_0(\boldsymbol{y}) = -\left(\frac{4}{3}\right)^s + \frac{1}{n^2}\boldsymbol{y}^T\boldsymbol{W}\boldsymbol{y} \\ s.t. \ \ \mathbf{1}_N^T\boldsymbol{y} = n, \ \ \boldsymbol{y} \in Z_+^N, \end{cases} \tag{4.3.14}$$

where $Z_+^N = Z_+ \times \cdots \times Z_+$, $Z_+ = \{0, 1, 2, \ldots\}$, $N = q_1 \ldots q_s$, and $\boldsymbol{y} = (y_1, \ldots, y_N)^T \in Z_+^N$ means $y_i \in Z_+$. Actually, from the constraint $\mathbf{1}_N^T\boldsymbol{y} = n$ and $\boldsymbol{y} \in Z_+^N$, we can reduce the range Z_+^N to Z_n^N, where $Z_n = \{0, 1, 2, \ldots, n\}$. Thus, the problem (4.3.14) is equivalent to the following optimization problem:

$$\begin{cases} \min \ f_1(\boldsymbol{y}) = \boldsymbol{y}^T\boldsymbol{W}\boldsymbol{y} \\ s.t. \ \ \mathbf{1}_N^T\boldsymbol{y} = n, \ \ \boldsymbol{y} \in Z_n^N. \end{cases} \tag{4.3.15}$$

The problem (4.3.15) is an integer programming problem and is not a convex optimization problem. It was shown that the integer programming problem is a NP-hard problem, while convex optimization problem can be solved in polynomial time (Boyd and Vandenberghe 2004). If the constraint $\boldsymbol{y} \in Z_n^N$ is relaxed, we have the following convex optimization problem

$$\begin{cases} \min \ f_2(\boldsymbol{y}) = \boldsymbol{y}^T\boldsymbol{W}\boldsymbol{y} \\ s.t. \ \ \mathbf{1}_N^T\boldsymbol{y} = n. \end{cases} \tag{4.3.16}$$

Problem (4.3.16) is a special convex quadratic programming, i.e., a semidefinite programming. Therefore, we can use the theory of convex optimization to solve problem (4.3.16). Especially, we have the following result.

Theorem 4.3.3 *The minimizer of problem* (4.3.16) *is*

$$\boldsymbol{y}^* = \frac{n}{N}\mathbf{1}_N, \tag{4.3.17}$$

where $N = q_1 \ldots q_s$.

Proof It is well known that the Lagrangian associated with the problem (4.3.16) is

$$L(\boldsymbol{y}, \lambda) = \boldsymbol{y}^T\boldsymbol{W}\boldsymbol{y} + \lambda(\mathbf{1}^T\boldsymbol{y} - n) = \boldsymbol{y}^T\boldsymbol{W}\boldsymbol{y} + \lambda\mathbf{1}^T\boldsymbol{y} - \lambda n, \tag{4.3.18}$$

where $\lambda \in R$ is the Lagrange multiplier associated with the constraint $\mathbf{1}^T\boldsymbol{y} = n$. The Lagrange dual function can be obtained as follows:

$$g(\lambda) = \inf_{\boldsymbol{y}} L(\boldsymbol{y}, \lambda) = \inf_{\boldsymbol{y}} \left(\boldsymbol{y}^T\boldsymbol{W}\boldsymbol{y} + \lambda\mathbf{1}^T\boldsymbol{y}\right) - \lambda n. \tag{4.3.19}$$

Derivate the Eq. (4.3.18) with respect to $\boldsymbol{y}$, and let the derivative be zero, and we have

$$\hat{\boldsymbol{y}} = -\frac{1}{2}\lambda\mathbf{W}^{-1}\mathbf{1}. \tag{4.3.20}$$

Thus

$$g(\lambda) = -\frac{1}{4}\lambda^2 \mathbf{1}^T \mathbf{W}^{-1}\mathbf{1} - \lambda n, \tag{4.3.21}$$

which gives a lower bound of the optimal value $\boldsymbol{y}^*$ of the optimization problem (4.3.16). And the Lagrange dual problem associated with the problem (4.3.16) becomes

$$\max \quad g(\lambda). \tag{4.3.22}$$

Let d^* and p^* be respective the optimal value of problem (4.3.16) and the Lagrange dual problem (4.3.22), respectively. Since the problem (4.3.16) is a semidefinite programming, the optimal duality gap is zero, which means $d^* = p^*$ (see Boyd and Vandenberghe 2004, p. 226) From (4.3.21), the maximizer of problem (4.3.22) is

$$\lambda^* = -\frac{2n}{\mathbf{1}^T \mathbf{W}^{-1}\mathbf{1}}. \tag{4.3.23}$$

Substituting (4.3.23) into (4.3.20), we have

$$\hat{\boldsymbol{y}} = -\frac{1}{2}(\mathbf{W}^{-1}\mathbf{1})\left(-\frac{2n}{\mathbf{1}^T \mathbf{W}^{-1}\mathbf{1}}\right). \tag{4.3.24}$$

Moreover, it can be easily checked that

$$\begin{cases} \mathbf{W}_k \mathbf{1}_{q_k} = \left(\frac{4q_k}{3} + \frac{1}{6q_k}\right)\mathbf{1}_{q_k}, & \mathbf{W}\mathbf{1}_N = \prod_{k=1}^s \left(\frac{4q_k}{3} + \frac{1}{6q_k}\right)\mathbf{1}_N, \\ \mathbf{W}_k^{-1}\mathbf{1}_{q_k} = \left(\frac{4q_k}{3} + \frac{1}{6q_k}\right)^{-1}\mathbf{1}_{q_k}, & \mathbf{W}^{-1}\mathbf{1}_N = \prod_{k=1}^s \left(\frac{4q_k}{3} + \frac{1}{6q_k}\right)^{-1}\mathbf{1}_N. \end{cases}$$

Then, we obtain the minimizer $\boldsymbol{y}^* = \frac{n}{N}\mathbf{1}_N$, which completes the proof.

The results in Theorem 4.3.3 for symmetrical and asymmetrical designs under WD are from Fang and Ma (2001a) and Zhou et al. (2012), respectively. According to Theorem 4.3.3, usually the optimal design $\mathcal{P}$ associated with $\boldsymbol{y}^*$ is not an exact design. However, a full design with $n = kN$ and $\boldsymbol{y} = k\mathbf{1}_N$ for some positive integer k is a uniform design under WD.

Corollary 4.3.3 *When $n = kN$, $N = q_1 \ldots q_s$ and k is a positive integer, the full design $\mathcal{P} \in \mathcal{U}(n; q_1, \ldots, q_s)$ is the uniform design under* WD *and its squared* WD *is*

$$WD^2(\mathcal{P}) = \prod_{i=1}^{s}\left(\frac{4}{3} + \frac{1}{6q_i^2}\right) - \left(\frac{4}{3}\right)^s.$$

Next consider the MD. For given n and $q_1, \ldots, q_s$, based on the formula (4.3.4), the problem of constructing uniform design can be formulated as the following optimization problem when the constant $\left(\frac{19}{12}\right)^s$ is ignored:

$$\begin{cases} \min -\frac{2}{n}\boldsymbol{m}^T\boldsymbol{y} + \frac{1}{n^2}\boldsymbol{y}^T\boldsymbol{M}\boldsymbol{y}, \\ s.t.\ \ \boldsymbol{1}_N^T\boldsymbol{y} = n,\ \boldsymbol{y} \in Z_n^N. \end{cases} \tag{4.3.25}$$

If the constraint $\boldsymbol{y} \in Z_n^N$ is relaxed and the objective function times n^2, we obtain the following convex optimization problem

$$\begin{cases} \min \boldsymbol{y}^T\boldsymbol{M}\boldsymbol{y} - 2n\boldsymbol{m}^T\boldsymbol{y}, \\ s.t.\ \ \boldsymbol{1}_N^T\boldsymbol{y} = n. \end{cases} \tag{4.3.26}$$

which is a special semidefinite programming problem. We have the following result.

Theorem 4.3.4 *For given n and $q_1, \ldots, q_s$, the minimizer of optimization problem (4.3.26) is*

$$\boldsymbol{y}^* = n\left(\boldsymbol{M}^{-1}\boldsymbol{m} + \frac{1 - \boldsymbol{m}^T\boldsymbol{M}^{-1}\boldsymbol{1}_N}{\boldsymbol{1}_N^T\boldsymbol{M}^{-1}\boldsymbol{1}_N}\boldsymbol{M}^{-1}\boldsymbol{1}_N\right). \tag{4.3.27}$$

where $\boldsymbol{m}$ and $\boldsymbol{M}$ are defined in Lemma 4.3.1.

The proof of Theorem 4.3.4 can be found in Chen et al. (2015). From Theorem 4.3.4, the elements of the optimal vector $\boldsymbol{y}^*$ seems not equal to each other. However, when $q_1 = \cdots = q_s = 2$, the elements $m_i^k = 305/192$, and $m_{ij}^k = 7/4$ if $i = j$, $3/2$ otherwise. Then,

$$\begin{aligned} \boldsymbol{m}_k &= (305/192, 305/192), \\ \boldsymbol{M}_k &= \begin{bmatrix} 7/4 & 3/2 \\ 3/2 & 7/4 \end{bmatrix},\ \boldsymbol{M}_k^{-1} = \begin{bmatrix} 28/13 & -24/13 \\ -24/13 & 28/13 \end{bmatrix}. \end{aligned}$$

Here, the notations $m_i^k, m_{ij}^k, \boldsymbol{m}_k$, and $\boldsymbol{M}_k$ are defined in Lemma 4.3.1. Moreover, according to the property of the operator Kronecker product

$$\begin{aligned} &(\boldsymbol{B} \otimes \boldsymbol{B})^{-1} = \boldsymbol{B}^{-1} \otimes \boldsymbol{B}^{-1},\ (\boldsymbol{B} \otimes \boldsymbol{B})(\boldsymbol{b} \otimes \boldsymbol{b}) = \boldsymbol{B}\boldsymbol{b} \otimes \boldsymbol{B}\boldsymbol{b}, \\ &(\boldsymbol{B} \otimes \boldsymbol{B})\boldsymbol{1}_{2k} = \boldsymbol{B}\boldsymbol{1}_k \otimes \boldsymbol{B}\boldsymbol{1}_k, \end{aligned}$$

then $\boldsymbol{M}^{-1}\boldsymbol{m} = \left(\frac{305}{634}\right)^s \boldsymbol{1}_N,\ \boldsymbol{M}^{-1}\boldsymbol{1}_N = \left(\frac{4}{13}\right)^s \boldsymbol{1}_N$ and $\boldsymbol{y}^* = \frac{n}{N}\boldsymbol{1}_N$. Then, we have the following result.

Corollary 4.3.4 *The uniform design under MD in $\mathcal{D}(n; 2^s)$ has the frequency vector $\boldsymbol{y}^* = \frac{n}{N}\boldsymbol{1}_N$, where $N = 2^s$. Moreover, the two-level full factorial design is the uniform design under MD.*

Ma et al. (2003) first showed some similar result under CD, and for details, one can refer to their paper.

4.3.4 Integer Programming Problem Method

The construction of a uniform design $U(n; q_1 \dots q_s)$ can be regarded as an optimization problem (4.3.15) that can be relaxed into (4.3.16). Theorem 4.3.3 gives the solution $\boldsymbol{y}^* = \frac{n}{N}\boldsymbol{1}_N$, where $N = q_1 \dots q_s$. However, the corresponding design with optimal frequency vector is not an exact design when $n < N$. It needs some algorithms to construct uniform designs when $n < N$. In this subsection, we consider the construction of uniform design by the integer programming problem method.

Without loss of any generality, consider the WD as the uniformity measure. Since $n < N$ and the uniform design requires the design points scatter on the domain as uniformity as possible, the constraint $\boldsymbol{1}_N^T \boldsymbol{y} = n$ can be changed as $\boldsymbol{y} \in \{0, 1\}^N$, where $\boldsymbol{y} \in \{0, 1\}^N$ means that each element of $\boldsymbol{y}$, $y_i \in \{0, 1\}$. Then, the quadratic integer programming problem (4.3.15) with a linear constraint can be transformed as the following unconstrained optimization problem (see Iasemidis et al. 2001):

$$\begin{cases} \min\ \boldsymbol{y}^T \boldsymbol{W} \boldsymbol{y} + K(\boldsymbol{W})(\boldsymbol{1}_N^T \boldsymbol{y} - n)^2 \\ s.t.\ \ \boldsymbol{y} \in \{0, 1\}^N, \end{cases} \tag{4.3.28}$$

where $K(\boldsymbol{W}) = 2\sum_{i=1}^N \sum_{j=1}^N |w_{ij}| + 1$, and $\boldsymbol{W} = (w_{ij})$ is defined in Lemma 4.3.1. Furthermore, because of the property $y_i^2 = y_i$, we can rewrite the problem (4.3.28) as follows:

$$\begin{cases} \min\ \boldsymbol{y}^T \boldsymbol{Q}_0 \boldsymbol{y} + K(\boldsymbol{W})n^2 \\ s.t.\ \ \boldsymbol{y} \in \{0, 1\}^N, \end{cases} \tag{4.3.29}$$

where $\boldsymbol{Q}_0 = \boldsymbol{W} + K(\boldsymbol{W})\boldsymbol{1}_N \boldsymbol{1}_N^T - 2K(\boldsymbol{W})N\boldsymbol{i}_N$ is a symmetric matrix. Note that the value of $K(\boldsymbol{W})$ is large in most cases and the elements of $\boldsymbol{Q}_0$ are also large. It is better to divide the objective function by $K(\boldsymbol{W})$ and remove the constant $K(\boldsymbol{W})n^2$, and problem (4.3.29) can be rewritten as the following unconstrained quadratic problem:

$$\begin{cases} \min\ f(\boldsymbol{y}) = \boldsymbol{y}^T \boldsymbol{Q} \boldsymbol{y} \\ s.t.\ \ \boldsymbol{y} \in \{0, 1\}^N, \end{cases} \tag{4.3.30}$$

where $\boldsymbol{Q} = \boldsymbol{Q}_0 / K(\boldsymbol{W})$ is a nonnegative definite matrix.

It is well known that the unconstrained zero-one quadratic programming problem (4.3.30) does not have analytic solution and is an NP-hard problem when N increases. For solving such problem, many methods were proposed in the literature, such as some trajectory methods including Tabu search (Beasley 1998; Palubeckis 2004) and simulated annealing (Beasley 1998; Katayama and Narihisa 2001), some population-based methods including scatter search (Amini et al. 1999) and evolutionary algorithms (Merz and Freisleben 1999; Merz and Katayama 2004), and other local search heuristics (Merz and Freisleben 2002). More details about these heuristics can refer to Gilli and Winkler (2009). Katayama and Narihisa (2001) presented an SA-based heuristic to test on publicly available benchmark instances of size ranging from 500 to 2500 variables and compared them with other heuristics. Computational results indicate that this SA leads to high-quality solutions with a short cpu times. For solving this problem, Merz and Freisleben (2002) proposed a greedy heuristic and two local search algorithms: 1-opt local search and k-opt local search.

Zhou et al. (2012) proposed an algorithm called *SA-based integer programming method* (SA-IPM) to solve the special zero-one quadratic programming problem for constructing uniform designs. SA-IPM combines SA-based heuristic and 1-opt local search with the best improvement move strategy. Its pseudo-code can be seen in Algorithm 4.3.1.

Algorithm 4.3.1 (*SA-based integer programming method*)

Initialize I, J, T_{init}, T_f, T_r;
Generate an initial random solution $\boldsymbol{y}_0 \in \{0, 1\}^N$;
for $i = 1 : I$
  Set $\boldsymbol{y} = \boldsymbol{y}_0$, $T = T_{init}$, ct$=$0;
  Calculate gains g_k of $\boldsymbol{y}$ for all k in $\{1, \ldots, N\}$;
  while ct$< J$;
    Set ct$=$ct+1;
    for t$=$1:N
      Find j with $g_j = \min_k g_k$;
      If $g_j < 0$, then set ct$=$0 and $y_j = 1 - y_j$ (and update all gains g_i);
      Otherwise, random choose $k \in \{1, \ldots, N\}$, set $y_k = 1 - y_k$ with probability $e^{-g_t/T}$ (and update all gains g_i);
    end
    Set $T = T_f \times T$;
  end
  If the design with respect to $\boldsymbol{y}$ reaches its lower bound, return $\boldsymbol{y}$;
  Otherwise, set $\boldsymbol{y}_0 = \boldsymbol{y}$, $T_{init} = T_r \times T_{init}$;
end
Return $\boldsymbol{y}$;

In the above pseudo-code parameters I, J, and T_{init} represent the number of time of annealing process, the termination conditional number at each iteration, and the initial temperature, respectively. Parameters $T_f, T_r \in (0, 1)$ are two temperature reduction rates. Usually, the initial values of the parameters in Algorithm 4.3.1 are set by some preliminary test, which is considered to have a trade-off between the quality of the resulted design and computer running time. Zhou et al. (2012) suggested that $T_{init} = 1/q$, where q is the number of levels, $J = 10$ and the ratio $T_f = 0.99$, $I = 10$, $T_r = 0.9$ if $N < 500$, and $I = 2$, $T_r = 0.8$ if $N \geqslant 500$.

Define a neighbor of current solution $\boldsymbol{y} = (y_1, \ldots, y_N)$ as

$$\{\boldsymbol{y}_i = (y_1, \ldots, y_{i-1}, 1 - y_i, y_{i+1}, \ldots, y_N), i = 1, \ldots, N\},$$

so the hamming distance between $\boldsymbol{y}_i$ and $\boldsymbol{y}$ is equal to 1. Define the gain $g_i = f(\boldsymbol{y}_i) - f(\boldsymbol{y})$, where $f(\cdot)$ is the objective function in problem (4.3.30), and $g_i < 0$ means $\boldsymbol{y}_i$ is a good neighbor; otherwise, it is a bad one. According to Merz and Freisleben (2002), the gain g_i can be calculated by

$$g_i = q_{ii}(\bar{y}_i - y_i) + 2 \sum_{k=1,k\neq i}^{m} q_{ki} y_k(\bar{y}_i - y_i), \qquad (4.3.31)$$

where $\bar{y}_i = 1 - y_i$, q_{ki} is the (k, i)-element of the matrix $\boldsymbol{Q}$ in problem (4.3.30). The gain g_i can be calculated in a linear time of N, but all gains of neighbors must be calculated in $O(N^2)$ times. However, the gains g_i do not have to be recalculated each time. Assuming that all g_i for a current solution have been calculated and the bit k is flipped, we can compute the new gain g_i^n efficiently with the formula:

$$g_i^n = \begin{cases} -g_i, & \text{if } i = k, \\ g_i + 2q_{ik}(\bar{y}_i - y_i)(\bar{y}_k - y_k), & \text{otherwise}, \end{cases} \qquad (4.3.32)$$

and the update gains can be performed in a linear time. Step (9:) includes a local search for the best improvement, which is different from the classical SA. It is possible in step (10:) that there are different j satisfied $g_j = \min_i g_i$. In this case, we randomly choose one bit to flip. In steps (10:) and (11:), all gains g_i can be updated by using (4.3.32). Moreover, if the lower bound in step (10:) is reached, the process is terminated. The lower bound of design under WD can be seen in Fang et al. (2005), Zhou et al. (2008).

Zhou et al. (2012) showed that the Algorithm 4.3.1 is powerful for searching (nearly) uniform designs when $n < N$. For example, when $N < 3200$, the algorithm can deliver designs with lower computational complexity and lower WD value compared with many existing construction methods. Therefore, SA-IPM is suitable to construct designs with a large number of runs and the total number of level-combinations to be below several thousands, e.g., $n > 200$ and $N < 3200$. Note that as N increases, the computational complexity of SA-IPM increases exponentially. For other uniformity criteria, one can use the SA-based integer programming problem method similarly.

Exercises

4.1

The Gauss–Newton method in optimization has been widely used in various fields. Let $f(\boldsymbol{x})$ be a function on $\mathcal{X}$. One wants to find a $\boldsymbol{x}^* \in \mathcal{X}$ such that $f(\boldsymbol{x}^*) = \max_{\boldsymbol{x}\in\mathcal{X}} f(\boldsymbol{x})$. List the conditions on $f(\boldsymbol{x})$ and $\mathcal{X}$ for applying the Gauss–Newton method for this optimal problem.

When the objective function $f(\boldsymbol{x})$ has several local maxima, how to apply the Gauss–Newton method to find the global maximum?

4.2

Give a brief introduction to the following in optimization methodology: Convex programming, linear programming, quadratic programming, integer programming, nonlinear programming, and stochastic programming.

4.3

Let $\mathcal{P}$ be a design below with eight runs and 4 two-level factors.

Row	1	2	3	4
1	1	1	1	1
2	2	2	2	2
3	1	1	2	2
4	2	2	1	1
5	1	2	1	2
6	2	1	2	1
7	1	2	2	1
8	2	1	1	2

Give the vector $\boldsymbol{y}_{\mathcal{P}}$ defined in (4.3.1).

4.4

Write a MATLAB code for the TA Algorithm 4.2.1. The code gives the choice of the discrepancy (WD, CD, or MD) to the user and provides the trace plot. Apply your own code to find nearly uniform designs $U_9(3^4)$, $U_{16}(4^5)$, $U_{18}(3^7)$, and $U_{25}(5^6)$.

4.5

Give the iteration formula for MD.

4.6

Write a MATLAB code for the SA-based integer programming method in Sect. 4.3.4. Apply your own code to find nearly uniform designs $U_9(3^4)$, $U_{45}(3^4)$, $U_{20}(5^4)$, $U_{200}(5^4)$, $U_{12}(4^5)$, and $U_{600}(4^5)$.

4.7

Suppose one wants to construct a uniform design $U_n(n^s)$. The so-called forward construction method is a natural idea. The first column can be chosen as $\boldsymbol{u}_1 = (1, 2 \ldots, n)^T$. The second column $\boldsymbol{u}_2$ is a permutation of $\{1, 2, \ldots, n\}$ such that $\boldsymbol{u}_2 \neq \boldsymbol{u}_1$ and $[\boldsymbol{u}_1, \boldsymbol{u}_2]$ have the minimum discrepancy among all $n! - 1$ permutations. The third column $\boldsymbol{u}_3$ is a permutation of $\{1, 2, \ldots, n\}$ such that $\boldsymbol{u}_3 \neq \boldsymbol{u}_1$, $\boldsymbol{u}_3 \neq \boldsymbol{u}_2$, and $[\boldsymbol{u}_1, \boldsymbol{u}_2, \boldsymbol{u}_3]$ have the minimum discrepancy among all $n! - 2$ permutations. Continue this procedure until we have s columns. A similar idea can be used for the construction of $U_n(q^s)$.

Write a MATLAB code for construction of $U_9(9^3)$, $U_9(3^4)$, and $U_8(2^7)$.

References

Amini, M.M., Alidaee, B., Kochenberger, G.A.: A scatter search approach to unconstrained quadratic binary programs. In: Corne, D., Dorigo, M., Glover, F. (eds.) New Ideas in Optimization, pp. 317–329. McGraw-Hill, New York (1999)

Beasley, J.E.: Heuristic algorithms for the unconstrained binary quadratic programming problem. Technical report, Management School, Imperial College, London, UK (1998)

Boyd, S., Vandenberghe, L.: Convex Optimization. Cambridge University Press, Cambridge (2004)

Chen, W., Qi, Z.F., Zhou, Y.D.: Constructing uniform designs under mixture discrepancy. Stat. Probab. Lett. **97**, 76–82 (2015)

Dueck, G., Scheuer, T.: Threshold accepting: a general purpose algorithm appearing superior to simulated annealing. J. Comput. Phys. **90**, 161–175 (1990)

Fang, K.T., Ma, C.X.: Orthogonal and Uniform Experimental Desigs. Science Press, Beijing (2001a)

Fang, K.T., Ma, C.X.: Wrap-around L_2-discrepancy of random sampling, Latin hypercube and uniform designs. J. Complex. **17**, 608–624 (2001b)

Fang, K.T., Qin, H.: A note on construction of nearly uniform designs with large number of runs. Stat. Probab. Lett. **61**, 215–224 (2003)

Fang, K.T., Lin, D.K.J., Winker, P., Zhang, Y.: Uniform design: theory and applications. Technometrics **42**, 237–248 (2000)

Fang, K.T., Ma, C.X., Winker, P.: Centered L_2-discrepancy of random sampling and Latin hypercube design, and construction of uniform designs. Math. Comput. **71**, 275–296 (2002)

Fang, K.T., Lu, X., Winker, P.: Lower bounds for centered and wrap-around L_2-discrepancies and construction of uniform. J. Complex. **20**, 268–272 (2003)

Fang, K.T., Tang, Y., Yin, J.X.: Lower bounds for wrap-around L_2-discrepancy and constructions of symmetrical uniform designs. J. Complex. **21**, 757–771 (2005)

Fang, K.T., Li, R., Sudjianto, A.: Design and Modeling for Computer Experiments. Chapman and Hall/CRC, New York (2006a)

Fang, K.T., Maringer, D., Tang, Y., Winker, P.: Lower bounds and stochastic optimization algorithms for uniform designs with three or four levels. Math. Comput. **75**, 859–878 (2006b)

Fang, K.T., Ke, X., Elsawah, A.M.: Construction of a new 27-run uniform orthogonal design. J. Complex. submitted (2016)

Gilli, M., Winkler, P.: Heuristic optimization methods in econometrics. In: Belsley, D., Kontoghiorghes, E. (eds.) Handbook of Computational Econometrics, pp. 81–119. Wiley, Chichester (2009)

Hiriart-Urruty, J.-B., Lemaréchal, C.: Convex Analysis and Minimization Algorithms. Springer, Berlin (1993) (Two volumes)

Iasemidis, L.D., Pardalos, P., Sackellares, J.C., Shiau, D.-S.: Quadratic binary programming and dynamical system approach to determine the predictability of epileptic seizures. J. Comb. Optim. **5**, 9–26 (2001)

Jiang, B.C., Ai, M.Y.: Construction of uniform designs without replications. J. Complex. **30**, 98–110 (2014)

Jin, R., Chen, W., Sudjianto, A.: An efficient algorithm for constructing optimal design of computer experiments. J. Stat. Plan. Inference **134**, 268–287 (2005)

Katayama, K., Narihisa, H.: Performance of simulated annealing-based heuristic for the unconstrained binary quadratic programming problem. Eur. J. Oper. Res. **134**(1), 103–119 (2001)

Kirkpatrick, S., Gelett, C., Vecchi, M.: Optimization by simulated annealing. Science **220**, 621–630 (1983)

Kozlov, M.K., Tarasov, S.P., Khachiyan, L.G.: Polynomial solvability of convex quadratic programming. Dokl. Akad. Nauk SSSR **248**, 1049–1051, Translated in: Sov. Math. - Dokl. **20**, 1108–1111 (1979)

Ma, C.X., Fang, K.T., Lin, D.K.J.: A note on uniformity and orthogonality. J. Stat. Plan. Inference **113**, 323–334 (2003)

Merz, P., Freisleben, B.: Genetic algorithms for binary quadratic programming. In: Proceedings of the 1999 Genetic and Evolutionary Computation Conference, pp. 417–424 (1999)

Merz, P., Freisleben, B.: Greedy and local search heuristics for unconstrained binary quadratic programming. J. Heuristics **8**, 197–213 (2002)

Merz, P., Katayama, K.: Memetic algorithms for the unconstrained binary quadratic programming problem. BioSystems **78**, 99–118 (2004)

Morris, M.D., Mitchell, T.J.: Exploratory design for computational experiments. J. Stat. Plan. Inference **43**, 381–402 (1995)
Palubeckis, G.: Multistart Tabu search strategies for the unconstrained binary quadratic optimization problem. Ann. Oper. Res. **131**, 259–282 (2004)
Pardalos, P.M., Vavasis, S.A.: Quadratic programming with one negative eigenvalue is np-hard. J. Global Optim. **1**, 15–22 (1991)
Shor, N.Z.: Minimization Methods for Non-differentiable Functions. Springer Series in Computational Mathematics. Springer, Berlin (1985)
Shor, N.Z.: The development of numerical methods for nonsmooth optimization in the USSR. History of Mathematical Programming. A Collection of Personal Reminiscences, pp. 135–139. North-Holland, Amsterdam (1991)
Winker, P., Fang, K.T.: Application of threshold accepting to the evaluation of the discrepancy of a set of points. SIAM Numer. Anal. **34**, 2038–2042 (1997)
Winker, P., Fang, K.T.: Optimal U-type design. In: Niederreiter, H., Zinterhof, P., Hellekalek, P. (eds.) Monte Carlo and Quasi-Monte Carlo Methods 1996, pp. 436–448. Springer, Berlin (1998)
Zhou, Y.D., Fang, K.T.: An efficient method for constructing uniform designs with large size. Comput. Stat. **28**(3), 1319–1331 (2013)
Zhou, Y.D., Ning, J.H., Song, X.B.: Lee discrepancy and its applications in experimental designs. Stat. Probab. Lett. **78**, 1933–1942 (2008)
Zhou, Y.D., Fang, K.T., Ning, J.H.: Constructing uniform designs: a heuristic integer programming method. J. Complex. **28**, 224–237 (2012)
Zhou, Y.D., Fang, K.T., Ning, J.H.: Mixture discrepancy for quasi-random point sets. J. Complex. **29**, 283–301 (2013)

Chapter 5
Modeling Techniques

Let $y = f(\boldsymbol{x}) = f(x_1, \ldots, x_s)$ be the true model of a system where $X_1, \ldots, X_s$ are factors that take values $(x_1, \ldots, x_s)$ on a domain $\mathcal{X}$, and y is response. For physical experiments with model unknown, the true model $f(\cdot)$ is unknown and for computer experiments the function $f(\cdot)$ is known, but it may have no an analytic formula. We are requested to find a metamodel $\hat{y} = g(\boldsymbol{x})$ with a high quality to approximate the true model in a certain sense. How to find a metamodel is a challenging problem. This chapter gives a brief introduction to various modeling techniques such as radial basis function, polynomial regression model, spline and Fourier model, wavelets basis and Kriging models in applications by the use of uniform designs. Readers can find more discussion on modeling methods in Eubank (1988), Wahba (1990), Hastie et al. (2001), and Fang et al. (2006).

When the true model is unknown, the statistician G.E.P. Box argued that "all models are wrong; but some are useful." A modeling method is a procedure for finding a high-quality metamodel that approximates the true model well over the domain $\mathcal{X}$. A good metamodel should have less computation complexity and be easy to explore relationship between the factors and the response or between the input variables and the output. Chen et al. (2006) said that "A mathematical model surrogate of system performance, to approximate the relationship between system performance and the design parameters." Our experiences also believe that there are only a few active factors in the model. There are various kinds of modeling techniques. Section 5.1 introduces the modeling technique by basis functions which can be used to analyze the data of physical experiments and computer experiments. Section 5.2 considers the Kriging models which are popularly used for data of computer experiments. Section 5.3 gives an example to show the usefulness of uniform designs.

K.-T. Fang et al., *Theory and Application of Uniform Experimental Designs*, Lecture Notes in Statistics 221,
https://doi.org/10.1007/978-981-13-2041-5_5

5.1 Basis Functions

Many modeling methods are based on a set of specific bases. Let

$$\{B_1(\boldsymbol{x}), B_2(\boldsymbol{x}), \ldots\}, \boldsymbol{x} \in \mathcal{X}$$

be a set of basis functions defined on the experimental domain $\mathcal{X}$. Here, the functions $B_i(\boldsymbol{x}), i = 1, 2, \ldots,$ are known. A metamodel g is a submodel of the maximal model of the interest,

$$g(\boldsymbol{x}) = \beta_1 B_1(\boldsymbol{x}) + \beta_2 B_2(\boldsymbol{x}) + \cdots + \beta_m B_m(\boldsymbol{x}), \tag{5.1.1}$$

where β_j's are unknown coefficients to be estimated and m can be finite or infinite. The polynomial basis, spline basis, and wavelets basis are popularly used in practical application. In this section, we introduce the application of these basis functions for modeling.

5.1.1 Polynomial Regression Models

The polynomial regression model has been widely used for modeling. The one-dimensional polynomial basis with order k is

$$1, x, x^2, \ldots, x^k,$$

and the centered polynomial basis is

$$1, (x - \bar{x}), (x - \bar{x})^2, \ldots, (x - \bar{x})^k,$$

where $\bar{x}$ is the sample mean of x. Sometimes, the centered polynomial basis is more useful than the polynomial basis. Consider the modeling for Example 1.1.1.

Example 5.1.1 (*Example* 1.1.1 *continued*) The experimenter wants to explore the relationship between the strength (y) and the amount of the chemical material (x) by an experiment. Obviously, it is not enough if one observes only at two or three different x-values, the latter are still called *levels* similar to that in factorial design. Suppose the number of levels is decided to be 12. The equal distance points on [0.4, 1] are $x = 0.4250, 0.4750, 0.5250, 0.5750, 0.6250, 0.6750, 0.7250, 0.7750, 0.8250, 0.8750, 0.9250, 0.9750$. The replication at each x-value is chosen as two times, and the corresponding strengths are denoted by $y_{k1}, y_{k2}, k = 1, \ldots, 12$ and are list in Table 5.1.

Table 5.1 One-factor UD and related responses

No	x	y_{k1}	y_{k2}
1	0.4250	1.2066	1.5996
2	0.4750	1.8433	0.9938
3	0.5250	1.4580	1.1150
4	0.5750	2.0226	1.9433
5	0.6250	2.1987	2.1855
6	0.6750	2.8344	3.3655
7	0.7250	3.4903	3.5215
8	0.7750	3.1198	3.6166
9	0.8250	3.7632	3.0041
10	0.8750	1.6193	2.1473
11	0.9250	0.8490	1.3500
12	0.9750	0.1136	0.8764

From the plot of y against x in Fig. 5.1, it is clear that the simple linear model $y = \beta_0 + \beta_1 x + \varepsilon$ is not suitable. There are many alternatives for the metamodels, for example, we can employ a polynomial regression model

$$y = \beta_0 + \beta_1 x + \beta_2 x^2 + \cdots + \beta_k x^k + \varepsilon \tag{5.1.2}$$

with some suitable order k. More often one considers centered polynomial regression model

$$y = \beta_0 + \beta_1(x - \bar{x}) + \beta_2(x - \bar{x})^2 + \cdots + \beta_k(x - \bar{x})^k + \varepsilon. \tag{5.1.3}$$

The fitting models by the centered polynomial regression models of order 5 and order 6 are shown in Fig. 5.1, respectively. The ANOVA table of the latter is given

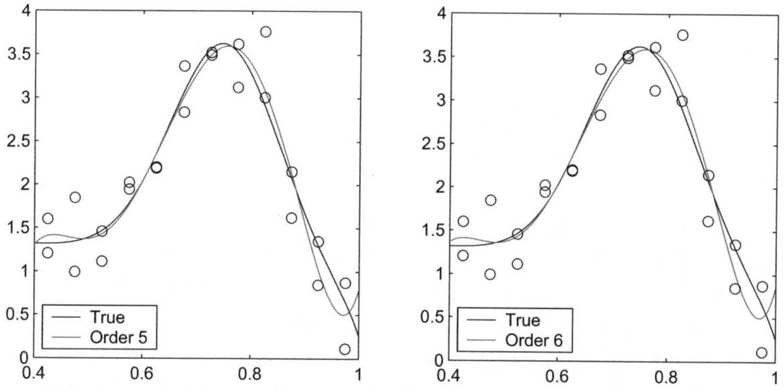

Fig. 5.1 Polynomial regressions of orders 5 and 6

Table 5.2 ANOVA of the polynomial regression of order 6

Source	Df	Sum of square	Mean square	F-value	Pr > F
Model	6	22.3939	3.7323	27.8322	<0.0001
Lack of fit	5	0.5861	0.1172	0.8740	0.5266
Error	12	1.6090	0.1341		
Total	23	24.5889			

in Table 5.2. In this table, the lack of fit test is implemented since there are 12 repeated points (For the lack of fit test, the reader can refer to some textbooks, such as Myers 1990.) If the lack of fit test is insignificant, the current model can be accepted as a metamodel; otherwise, we should try another model. For our case the test is insignificant, and so the polynomial model of order 6 can be used as a metamodel. The estimator of $\hat{\sigma}^2$ now can combine the two sums of squares "lack of fit" and "error": $\hat{\sigma}^2 = (0.5861 + 1.6090)/(5 + 12) = 0.1291$. The model test is significant, which means the model can fit the data very well. However, from the plots in Fig. 5.1 it is clear that this metamodel can be further improved as the fitting is not so good in the neighborhoods of 0.4 and 1, respectively.

For multi-dimensional cases, a polynomial regression model is based on special basis functions $x_1^{r_1} \ldots x_s^{r_s}$, where $r_1, \ldots, r_s$ are nonnegative integers. Low-order polynomials such as the first-order regression model

$$E(y) = \beta_0 + \beta_1 x_1 + \cdots + \beta_s x_s, \tag{5.1.4}$$

the quadratic regression model

$$E(y) = \beta_0 + \sum_{i=1}^{s} \beta_i x_i + \sum_{i \leqslant j}^{s} \beta_{ij} x_i x_j, \tag{5.1.5}$$

and the centered quadratic regression model

$$E(y) = \beta_0 + \sum_{i=1}^{s} \beta_i (x_i - \bar{x}_i) + \sum_{i \leqslant j}^{s} \beta_{ij} (x_i - \bar{x}_i)(x_j - \bar{x}_j) \tag{5.1.6}$$

are recommended, where $\bar{x}_i$ is the sample mean of x_i.

In the presence of high-order polynomials, we face two problems: (1) the number of polynomial basis functions dramatically increases with the number of factors and the degree of polynomial; (2) the polynomial basis may cause the collinearity problem, i.e., there are high correlations among regressors. In such situations, *orthogonal polynomial models* are recommended to overcome the difficulty arising out of the presence of multi-collinearity. The orthogonal regression model for one-factor experiment is well known. For multi-factor case, the corresponding model can be

$$E(y) = \alpha_0 + \sum_{i=1}^{s} \alpha_i \phi_i(x_i) + \sum_{i \leqslant j}^{s} \alpha_{ij} \phi_i(x_i) \phi_j(x_j), \tag{5.1.7}$$

where $\phi_j(u)$'s satisfy

$$\int_0^1 \phi_j(u)\, du = 0, \quad \int_0^1 \phi_j^2(u)\, du = 1, \quad \text{and} \quad \int_0^1 \phi_j(u)\phi_k(u)\, du = 0, \ \text{for } j \neq k.$$

The univariate orthogonal polynomials over [0,1] can be constructed using Legendre polynomials over $[-1/2, 1/2]$ by a location transformation. The first few Legendre polynomials are:

$$\begin{aligned}
\phi_0(u) &= 1, \\
\phi_1(u) &= \sqrt{12}(u - 1/2), \\
\phi_2(u) &= \sqrt{180}\left\{(u - 1/2)^2 - \frac{1}{12}\right\}, \\
\phi_3(u) &= \sqrt{2800}\left\{(u - 1/2)^3 - \frac{3}{20}(u - 1/2)\right\}, \\
\phi_4(u) &= 210\left\{(u - 1/2)^4 - \frac{3}{14}(u - 1/2)^2 + \frac{3}{560}\right\}, \\
\phi_5(u) &= 252\sqrt{11}\left\{(u - 1/2)^5 - \frac{5}{18}(u - 1/2)^3 + \frac{5}{336}(u - 1/2)\right\}, \\
\phi_6(u) &= 924\sqrt{13}\left\{(u - 1/2)^6 - \frac{15}{44}(u - 1/2)^4 + \frac{5}{176}(u - 1/2)^2 - \frac{5}{14{,}784}\right\}.
\end{aligned}$$

These orthogonal polynomials together with their tensor products

$$\Phi_{r_1,\ldots,r_s}(\boldsymbol{x}) = \prod_{j=1}^{s} \phi_{r_j}(x_j),$$

can be easily used to construct an orthogonal polynomial basis in an experimental domain $\mathcal{X}$. Let $B_0(\boldsymbol{x}) \equiv \Phi_{0,\ldots,0}(\boldsymbol{x}) = 1$, and any finite set of function $\Phi_{r_1,\ldots,r_s}(\boldsymbol{x})$ can serve to define the basis function.

Fourier basis functions are well-known orthogonal basis functions. When the response y is a periodic function of the factors, Fourier regression models are useful. Riccomango et al. (1997) gave a comprehensive study on this basis. It is well known that

$$1, \cos(2\pi x), \sin(2\pi x), \ldots, \cos(2k\pi x), \sin(2k\pi x), \ldots,$$

form an orthogonal basis for a functional space over [0, 1]. In practice, the following Fourier regression model is recommended:

$$E(y) = \beta_0 + \sum_{i=1}^{s} \sum_{j=1}^{m} \{\alpha_{ij} \cos(2j\pi x_i) + \beta_{ij} \sin(2j\pi x_i)\}.$$

When there are some interactions, the two-factor complete model is

$$\begin{aligned} E(y) = \theta_0 &+ \sqrt{2} \sum_{r=1}^{m_1} [\sin(2\pi r x_1)\beta_{1,r} + \cos(2\pi r x_1)\gamma_{1,r}] \\ &+ \sqrt{2} \sum_{s=1}^{m_2} [\sin(2\pi s x_2)\beta_{2,s} + \cos(2\pi s x_2)\gamma_{2,s}] \\ &+ 2 \sum_{r=1}^{m_1} \sum_{s=1}^{m_2} [\sin(2\pi r x_1)\sin(2\pi s x_2)\theta_{r,s} + \sin(2\pi r x_1)\cos(2\pi s x_2)\lambda_{r,s} \\ &+ \cos(2\pi r x_1)\sin(2\pi s x_2)\tau_{r,s} + \cos(2\pi r x_1)\cos(2\pi s x_2)\phi_{r,s}], \end{aligned}$$

where m_1 and m_2 are orders of the Fourier regression model, $\beta_{1,r}$, $\gamma_{1,r}$, $\beta_{2,s}$, $\gamma_{2,s}$, $\theta_{r,s}$, $\lambda_{r,s}$, $\tau_{r,s}$, $\phi_{r,s}$'s are unknown coefficients. Riccomango et al. (1997) pointed out that if you choose a suitable design, it will be D-optimal under the above models. Shi et al. (2001) proposed the so-called FOUND algorithm, by which we can find many D-optimal designs. For more discussions, the reader can refer to Xie et al. (2007).

5.1.2 *Spline Basis*

Each item in the polynomial basis gives inference to the whole domain $\mathcal{X}$. When the function $g(\boldsymbol{x})$ is flat in some area of $\mathcal{X}$, but fluctuates in other area of $\mathcal{X}$, it may be difficult for the polynomial regression model to model such a function. We wish to have a basis in which some items appear only in a special area of $\mathcal{X}$. For example, the term $I(a \leqslant x \leqslant b)$ disappears outside of $[a, b]$, where $I(\cdot)$ is the indicator function and the term $(x - a)_+$ disappear in $[-\infty, a]$, where

$$x_+ = \begin{cases} x, & \text{if } x > 0; \\ 0, & \text{otherwise.} \end{cases}$$

For one-factor case, if we can choose predecided knots $\kappa_1, \ldots, \kappa_m$, a *power spline basis* has the following general form:

$$1, x, x^2, \ldots, x^p, (x - \kappa_1)_+^p, \ldots, (x - \kappa_m)_+^p. \tag{5.1.8}$$

The most widely used orders are $p \leqslant 5$. Hastie et al. (2001) said that "It is claimed that cubic spline are the lowest-order spline for which the knot-discontinuity is not visible to the human eye." Certainly, the cubic model with $p = 3$ is useful in practical applications.

Spline basis functions for multi-factor experiments are constructed by using the method of tensor product. It is common to standardize the x-variables first such that the ranges for all x-variables are the same, for example [0, 1]. Then take the same knots for each x-variables. For given fixed knots $\kappa_1, \ldots, \kappa_m$, denote

$$S_0(u) = 1,\ S_1(u) = u, \ldots, S_p(u) = u^p,$$
$$S_{p+1}(u) = (u - \kappa_1)_+^p, \ldots, S_{p+m}(u) = (u - \kappa_m)_+^p.$$

Then an s-dimensional tensor product basis function over $\boldsymbol{x} = (x_1, \ldots, x_s)'$ is

$$B_{r_1,\ldots,r_s}(\boldsymbol{x}) = \prod_{j=0}^{s} S_{r_j}(x_j). \tag{5.1.9}$$

It is easy to see that $B_{0,\ldots,0}(\boldsymbol{x}) = 1$, and any finite set of functions $B_{r_1,\ldots,r_s}(\boldsymbol{x})$, $0 \leqslant r_1, \ldots, r_s \leqslant m + p$ can form a basis function. However, the tensor product method increases the number of basis functions exponentially as the dimension s increases. Friedman (1991) proposed the so-called *multivariate adaptive regression splines* (MARS). In MARS, the number of basis functions and the knot locations are adaptively determined by the data. Friedman (1991) presented a thorough discussions on MARS and gave an algorithm to create spline basis functions for MARS. Readers can refer to this paper for details.

5.1.3 Wavelets Basis

Wavelets bases have been widely used in numerical analysis and wavelet shrinkage methods in nonparametric regression (see Daubechies 1992; Chui 1992; Antoniadis and Oppenheim 1995). The latter has been often used in modeling for experimental data. In practice, the most widely used wavelet shrinkage method is the Donoho–Johnstone's VisuShrink procedure (see Donoho and Johnstone 1994; Donoho et al. 1995).

The definition of the wavelets basis is as follows. Let ϕ and ψ denote the orthogonal father and mother wavelet functions. Assume that ψ has r vanishing moments and ϕ satisfies $\int \phi(x)dx = 1$. Define

$$\phi_{jk}(x) = 2^{j/2}\phi(2^j x - k),\ \ \psi_{jk}(x) = 2^{j/2}\psi(2^j x - k). \tag{5.1.10}$$

The set $\{\phi_{jk}(x), \psi_{jk}(x)\}$ constitutes an orthogonal basis on $L_2([0, 1])$, the set of square-integrable functions on [0, 1]. This basis has an associated exact orthogonal discrete wavelet transform that transforms data into wavelet coefficient domain. For a given $g \in L_2([0, 1]))$, it can be expanded into a wavelet series

$$g(x) = \sum_{k=1}^{2^{j_0}} a_{j_0k}\phi_{j_0k}(x) + \sum_{j=j_0}^{\infty}\sum_{k=1}^{2^j} b_{jk}\psi_{jk}(x), \tag{5.1.11}$$

where

$$a_{jk} = \int g(x)\phi_{jk}(x)dx, \; b_{jk} = \int g(x)\psi_{jk}(x)dx.$$

Wavelet transform decomposes a function into different resolution components. In (5.1.11), a_{j_0k} are the coefficients at the coarsest level. They represent the gross structure of the function g, while b_{jk} are the wavelet coefficients that represent finer and finer structures of the function g as the resolution level j increases. The wavelet basis transforms i.i.d. Gaussian noise to i.i.d. Gaussian noise and is norm-preserving. This nice property allows us to transform the problem in the function domain into a problem in the sequence domain of the wavelet coefficients. The reader can refer to Brown and Cai (1997) and Cai and Brown (1998) for a systematic introduction.

The so-called *least interpolating polynomials* proposed by De Boor and Bon (1990) are other examples of metamodels in base function (5.1.1), where the $B_j(\boldsymbol{x})$ terms are polynomial, determined by the locations of the design points. The method favors low-degree terms over higher-degree terms. Several authors pointed out poor behavior of the method, and therefore, we omit the details.

5.1.4 Radial Basis Functions

A *radial symmetrical function* has the following form

$$\phi(\boldsymbol{x}) = \phi(x_1, \ldots, x_s) = r(||\boldsymbol{x}||), \tag{5.1.12}$$

where $||\boldsymbol{x}||$ is the Euclidean norm of $\boldsymbol{x}$ and function r is given. Various spherical density distributions, like multivariate normal, mixtures of normal, uniform distribution on a ball in R^s, symmetrical Kotz-type, symmetrical multivariate types II and VII including multivariate t- and multivariate Cauchy distributions (Fang et al. 1990), give us a large range of model candidates. Radial basis functions were proposed by Hardy (1971) and developed by Dyn et al. (1986) and Powell (1987). A radial basis function is a linear combination of radially symmetrical functions. An extension of (5.1.12) is

$$K(||\boldsymbol{x} - \boldsymbol{x}_i||/\theta), i = 1, \ldots, n, \tag{5.1.13}$$

where $\boldsymbol{x}_1, \ldots, \boldsymbol{x}_n$ are design points, $K(\cdot)$ is a kernel function, and θ is a smoothing parameter. A metamodel is suggested to have the form of

$$g(\mathbf{x}) = \mu + \sum_{j=1}^{n} \beta_j K(||\mathbf{x} - \mathbf{x}_i||/\theta). \tag{5.1.14}$$

There are several popular choices of basis functions such as

Linear	$K(z) = z$
Cubic	$K(z) = z^3$
Thin plate spline	$K(z) = z^2 \log z$
Gaussian	$K(z) = \exp(-z^2)$

A number of modifications to the exact interpolation have been proposed. For a detailed introduction and discussion, one can refer to Section 5.6.3 of Fang et al. (2006).

5.1.5 *Selection of Variables*

There are so many functions in a function base as the number of factors increases. When a metamodel involves so many items, it will cost

- *Collinearity*: There are high correlations among variables.
- *Sparsity*: The number of observations is relatively small compared with the number of variables.
- *The curse of dimensionality*: Many methods cannot be implemented due to computational complexity.

Therefore, we have to choose a subset of the basis functions and need techniques of variable selections. There are a lot of methods in this direction, such as

- forward selection;
- backward elimination;
- stepwise regression;
- the best subset;
- principal component regression;
- partial least squares regression;
- penalized likelihood approaches;
- Bayesian approach.

This is a hot research area, and we do not want to give any discussion here. The reader can easily find many references in the literature.

5.2 Modeling Techniques: Kriging Models

In the previous section, we consider the maximal model as of the form

$$y(\boldsymbol{x}_k) = \sum_{i=1}^{m} \beta_i B_i(\boldsymbol{x}_k) + \varepsilon_k, \tag{5.2.1}$$

where $\varepsilon_k, k = 1, \ldots, n$, are i.i.d. with mean zero and variance $\hat{\sigma}^2$. In many case studies, the i.i.d. assumption on the error cannot be accepted. A natural modification to the above model is given by

$$y(\boldsymbol{x}) = \sum_{i=1}^{m} \beta_i B_i(\boldsymbol{x}) + z(\boldsymbol{x}), \ \ \boldsymbol{x} \in \mathcal{X}, \tag{5.2.2}$$

where $z(\boldsymbol{x})$ is a random function of $\boldsymbol{x}$. For physical experiments, we treat z as a white noise (see Definition 5.2.1 below) corresponding to measurement error. In computer experiments, however, there is no measurement error, and $z(\boldsymbol{x})$ is then systematic departure from the given linear model.

There are many choices of $z(\boldsymbol{x})$, for example, assume that $z(\boldsymbol{x})$ is a white noise or $z(\boldsymbol{x})$ is a Gaussian random function. On the other hand, from the characteristics of computer experiments there is no random error in the experiments. We should have $\hat{y}(\boldsymbol{x}_k) = y(\boldsymbol{x}_k), k = 1, \ldots, n$, where $\hat{y}(\boldsymbol{x}_k)$ is the estimator of $y(\boldsymbol{x}_k)$ by the chosen metamodel. This requires that the metamodel predicts the response at any $\boldsymbol{x} \in \mathcal{X}$ essentially by an *interpolation* among the experimental (training) data.

The so-called *Kriging model* or *spatial correlation model*, motivated in the field of geostatistics, is a modeling method that can fit the above requirements. The word "Kriging" is synonymous with optimal spatial prediction. It has been named after a South African mining engineer with the name Krige, who first popularized stochastic methods for spatial predictions (Krige 1951). His work was furthered in the early 1970s by authors such as Matheron (1971) and formed the foundation for an entire field of study now known as geostatistics, see for example, Cressie (1993, 1997) and Goovaerts (1997). The Kriging models have been utilized in modeling for computer experiments. There are a lot of studies on this direction, for example, the reader can refer to Sacks et al. (1989a, b) and Welch et al. (1992) for a comprehensive review. Stein (1999) gave a more general theory on Kriging model.

5.2.1 Models

Definition 5.2.1 Let $y(\boldsymbol{x})$ be a stochastic process on the domain $\mathcal{X} \subset R^s$. We say that $y(\boldsymbol{x})$ is a *Gaussian random function* if for any $m > 1$ and any choice of $\boldsymbol{x}_1, \ldots, \boldsymbol{x}_m$ in $\mathcal{X}$, the vector $(y(\boldsymbol{x}_1), \ldots, y(\boldsymbol{x}_m))$ follows a multivariate normal distribution with the covariance function

$$\mathrm{Cov}(y(\boldsymbol{x}_i), y(\boldsymbol{x}_j)) = \hat{\sigma}^2 R(\boldsymbol{x}_i, \boldsymbol{x}_j),$$

where $\hat{\sigma}^2 > 0$ is the unknown variance and $R(\cdot, \cdot)$ is the correlation function. When the latter depends on $\boldsymbol{x}_i$ and $\boldsymbol{x}_j$ only through $\boldsymbol{x}_i - \boldsymbol{x}_j$, the Gaussian random function $y(\boldsymbol{x})$ is called stationary. In this case, the correlation function is denoted by $R(\cdot)$. For a stationary Gaussian random function, if the correlation function satisfies $R(0) = 1$ and $R(a) = 0, a \neq 0$, the corresponding Gaussian random function is called a white noise.

Gaussian random functions are determined by their mean function, $\boldsymbol{\mu}(\boldsymbol{x}) = E(y(\boldsymbol{x}))$ and their covariance function $\text{Cov}(\boldsymbol{x}_1, \boldsymbol{x}_2) = \text{Cov}(y(\boldsymbol{x}_1), y(\boldsymbol{x}_2))$.

Definition 5.2.2 A Kriging model is expressed as

$$y(\boldsymbol{x}) = \sum_{i=1}^{m} \beta_i h_i(\boldsymbol{x}) + z(\boldsymbol{x}), \tag{5.2.3}$$

where $h_j(\boldsymbol{x})$'s are known functions, β_j's are unknown coefficients to be estimated, and $z(\boldsymbol{x})$ is a stationary random function with mean $E(z(\boldsymbol{x})) = 0$ and covariance

$$\text{Cov}(z(\boldsymbol{x}_i), z(\boldsymbol{x}_j)) = \hat{\sigma}_z^2 R(\boldsymbol{x}_i - \boldsymbol{x}_j),$$

where $\hat{\sigma}_z^2$ is the unknown variance and the correlation function R is given. The part of $\sum_{i=1}^{m} \beta_i h_i(\boldsymbol{x})$ is called the *parametric item*. When the parametric item is a constant, say μ, the model reduces to

$$y(\boldsymbol{x}) = \mu + z(\boldsymbol{x}) \tag{5.2.4}$$

and is said to be the *ordinary Kriging model*, which has been widely used in modeling.

In the literature, the power exponential correlation function

$$R(\boldsymbol{x}_i, \boldsymbol{x}_k) = \prod_{j=1}^{s} R_j(x_{ij}, x_{kj}) = \prod_{j=1}^{s} \exp(-\theta_j |x_{ij} - x_{kj}|^{\gamma_j}), \tag{5.2.5}$$

has been widely used, where $\theta_j \geq 0$ are unknown parameters and $0 \leqslant \gamma_j \leqslant 2$ are given. Let $d_j = x_{ij} - x_{kj}$. More choices of the function $R_j(d_j)$ are

- EXP: $\exp(-\theta_j |d_j|)$;
- Guass: $\exp(-\theta_j d_j^2)$;
- LIN: $\max\{0, 1 - \theta_j |d_j|\}$;
- Spline: $1 - 3\xi_j^2 + 2\xi_j^3, \; \xi_j = \min\{1, \theta_j |d_j|\}$;
- Matérn family:

$$\frac{1}{\Gamma(\nu) 2^{\nu-1}} \left(\frac{2\sqrt{\nu}|d_j|}{\theta_j} \right)^{\nu} K_{\nu} \left(\frac{2\sqrt{\nu}|d_j|}{\theta_j} \right), \tag{5.2.6}$$

where θ_j are unknown parameters, $\nu > 0$ and $K_\nu(\cdot)$ is the modified Bessel function of order ν whose definition can be found in Wikipedia, the free encyclopedia. Let $\boldsymbol{\beta} = (\beta_1, \beta_2, \ldots, \beta_m)'$ and covariance's parameters σ_z^2, $\boldsymbol{\theta} = (\theta_1, \theta_2, \ldots, \theta_s)$, the smoothing parameter vector. As with all choices of the correlation function, the function goes to zero as the distance between $\boldsymbol{x}_i$ and $\boldsymbol{x}_k$ increases. This shows that the influence of a sampled data point on the point to be predicted becomes weaker as their separation distance increases. The magnitude of $\boldsymbol{\theta}$ dictates how quickly the influence deteriorates. For large values of $\boldsymbol{\theta}$, only the data points that are very near to each other are well correlated. For small values of $\boldsymbol{\theta}$, the points further away still influence the point to be predicted because they are still well correlated. Very often we choose $\theta_1 = \cdots = \theta_s = \theta$ that is called the *width of the Gaussian kernel*. The parameter γ_j is a measure of smoothness. The response becomes increasingly smooth as the values of γ_j's increase. One should also note the interpolating behavior of the correlation function. Because $R(\boldsymbol{x}_i, \boldsymbol{x}_i) = 1$, the predictor will go exactly through any measured data point.

5.2.2 Estimation

For a given data set $\{(y_i, \boldsymbol{x}_i), i = 1, \ldots, n\}$, we need to estimate all the unknown parameters $\boldsymbol{\beta}$, $\hat{\sigma}_z^2$ and $\boldsymbol{\theta}$ in the model. For estimating these unknown parameters, the unbiased linear predictor $\hat{y}(\boldsymbol{x}) = \boldsymbol{c}(\boldsymbol{x})^T \boldsymbol{y}$ is favorable, where $\boldsymbol{c}(\boldsymbol{x}) = (c_1(\boldsymbol{x}), \ldots, c_n(\boldsymbol{x}))^T$ is a constant vector, $\boldsymbol{y} = (y_1, \ldots, y_n)^T$ is the response vector of the data and $\hat{y}$ is the fitting result. In the literature, it is preferred to use the following linear predictor:

Definition 5.2.3 (a) A predictor $\hat{y}(\boldsymbol{x})$ of $y(\boldsymbol{x})$ is a *linear predictor* if it has the form $\hat{y}(\boldsymbol{x}) = \sum_{i=1}^n c_i(\boldsymbol{x}) y_i$.

(b) A predictor $\hat{y}(\boldsymbol{x})$ is an *unbiased predictor* if $E\{\hat{y}(\boldsymbol{x})\} = E\{y(\boldsymbol{x})\}$.

(c) A predictor $\hat{y}(\boldsymbol{x})$ is the *best linear unbiased predictor* (*BLUP*) if it has the minimum mean square prediction error (MSPE), $\text{MSPE}(\hat{y}) = E\{\hat{y}(\boldsymbol{x}) - y(\boldsymbol{x})\}^2$, among all linear unbiased predictors.

The BLUP is given by

$$\hat{y}(\boldsymbol{x}) = \boldsymbol{h}^T(\boldsymbol{x})\hat{\boldsymbol{\beta}} + \mathbf{v}^T(\boldsymbol{x})\boldsymbol{V}^{-1}(\boldsymbol{y} - \boldsymbol{H}\hat{\boldsymbol{\beta}}), \tag{5.2.7}$$

where

$$\begin{aligned}
\boldsymbol{h}(\boldsymbol{x}) &= (h_1(\boldsymbol{x}), \ldots, h_m(\boldsymbol{x}))^T : m \times 1, \\
\boldsymbol{H} &= (h_j(\boldsymbol{x}_i)) : n \times m, \\
\boldsymbol{V} &= (\text{Cov}(z(\boldsymbol{x}_i), z(\boldsymbol{x}_j))) : n \times n, \\
\mathbf{v}(\boldsymbol{x}) &= (\text{Cov}(z(\boldsymbol{x}), z(\boldsymbol{x}_1)), \ldots, \text{Cov}(z(\boldsymbol{x}), z(\boldsymbol{x}_n)))^T : n \times 1,
\end{aligned}$$

and

$$\hat{\boldsymbol{\beta}} = [\boldsymbol{H}^T \boldsymbol{V}^{-1} \boldsymbol{H}]^{-1} \boldsymbol{H} \boldsymbol{V}^{-1} \boldsymbol{y},$$

which is the generalized least squares estimate of $\boldsymbol{\beta}$. The MSPE of $\hat{y}(\boldsymbol{x})$ is

$$\mathrm{MSPE}(\hat{y}(\boldsymbol{x})) = \hat{\sigma}_z^2 - (\boldsymbol{h}^T(\boldsymbol{x}), \mathbf{v}^T(\boldsymbol{x})) \begin{pmatrix} \mathbf{0} & \boldsymbol{H}^T \\ \boldsymbol{H} & \boldsymbol{V} \end{pmatrix}^{-1} \begin{pmatrix} \boldsymbol{h}(\boldsymbol{x}) \\ \mathbf{v}(\boldsymbol{x}) \end{pmatrix}.$$

Let $\mathbf{r}(\boldsymbol{x}) = \mathbf{v}(\boldsymbol{x})/\hat{\sigma}_z^2$ and $\boldsymbol{R} = \boldsymbol{V}/\hat{\sigma}_z^2$. Now the MSPE can be expressed as

$$\mathrm{MSPE}(\hat{y}(\boldsymbol{x})) = \hat{\sigma}_z^2 \left[1 - (\boldsymbol{h}^T(\boldsymbol{x}), \mathbf{r}^T(\boldsymbol{x})) \begin{pmatrix} \mathbf{0} & \boldsymbol{H}^T \\ \boldsymbol{H} & \boldsymbol{R} \end{pmatrix}^{-1} \begin{pmatrix} \boldsymbol{h}(\boldsymbol{x}) \\ \mathbf{r}(\boldsymbol{x}) \end{pmatrix} \right]. \quad (5.2.8)$$

The maximum likelihood estimator of $\hat{\sigma}_z^2$ is

$$\hat{\sigma}_z^2 = \frac{1}{n}(\boldsymbol{y} - \boldsymbol{H}\hat{\boldsymbol{\beta}})^T \boldsymbol{R}^{-1} (\boldsymbol{y} - \boldsymbol{H}\hat{\boldsymbol{\beta}}).$$

Note that the above estimates involve the unknown parameters $\boldsymbol{\theta}$. One way for estimating $\boldsymbol{\theta}$ is to minimize $\mathrm{MSPE}(\hat{y}(\boldsymbol{x}))$ with respect to $\boldsymbol{\theta}$. The reader can refer to Sacks et al. (1989a), Miller and Frenklach (1983) and Santner et al. (2003) for a detailed discussion and some examples.

5.2.3 *Maximum Likelihood Estimation*

If the Kriging model is considered, the maximum likelihood method could be an alternative approach for the estimation of $\boldsymbol{\beta}$, $\hat{\sigma}_z^2$ and $\boldsymbol{\theta}$. For emphasizing the parameter $\boldsymbol{\theta}$, denote $\boldsymbol{R} = \boldsymbol{R}(\boldsymbol{\theta})$. In this case, the density of $\boldsymbol{y}$ is given by

$$(2\pi\hat{\sigma}_z^2)^{-n/2} |\boldsymbol{R}(\boldsymbol{\theta})|^{-1/2} \exp\left\{ -\frac{1}{2\hat{\sigma}_z^2} (\boldsymbol{y} - \boldsymbol{H}\boldsymbol{\beta})^T \boldsymbol{R}(\boldsymbol{\theta})^{-1} (\boldsymbol{y} - \boldsymbol{H}\boldsymbol{\beta}) \right\}. \quad (5.2.9)$$

The log-likelihood function of the data, after dropping a constant, equals

$$l(\boldsymbol{\beta}, \sigma_z^2, \boldsymbol{\theta}) = -\frac{n}{2}\log(\sigma_z^2) - \frac{1}{2}\log|R(\boldsymbol{\theta})| - \frac{1}{2\sigma_z^2}(\boldsymbol{y} - \boldsymbol{H}\boldsymbol{\beta})^T \boldsymbol{R}(\boldsymbol{\theta})^{-1}(\boldsymbol{y} - \boldsymbol{H}\boldsymbol{\beta}).$$

Maximizing the log-likelihood function yields the maximum likelihood estimators of $(\boldsymbol{\beta}, \sigma_z^2, \boldsymbol{\theta})$. In practice, simultaneous maximization over $(\boldsymbol{\beta}, \sigma_z^2, \boldsymbol{\theta})$ has some technical difficulty. It has been empirically observed that the prediction based on the simultaneous maximization over $(\boldsymbol{\beta}, \sigma_z^2, \boldsymbol{\theta})$ performs almost the same as that relying on the estimation of $\boldsymbol{\beta}$ and $(\sigma_z^2, \boldsymbol{\theta})$ separately. This is consistent with the above theoretical analysis. In practice, we may estimate $\boldsymbol{\beta}$ and $(\sigma_z^2, \boldsymbol{\theta})$ iteratively in the following way:

Step 1: Initial value of $\boldsymbol{\theta}$. For choosing a good initial value of $\boldsymbol{\theta} = \boldsymbol{\theta}_0$, it needs some additional information. In fact, it is clear that a good initial value for $\boldsymbol{\beta}$ is the least squares estimator. This is equivalent to setting the initial value of $\boldsymbol{\theta}$ to be $\mathbf{0}$. We might to choose $\boldsymbol{\theta}_0 = \mathbf{0}$.

Step 2: The maximum likelihood estimator of $\boldsymbol{\beta}$. For a given $\boldsymbol{\theta}_0$, the maximum likelihood estimator of $\boldsymbol{\beta}$ is

$$\hat{\boldsymbol{\beta}}_0 = (\boldsymbol{H}^T \boldsymbol{R}^{-1}(\boldsymbol{\theta}_0)\boldsymbol{H})^{-1}\boldsymbol{H}^T \boldsymbol{R}^{-1}(\boldsymbol{\theta}_0)\boldsymbol{y}. \tag{5.2.10}$$

Step 3: Estimation of $\hat{\sigma}_z^2$. The maximum likelihood estimator of $\hat{\sigma}_z^2$ for a given $\boldsymbol{\theta}_0$ is given by

$$\hat{\sigma}_z^2 = n^{-1}(\boldsymbol{y} - \boldsymbol{H}\hat{\boldsymbol{\beta}}_0)^T \boldsymbol{R}^{-1}(\boldsymbol{\theta}_0)(\boldsymbol{y} - \boldsymbol{H}\hat{\boldsymbol{\beta}}_0), \tag{5.2.11}$$

which is a biased estimator for $\hat{\sigma}_z^2$.

Step 4: Estimation of $\boldsymbol{\theta}$. The maximum likelihood estimator for $\boldsymbol{\theta}$ does not have a closed form. Newton–Raphson algorithm or Fisher score algorithm may be used to search for the solution. Find a solution as a new $\boldsymbol{\theta}_0$ and go back to *Step 2*. Iterate *Step 2* to *Step 4* until it converges.

5.2.4 Parametric Empirical Kriging

Let $\boldsymbol{e}_i$ be the unit column vector with 1 at the ith coordinate and 0 otherwise. When $\boldsymbol{x}$ is the ith observation $\boldsymbol{x}_i$, from (5.2.7) we have

$$\begin{aligned}\hat{y}(\boldsymbol{x}_i) &= \boldsymbol{h}^T(\boldsymbol{x}_i)\boldsymbol{\beta} + \mathbf{v}^T(\boldsymbol{x}_i)\boldsymbol{V}^{-1}(\boldsymbol{y} - \boldsymbol{H}\boldsymbol{\beta}) \\ &= \boldsymbol{h}^T(\boldsymbol{x}_i)\boldsymbol{\beta} + \boldsymbol{e}_i^T(\boldsymbol{y} - \boldsymbol{H}\boldsymbol{\beta}) = y_i,\end{aligned}$$

which indicates that Kriging predictor interpolates the data. This property is suitable for modeling data from computer experiments where there is no random error, but is not good for data from physical experiments. Therefore, the so-called parametric empirical Kriging model is suggested for physical experiment data analysis (Sacks et al. 1989a). Some details can be found in Appendix C of Santner et al. (2003). Let

$$y(\boldsymbol{x}) = \sum_{i=1}^{m} \beta_i h_i(\boldsymbol{x}) + z(\boldsymbol{x}) + \epsilon(\boldsymbol{x}) = \boldsymbol{h}^T(\boldsymbol{x})\boldsymbol{\beta} + z(\boldsymbol{x}) + \epsilon(\boldsymbol{x}), \qquad (5.2.12)$$

where $\boldsymbol{h}(\boldsymbol{x})$, $\boldsymbol{\beta}$ and $z(\boldsymbol{x})$ have the same meaning and assumptions as before, $\epsilon(\boldsymbol{x})$ denotes the random error at $\boldsymbol{x}$ with mean zero and variance $\hat{\sigma}^2$. We always assume that $\epsilon(\boldsymbol{x})$ and $z(\boldsymbol{x})$ are uncorrelated and $\epsilon(\boldsymbol{x})$ is a white noise (see Definition 5.2.1). Let $v(\boldsymbol{x}) = z(\boldsymbol{x}) + \epsilon(\boldsymbol{x})$. Obviously, $v(\boldsymbol{x})$ is a Gaussian random function with mean zero and covariance function

$$\boldsymbol{\Sigma}_v = \hat{\sigma}_z^2 \boldsymbol{R}(\boldsymbol{\theta}) + \hat{\sigma}^2 \boldsymbol{I} = \hat{\sigma}_z^2 \left[\boldsymbol{R}(\boldsymbol{\theta}) + \frac{\hat{\sigma}^2}{\hat{\sigma}_z^2} \boldsymbol{I} \right] = \hat{\sigma}_z^2 \tilde{\boldsymbol{R}}(\boldsymbol{\theta}, \alpha),$$

where $\alpha = \hat{\sigma}^2/\hat{\sigma}_z^2$, $\boldsymbol{R}(\boldsymbol{\theta})$ is the correlation function of $z(\boldsymbol{x})$ with unknown parameter vector $\boldsymbol{\theta}$, and

$$\tilde{\boldsymbol{R}}(\boldsymbol{\theta}, \alpha) = \boldsymbol{R}(\boldsymbol{\theta}) + \frac{\hat{\sigma}^2}{\hat{\sigma}_z^2} \boldsymbol{I} = \boldsymbol{R}(\boldsymbol{\theta}) + \alpha \boldsymbol{I},$$

with $\boldsymbol{I}$ being the identity matrix. Use $\tilde{\boldsymbol{R}}(\boldsymbol{\theta}, \alpha)$ instead of $\boldsymbol{R}(\boldsymbol{\theta})$ in (5.2.9), although $\tilde{\boldsymbol{R}}(\boldsymbol{\theta}, \alpha)$ is not a correlation matrix now, $\boldsymbol{\Sigma}_v = \hat{\sigma}_z^2 \tilde{\boldsymbol{R}}(\boldsymbol{\theta}, \alpha)$ is a covariance matrix. We can still apply the above procedure for estimating parameters $(\boldsymbol{\beta}, \hat{\sigma}_z^2, \boldsymbol{\theta}, \alpha)$, where $\tilde{\boldsymbol{R}}(\boldsymbol{\theta}, \alpha)$ involves unknown parameters $\boldsymbol{\theta}$ and α. Then maximum likelihood estimator of $\hat{\sigma}^2$ can be obtained by $\hat{\sigma}^2 = \hat{\alpha}\hat{\sigma}_z^2$.

There is a MATLAB Kriging toolbox, called DACE, that can help us to carry out the above calculations.

5.2.5 Examples and Discussion

This subsection applies the Kriging method to some examples that were studied before. First, let us continue to discuss the modeling for Example 1.1.1 by the parametric empirical Kriging.

Example 5.2.1 (*Examples* 1.1.1 *and* 5.1.1 *continued*) In Sect. 1.1, we have employed polynomial regression models for fitting the data. Consider a simple model

$$y(x) = \beta_0 + \beta_1 x + z(x) + \epsilon(x), \qquad (5.2.13)$$

where $\epsilon(x)$ is a white noise with mean zero and variance $\hat{\sigma}^2$, $z(x)$ is a Gaussian random function with zero mean function and covariance function $\hat{\sigma}_z^2 R(\theta)$, and

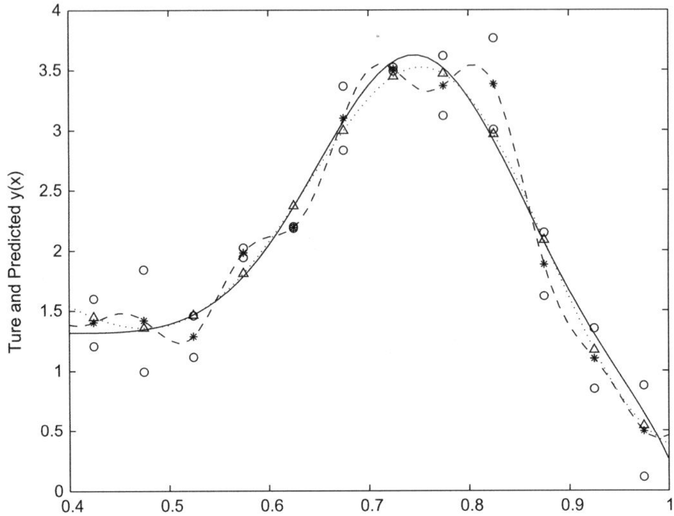

Fig. 5.2 Data in Example 1.1.1 fitted by Kriging model and parametric empirical Kriging model

$$R(d) = \exp(-\theta d^2).$$

The maximum likelihood estimates are $\hat{\boldsymbol{\beta}}_0 = -0.3411$, $\hat{\boldsymbol{\beta}}_1 = -0.2599$, $\hat{\theta} = 1.000$, $\hat{\alpha} = 0.1768$, $\hat{\sigma}_z^2 = 0.6626$, and $\hat{\sigma}^2 = \alpha\hat{\sigma}_z^2 = 0.1171$. The latter is consistent with our previous estimation, $\hat{\sigma}^2 = 0.1291$ under the ANOVA model discussed in Example 5.1.1. Figure 5.2 shows that the parametric empirical Kriging model (the dashed line) has a perfect fitting. Figure 5.2 also shows the fitting by the Kriging model

$$y(x) = \boldsymbol{\beta}_0 + \boldsymbol{\beta}_1 x + z(x),$$

Obviously, the latter (dashed line with star) is not good as the empirical Kriging model.

Example 5.2.2 (*Example* 1.5.1 *continued*) Apply the parametric empirical Kriging method to Example 1.5.1. According to the discussions in Example 1.5.1, we consider the following three models:

$$y(\boldsymbol{x}) = \beta_0 + \beta_1 x_1 + \beta_2 x_2 + \beta_3 x_3 + \beta_4 x_4 + z(\boldsymbol{x}) + \varepsilon(\boldsymbol{x}), \tag{5.2.14}$$

$$\begin{aligned} y(\boldsymbol{x}) = {} & \beta_0 + \beta_1 x_2 + \beta_2 x_3 + \beta_3 x_1 x_3 + \beta_4 x_2 x_4 \\ & + \beta_5 x_2^2 + z(\boldsymbol{x}) + \varepsilon(\boldsymbol{x}), \end{aligned} \tag{5.2.15}$$

$$\begin{aligned} y(\boldsymbol{x}) = {} & \beta_0 + \beta_1 (x_1 - \bar{x}_1) + \beta_2 (x_2 - \bar{x}_2) + \beta_3 (x_4 - \bar{x}_4) \\ & + \beta_4 (x_3 - \bar{x}_3)(x_4 - \bar{x}_4) + \beta_5 (x_2 - \bar{x}_2)^2 + z(\boldsymbol{x}) + \varepsilon(\boldsymbol{x}). \end{aligned} \tag{5.2.16}$$

The estimates of the parameters are given as follows:

Model (5.2.14):
$\hat{\beta} = (-0.0019, 0.7378, 0.3147, -0.1118, 0.3737)$,
$\hat{\boldsymbol{\theta}} = (0.8620, 0.4418, 1.5232, 0.3715)$,
$\hat{\sigma}_z^2 = 0.00024831$, and $\hat{\sigma}^2 = 0.00023053$.

Model (5.2.15):
$\hat{\beta} = (-0.0043, 0.9900, -0.8630, 1.0699, 0.6673, -1.2060)$,
$\hat{\boldsymbol{\theta}} = (0.6771, 0.8524, 1.000, 1.000, 1.000)$,
$\hat{\sigma}_z^2 = 0.000037869$, and $\hat{\sigma}^2 = 0.000037869$.

Model (5.2.16):
$\hat{\beta} = (0.0011, 0.7354, 0.3092, 0.3697, 0.00058, -0.000082)$,
$\hat{\boldsymbol{\theta}} = (0.2726, 0.2500, 1.2968, 0.2838, -0.4125)$,
$\hat{\sigma}_z^2 = 0.000043704$, and $\hat{\sigma}^2 = 0.00003675$.

Table 5.3, where Model (1), Model (2), and Model (3), respectively, represent Model (5.2.14), Model (5.2.15), and Model (5.2.16), shows the fitting values of the responses and related residuals. From these results, we observe:

• Model (5.2.14) has larger absolute values of residuals and the largest variances of $\hat{\sigma}_z^2$ and $\hat{\sigma}^2$ among the three models.

• The last two models provide estimates, $\hat{\sigma}_z^2$ and $\hat{\sigma}^2$, at the same level, respectively, and are smaller than that under model (5.2.14). These results are consisted with the results obtained in Example 1.5.1.

Table 5.3 Fitting values of y and related residuals in Example 1.5.1 under the three models

Observations	Model (1) $\hat{y}$	Model (1) e_i	Model (2) $\hat{y}$	Model (2) e_i	Model (3) $\hat{y}$	Model (3) e_i
0.1836	0.1859	−0.0023	0.1790	0.0046	0.1810	0.0026
0.1739	0.1540	0.0199	0.1707	0.0032	0.1758	−0.0019
0.0900	0.0850	0.0050	0.0827	0.0073	0.0851	0.0049
0.1176	0.1358	−0.0182	0.1154	0.0022	0.1180	−0.0004
0.0795	0.0731	0.0064	0.0850	−0.0055	0.0699	0.0096
0.0118	0.0145	−0.0027	0.0101	0.0017	0.0168	−0.0050
0.0991	0.0873	0.0118	0.1056	−0.0065	0.0995	−0.0004
0.1319	0.1301	0.0018	0.1289	0.0030	0.1303	0.0016
0.0717	0.0725	−0.0008	0.0706	0.0011	0.0743	−0.0026
0.0109	0.0303	−0.0194	0.0131	−0.0022	0.0134	−0.0025
0.1266	0.1279	−0.0013	0.1276	−0.0010	0.1322	−0.0056
0.1424	0.1426	−0.0002	0.1504	−0.0080	0.1429	−0.0005

• Delete $z(\boldsymbol{x})$ from Model (5.2.14) and denote the resulting model by Model ($5.2.14_0$). Similarly, we have Model ($5.2.15_0$) and Model ($5.2.16_0$). We find the parametric empirical Kriging Model (5.2.14) has a smaller value of the estimated error variance, $\hat{\sigma}^2$, than Model ($5.2.14_0$) has. It shows that the parametric empirical Kriging model can improve the fitting. Similar results can be found for other two models (5.2.15) and (5.2.16).

• The parametric item in model (5.2.12) is important. Many existing modeling techniques can be used for selecting the parametric item.

5.3 A Case Study on Environmental Data—Model Selection

Example 1.1.4 introduces an environmental problem in order to determine the amount of concentration of the six metals affects toxicity. Environmentalists believe that contents of some metal elements in water would directly affect human health. It is of interest to study the association between the mortality of some kind cell of mice and contents of six metals: Cadmium (Cd), Copper (Cu), Zinc (Zn), Nickel (Ni), Chromium (Cr), and Lead (Pb). An experiment was conducted by a uniform design $U_{17}(17^6)$ (see the left portion of Table 5.4) as all the metals vary 17 levels. For each design point, three experiments were conducted, the outputs (mortalities) are depicted in Table 5.4 (the data are adopted from Fang and Wang 1994), from which it can be seen that the mortality in the last row corresponding to the highest level-combination of the contents of metals is higher than the others. This implies that the contents of metals may affect the mortality. After conducting the experiments and collecting the data, the investigator has to analyze the data in order to understand how the mortality associates with the levels of metal contents.

Note that the ratio of the maximum to minimum of contents of six metals is 2000. The standardized procedure is necessary. Let $x_1, \ldots, x_6$ denote the standardized variables of Cd, Cu, Zn, Ni, Cr, and Pb, respectively. Thus, $x_1 = (\text{Cd} - 6.5624)/7.0656$, where 6.5624 and 7.0656, respectively, are the sample mean and sample standard deviation for the factor Cd and similar for other five variables. Moreover, for the three outputs of each run, let y be the mean of the three responses.

Fang and Wang (1994) employed stepwise regression for modeling and gave the following metamodel

$$\begin{aligned}\hat{y} = {} & 32.68 + 5.03\log(x_1) + 3.48\log(x_2) + 2.03\log(x_4) + 0.55(\log(x_2))^2 \\ & -0.63(\log(x_3))^2 + 0.94(\log(x_4))^2 + 0.53\log(x_1)\log(x_2) \\ & -0.70\log(x_1)\log(x_5) + 0.92\log(x_2)\log(x_6),\end{aligned}$$

where they took the logarithm for all the six variables as the ranges they varied are too large. We found that the stepwise regression for this data is very unstable; that is,

Table 5.4 Environmental experiments and related mortalities by a uniform design

Cd	Cu	Zn	Ni	Cr	Pb	Y_1	Y_2	Y_3
0.01	0.2	0.8	5.0	14.0	16.0	19.95	17.6	18.22
0.05	2.0	10.0	0.1	8.0	12.0	22.09	22.85	22.62
0.1	10.0	0.01	12.0	2.0	8.0	31.74	32.79	32.87
0.2	18.0	1.0	0.8	0.4	4.0	39.37	40.65	37.87
0.4	0.1	12.0	18.0	0.05	1.0	31.90	31.18	33.75
0.8	1.0	0.05	4.0	18.0	0.4	31.14	30.66	31.18
1.0	8.0	2.0	0.05	12.0	0.1	39.81	39.61	40.80
2.0	16.0	14.0	10.0	5.0	0.01	42.48	41.86	43.79
4.0	0.05	0.1	0.4	1.0	18.0	24.97	24.65	25.05
5.0	0.8	4.0	16.0	0.2	14.0	50.29	51.22	50.54
8.0	5.0	16.0	2.0	0.01	10.0	60.71	60.43	59.69
10.0	14.0	0.2	0.01	16.0	5.0	67.01	71.99	67.12
12.0	0.01	5.0	8.0	10.0	2.0	32.77	30.86	33.70
14.0	0.4	18.0	0.2	4.0	0.8	29.94	28.68	30.66
16.0	4.0	0.4	14.0	0.8	0.2	67.87	69.25	67.04
18.0	12.0	8.0	1.0	0.1	0.05	55.56	55.28	56.52
20.0	20.0	20.0	20.0	20.0	20.0	79.57	79.43	78.48

the final model is very sensitive to F-value that is for choosing variables or deleting variables. This phenomena have often happened in the use of stepwise regression. Another technique via penalized least squares, called SCAD, was proposed by Fan and Li (2001). The AIC and BIC are a special case of their approach. The reader can find the details of the theory and algorithm of the method in Fan and Li (2001).

All linear terms, quadratic terms, and interactions between the linear terms are chosen as x variables in the linear model. Thus, including an intercept term, there are 28 predictors in the model. The model is over-fitted because totally there are only 17 level-combinations in this experiment. Thus, variable selection is necessary. The SCAD variable selection procedure is applied for this data set. The estimated coefficients of x variables (rather than the original scale of contents) are depicted in Table 5.5. A total of 12 variables are included in the final model (comparing with $t_{0.005}(38) = 2.7116$). All selected variables are very statistically significant. From Table 5.5, it can be seen that the effect of Cd is quadratic, the positive effect of Cu and Ni is linear, in addition, Cu & Cr and Ni & Cr have negative interactions. Cr has a negative effect; further, Cr has negative interactions with Cu, Zn, and Ni, respectively. The effect of Zn is quadratic. Finally, Pb has a positive linear and quadratic effect. Moreover, Pb and Zn have a positive interaction.

We now try to apply parametric empirical Kriging models to this data set. Considered four models, where the first one is the simply linear model, the second one is parametric empirical Kriging model with the simply linear parametric item, the

Table 5.5 Estimates, standard errors and t-value

X-variable	Estimate	Standard error	$\|t\|$
Intercept	36.4539	0.5841	62.4086
Cd	14.9491	0.2944	50.7713
Cu	12.8761	0.2411	53.4060
Ni	0.9776	0.2510	3.8950
Cr	−7.2696	0.2474	29.3900
Pb	4.0646	0.2832	14.3536
Cd^2	−6.2869	0.3624	17.3480
Zn^2	2.8666	0.3274	8.7554
Pb^2	9.2251	0.4158	22.1856
Cu*Cr	−1.6788	0.3171	5.2945
Zn*Cr	−6.2955	0.3306	19.0401
Zn*Pb	11.9110	0.2672	44.5708
Ni*Cr	−11.3896	0.4303	26.4680

third one has been recommended by Li (2002), and the last one combines the third and parametric empirical Kriging model.

$$y(\boldsymbol{x}) = \beta_0 + \beta_1 x_1 + \beta_2 x_2 + \beta_3 x_3 + \beta_4 x_4 + \beta_5 x_5 + \beta_6 x_6 + \epsilon(\boldsymbol{x}), \tag{5.3.1}$$

$$\begin{aligned} y(\boldsymbol{x}) = {} & \beta_0 + \beta_1 x_1 + \beta_2 x_2 + \beta_3 x_3 + \beta_4 x_4 + \beta_5 x_5 \\ & + \beta_6 x_6 + z(\boldsymbol{x}) + \epsilon(\boldsymbol{x}), \end{aligned} \tag{5.3.2}$$

$$\begin{aligned} y(\boldsymbol{x}) = {} & \beta_0 + \beta_1 x_1 + \beta_2 x_2 + \beta_3 x_4 + \beta_4 x_5 + \beta_5 x_6 + \beta_6 x_1^2 + \beta_7 x_3^2 \\ & + \beta_8 x_6^2 + \beta_9 x_2 x_5 + \beta_{10} x_3 x_5 + \beta_{11} x_3 x_6 + \beta_{12} x_4 x_5 + \epsilon(\boldsymbol{x}), \end{aligned} \tag{5.3.3}$$

$$\begin{aligned} y(\boldsymbol{x}) = {} & \beta_0 + \beta_1 x_1 + \beta_2 x_2 + \beta_3 x_4 + \beta_4 x_5 + \beta_5 x_6 + \beta_6 x_1^2 + \beta_7 x_3^2 + \beta_8 x_6^2 \\ & + \beta_9 x_2 x_5 + \beta_{10} x_3 x_5 + \beta_{11} x_3 x_6 + \beta_{12} x_4 x_5 + z(\boldsymbol{x}) + \epsilon(\boldsymbol{x}), \end{aligned} \tag{5.3.4}$$

where $x_1 \sim x_6$ are standardized. Estimates of the unknown parameters in models (5.3.1) and (5.3.2) are given in Table 5.6, and the fitting values of y and related residuals are listed in Table 5.7. For models (5.3.3) and (5.3.4), the results are presented in Tables 5.8 and 5.9, respectively. From the above results, we might to address some conclusions:

The experiment was duplicated three times, and an estimate of the variance the pure random error can be obtained by

$$\hat{\sigma}^2 = \frac{1}{2 \times 17} \sum_{i=1}^{17} \sum_{j=1}^{3} (y_{ij} - \bar{y}_i)^2 = 1.2324.$$

Table 5.6 Estimates of models (5.3.1) and (5.3.2)

	Model (5.3.1)	Model (5.3.2)	Model (5.3.2)
X-variable	$\hat{\beta}$	$\hat{\beta}$	$\hat{\theta}$
constant	42.8639	−0.0394	
x_1	10.7108	0.6758	0.0010
x_2	8.1125	0.3747	0.3200
x_3	−2.1168	−0.1278	4.0637
x_4	3.8514	0.2418	2.3702
x_5	−0.3443	0.0111	0.0864
x_6	0.3579	0.0562	0.9406
σ_z^2			123.2917
σ^2			1.1071

Table 5.7 Fitting values of y and related residuals under models (5.3.1) and (5.3.2)

$\bar{y}$	Model (5.3.1) $\hat{y}$	Model (5.3.1) e_i	Model (5.3.2) $\hat{y}$	Model (5.3.2) e_i
18.5900	26.6164	−8.0264	18.6129	−0.0229
22.5200	23.4063	−0.8863	22.5214	−0.0014
32.4667	42.2368	−9.7701	32.4939	−0.0273
39.2967	45.0475	−5.7509	39.3332	−0.0365
32.2767	30.7434	1.5333	32.2691	0.0076
30.9933	27.4271	3.5662	30.9871	0.0062
40.0733	33.3073	6.7660	40.0573	0.0161
42.7100	46.1736	−3.4636	42.7081	0.0019
24.8900	30.9297	−6.0397	24.9044	−0.0144
50.6833	40.4781	10.2052	50.6551	0.0282
60.2767	38.4284	21.8483	60.2051	0.0715
68.7067	54.4104	14.2963	68.6476	0.0591
32.4433	44.4368	−11.9935	32.4846	−0.0412
29.7600	40.0015	−10.2415	29.7992	−0.0392
68.0533	60.0874	7.9659	68.0277	0.0256
55.7867	62.9681	−7.1814	55.8050	−0.0183
79.1600	81.9878	−2.8278	79.1751	−0.0151
MSE		9.2695		0.0316

Estimator of σ^2 is 1.1071 under the model (5.3.2) and is 0.9140 under the model (5.3.4), respectively. Both are close to the estimator of the variance of the pure random error, 1.2324.

The parametric empirical models (5.3.2) have the smallest MSE-value among the four models. It is sure that the parametric empirical model may improve fitting if the parametric item is good. Both models of (5.3.3) and (5.3.4) have the same level of MSE-value, but they have larger MSE values than model (5.3.2). It indicates that how to choose a good parametric item is not easy and is an open problem.

Table 5.8 Estimates of models (5.3.3) and (5.3.4)

	Model (5.3.3)	Model (5.3.4)	Model (5.3.4)
X-variable	$\hat{\beta}$	$\hat{\beta}$	$\hat{\theta}$
$constant$	36.4538	−0.0006	
x_1	14.9491	0.8330	0.0064
x_2	12.8760	0.7175	0.0043
x_4	0.9776	0.0545	0.0029
x_5	−7.2695	−0.4051	0.0019
x_6	4.0644	0.2265	0.0013
x_1^2	−6.2869	−0.3292	0.0010
x_3^2	2.8666	0.1501	0.0010
x_6^2	9.2251	0.4830	0.0010
x_2x_5	−1.6788	−0.1098	0.0010
x_3x_5	−6.2954	−0.4131	0.0010
x_3x_6	11.9109	0.7777	0.0010
x_4x_5	−11.3897	−0.7473	0.0010
σ_z^2			0.7192
σ^2			0.9140

Table 5.9 Fitting values of y and related residuals under models (5.3.3) and (5.3.4)

$\bar{y}$	Model (5.3.3) $\hat{y}$	Model (5.3.3) e_i	Model (5.3.4) $\hat{y}$	Model (5.3.4) e_i
18.5900	18.1872	0.4028	18.1876	0.4024
22.5200	22.0938	0.4262	22.0941	0.4259
32.4667	33.0771	−0.6104	33.0767	−0.6100
39.2967	39.1439	0.1528	39.1439	0.1527
32.2767	32.1827	0.0939	32.1828	0.0938
30.9933	31.5229	−0.5296	31.5223	−0.5290
40.0733	40.0871	−0.0138	40.0872	−0.0139
42.7100	42.3616	0.3484	42.3619	0.3481
24.8900	25.1529	−0.2629	25.1528	−0.2628
50.6833	50.9327	−0.2494	50.9322	−0.2489
60.2767	60.1480	0.1287	60.1482	0.1285
68.7067	68.7445	−0.0378	68.7445	−0.0378
32.4433	32.3263	0.1170	32.3265	0.1168
29.7600	29.8031	−0.0431	29.8030	−0.0430
68.0533	67.6520	0.4013	67.6525	0.4008
55.7867	56.0953	−0.3087	56.0950	−0.3083
79.1600	79.1755	−0.0155	79.1754	−0.0154
MSE		0.3041		0.3038

There are many other modeling techniques, such as Bayesian approach including Bayesian Gaussian Kriging model, neural networks, local polynomial regression. The reader can refer to Fang et al. (2006).

Exercises

5.1

Suppose the response y and factor x have the following underlying relationship

$$y = f(x) + \boldsymbol{e} = 1 - e^{-x^2} + \varepsilon, \quad \varepsilon \sim N(0, 0.1^2), x \in [0, 3],$$

but the experimenter does not know this model and he/she wants to find an approximation model to the real one by experiments. Therefore, he/she considers four designs as follows:

$$D_3 = \{0, 0, 0, 0, 1, 1, 1, 1, 2, 2, 2, 2\},$$
$$D_4 = \left\{\frac{1}{4}, \frac{1}{4}, \frac{1}{4}, \frac{3}{4}, \frac{3}{4}, \frac{3}{4}, \frac{5}{4}, \frac{5}{4}, \frac{5}{4}, \frac{7}{4}, \frac{7}{4}, \frac{7}{4}\right\},$$
$$D_6 = \left\{\frac{1}{6}, \frac{1}{6}, \frac{3}{6}, \frac{3}{6}, \frac{5}{6}, \frac{5}{6}, \frac{7}{6}, \frac{7}{6}, \frac{9}{6}, \frac{9}{6}, \frac{11}{6}, \frac{11}{6}\right\},$$
$$D_{12} = \left\{\frac{1}{12}, \frac{3}{12}, \frac{5}{12}, \frac{7}{12}, \frac{9}{12}, \frac{11}{12}, \frac{13}{12}, \frac{15}{12}, \frac{17}{12}, \frac{19}{12}, \frac{21}{12}, \frac{23}{12}\right\}.$$

Implement the following steps:

1. Plot the function
$$y = f(x) = 1 - e^{-x^2}, \quad x \in [0, 3].$$
2. Generate a data set for each design by the statistical simulation.
3. Find a suitable regression model and related ANOVA table for each data set. Plot the fitting models.
4. Randomly generate $N = 1000$ points $x_1, \ldots, x_{1000}$ and calculate the mean square error (MSE) defined by
$$\text{MSE} = \frac{1}{N}\sum_{i=1}^{N}(y_i - \hat{y}_i)^2$$
for each model, where $\hat{y}_i$ is the estimated value of y_i under the model.
5. According to the plots, MSE and SS_E, give your conclusions based on your comparisons among the above models.

5.2

For comparing different kinds of designs and modeling techniques in computer experiments, there is a popular way to consider several case studies. The models are known. Choose several designs like the orthogonal design (OD), Latin hypercube sampling (LHS), uniform design (UD), and modeling techniques. Then compare all design-modeling combinations.

Suppose that the following models are given. Please consider three kinds of designs OD, UD, and LHS (with $n = 16, 25, 29, 64$) and modeling techniques: the quadratic regression models, a power spline basis with the following general form of $1, x, x^2, \ldots, x^p, (x - \kappa_1)_+^p, \ldots, (x - \kappa_m)_+^p$ in (5.1.8), a Kriging model $y(\boldsymbol{x}) = \sum_{i=1}^m \beta_i h_i(\boldsymbol{x}) + z(\boldsymbol{x})$ defined in Definition 5.2.2 and artificial neural network. Give your comparisons for all possible design-modeling combinations.

Model 1:

$$Y = \frac{\ln(x_1) \times (\sin(x_2) + 4)}{e^{x_3}} + \ln(x_1)e^{x_3} \tag{5.3.5}$$

where the ranges of the independent variables are $x_1 : [0.1, 10]$, $x_2 : [-\pi/2, \pi/2]$, and $x_3 : [0, 1]$, respectively.

Model 2:

$$\begin{aligned} Y = -\Bigg[& 2\exp\left\{-\frac{1}{2}(x_1^2 + (x_2 - 4))^2\right\} \\ & + \exp\left\{-\frac{1}{2}((x_1 - 4)^2 + \frac{x_2^2}{4}\right\} + \exp\left\{-\frac{1}{2}\left(\frac{(x_1 + 4)^2}{4} + x_2^2\right)\right\}\Bigg] \end{aligned} \tag{5.3.6}$$

where the ranges of the independent variables are $x_1 : [-10, 7]$, $x_2 : [-6, 7]$, respectively.

Model 3:

$$\begin{aligned} Y = & \ 10(x_2 - x_1^2)^2 + (1 - x_1)^2 + 9(x_4 - x_3^2) + (1 - x_3)^2 \\ & + 1.01[(x_2 - 1)^2 + (x_4 - 1)^2] + 1.98(x_2 - 1)(x_4 - 1)^2 \end{aligned} \tag{5.3.7}$$

where the ranges of the independent variables are $x_i : [-2, 2], i = 1, 2, 3, 4$.

Model 4:

$$Y = \sum_{k=1}^{4} [100(x_{k+1} - x_k^2)^2 + (1 - x_k)^2], \tag{5.3.8}$$

where the ranges of the independent variables are $x_i : [-2, 2], i = 1, \ldots, 5$.

5.3

Example 1.1.7 is a good platform for comparing various design-modeling combinations. Let y be the distance from the end of the arm to the origin expressed as a function of $2m$ variables $\theta_j \in [0, 2\pi]$ and $L_j \in [0, 1]$, where $y = \sqrt{u^2 + v^2}$ and (u, v) are defined in (1.1.3). Consider three kinds of designs OD, UD, and LHS and three kinds of metamodel: polynomial regression model, Kriging model, and empirical Kriging model. Give your comparisons for possible design-modeling combinations with $m = 2$ and $m = 3$, respectively.

References

Antoniadis, A., Oppenheim, G.: Wavelets and Statistics. Springer, New York (1995)

Brown, L.D., Cai, T.: Wavelet regression for random uniform design, Technical report, no. 97-15, Department of Statistics, Purdue University (1997)

Cai, T., Brown, L.D.: Wavelet shrinkage for nonequispaced samples. Ann. Stat. **26**, 425–455 (1998)

Chen, V.C.P., Tsui, K.L., Barton, R.R., Meckesheimer, M.: A review on design, modeling and applications of computer experiments. IIE Trans. **38**, 273–291 (2006)

Chui, C.K.: Wavelets: A Tutorial in Theory and Applications. Academic, Boston (1992)

Cressie, N.: Spatial prediction and ordinary kriging. Math. Geol. **20**, 405–421 (1997)

Cressie, N.A.: Statistics for Spatial Data. Wiley, New York (1993)

Daubechies, I.: Ten Lectures on Wavelets. SIAM, Philadelphia (1992)

De Boor, C., Bon, A.: On multivariate polynomial interpolation. Constr. Approx. **6**, 287–302 (1990)

Donoho, D.L., Johnstone, I.M.: Ideal spatial adaptation by wavelet shrinkage. Biometrika **81**, 425–455 (1994)

Donoho, D.L., Johnstone, I.M., Kerkyacharian, G., Pieard, D.: Wavelet shrinkage: asymptopia? J. R. Stat. Soc. Ser. B **57**, 301–369 (1995)

Dyn, N., Levin, D., Rippa, S.: Numerical procedures for surface fitting of scattered data by radial basis functions. SIAM J. Sci. Stat. Comput. **7**, 639–659 (1986)

Eubank, R.L.: Spline Smoothing and Nonparametric Regression. Marcel Dekker, New York (1988)

Fan, J., Li, R.: Variable selection via nonconcave penalized likelihood and its oracle properties. J. Am. Stat. Assoc. **96**, 1348–1360 (2001)

Fang, K.T., Wang, Y.: Number-Theoretic Methods in Statistics. Chapman and Hall, London (1994)

Fang, K.T., Kotz, S., Ng, K.: Symmetric Multivariate and Related Distributions. Chapman and Hall, London (1990)

Fang, K.T., Li, R., Sudjianto, A.: Design and Modeling for Computer Experiments. Chapman and Hall/CRC, New York (2006)

Friedman, J.H.: Multivariate adaptive regression splines. Ann. Stat. **19**, 1–141 (1991)

Goovaerts, P.: Geostatistics for Natural Resources Evaluation. Oxford University Press, New York (1997)

Hardy, R.L.: Multiquadratic equations of topography and other irregular surfaces. J. Geophys. Res. **76**, 1905–1915 (1971)

Hastie, T., Tibshirani, R., Friedman, J.: The Elements of Statistical Learninig, Data Mining, Inference, and Prediction. Springer, New York (2001)

Krige, D.G.: A statistical approach to some mine valuations and allied problems at the witwatersrand, Master's thesis, University of Witwatersrand (1951)

Li, R.: Model selection for analysis of uniform design and computer experiment. Int. J. Reliab. Qual. Saf. Eng. **9**, 305–315 (2002)

Matheron, G.: The theory of regionalized variables and its applications, mathematiques de fontainebleau, 5th edn. Les Cahiers du Centre de Morphologie, Fontainebleau (1971)

Miller, D., Frenklach, M.: Sensitivity analysis and parameter estimation in dynamic modeling of chemical kinetics. Int. J. Chem. Kinet. **15**, 677–696 (1983)

Myers, R.H.: Classical and Modern Regression with Applications, 2nd edn. Duxbury Press, Belmont (1990)

Powell, M.J.D.: Radial basis functions for multivariable interpolation: a review. In: Mason, J., Cox, M. (eds.) Algorithms for Approximation, pp. 143–167. Oxford University Press, London (1987)

Riccomango, E., Schwabe, R., Wynn, H.P.: Lattice-based *D*-optimal design for Fourier regression. Ann. Statist. **25**, 2313–2317 (1997)

Sacks, J., Schiller, S.B., Welch, W.J.: Designs for computer experiments. Technometrics **31**, 41–47 (1989a)

Sacks, J., Welch, W.J., Mitchell, T.J., Wynn, H.P.: Design and analysis of computer experiments (with discussion). Stat. Sin. **4**, 409–435 (1989b)

Santner, T.J., Williams, B.J., Notz, W.I.: The Design and Analysis of Computer Experiments. Springer, New York (2003)

Shi, P., Fang, K.T., Tsai, C.L.: Optimal multi-criteria designs for fourier regression model. J. Stat. Plan. Inference **96**, 387–401 (2001)

Stein, M.L.: Interpolation of Spatial Data. Some Theory for Kriging. Springer, New York (1999)

Wahba, G.: Spline Models for Observational Data. SIAM, Philadelphia (1990)

Welch, W.J., Buck, R.J., Sacks, J., Wynn, H.P., Mitchell, T.J., Morris, M.D.: Screening, predicting and computer experiments. Technometrics **34**, 15–25 (1992)

Xie, M.Y., Ning, J.H., Fang, K.T.: Orthogonality and *D*-optimality of the U-type design under general Fourier regression models. Stat. Probab. Lett. **77**, 1377–1384 (2007)

Chapter 6
Connections Between Uniformity and Other Design Criteria

Most experimental designs, such as simple random design, random block design, Latin square design, fractional factorial design (FFD, for short), optimal design, and robust design are concerned with *randomness, balance between factors and levels of each factor, orthogonality, efficiency, and robustness*. From the previous chapters, we see that the *uniformity* has played an important role in the evaluation and construction of uniform designs. In this chapter, we shall show that uniformity is intimately connected with many other design criteria.

6.1 Uniformity and Isomorphism

Let $D(n; q^s)$ be a factorial design of n runs and s factors each having q levels. The orthogonal designs $L_n(q^s)$ are special cases of factorial designs. See Sect. 1.3.2 for the basic knowledge of factorial designs.

Definition 6.1.1 Two factorial designs are called isomorphic if one can be obtained from the other by relabeling the factors, reordering the runs, or switching the levels of factors.

Two isomorphic designs are considered to be equivalent because they share the same statistical properties in a classical ANOVA model. It is important to identify design isomorphism in practice.

Example 6.1.1 Suppose that there are 5 two-level factors in an experiment and the experimenter wants to arrange this experiment by the Plackett and Burman design $L_{12}(2^{11})$. We need to choose a subdesign of five columns from $L_{12}(2^{11})$ such that it has the best statistical property in a certain sense. Unfortunately, the Plackett and

K.-T. Fang et al., *Theory and Application of Uniform Experimental Designs*, Lecture Notes in Statistics 221,
https://doi.org/10.1007/978-981-13-2041-5_6

Table 6.1 Two non-isomorphic $L_{12}(2^{11})$ designs

No	$L12-5.1$					$L12-5.2$				
1	1	1	2	1	1	1	1	2	1	1
2	2	1	1	2	2	2	1	1	2	1
3	1	2	1	1	2	1	2	1	1	2
4	2	1	2	1	2	2	1	2	1	1
5	2	2	1	2	1	2	2	1	2	1
6	2	2	2	1	1	2	2	2	1	2
7	1	2	2	2	1	1	2	2	2	1
8	1	1	2	2	2	1	1	2	2	2
9	1	1	1	2	1	1	1	1	2	2
10	2	1	1	1	1	2	1	1	1	2
11	1	2	1	1	2	1	2	1	1	1
12	2	2	2	2	2	2	2	2	2	2

Burman design is non-regular and we cannot use the minimum aberration criterion (see Sect. 1.4.4) to choose such a design. Therefore, Lin and Draper (1992) sorted the $\binom{11}{5} = 462$ subdesigns into two non-isomorphic subclass, denoted by $L12-5.1$ and $L12-5.2$, respectively. There are 66 designs belonging to the group $L12-5.1$, while there are 396 designs belonging to the group $L12-5.2$. Table 6.1 shows two representative non-isomorphic subdesigns of $L_{12}(2^5)$, one with a repeat-run pair and the other without any repeat-run pair. From the geometric viewpoint, they prefer the $L12-5.1$ as it has one more degree of freedom than the $L12-5.2$. For identifying two $D(n; q^s)$ designs, a complete search compares $n!(q!)^s s!$ designs from the definition of isomorphism. For example, to see if two orthogonal $L_{12}(2^5)$ designs are isomorphic requires $12!5!2^5 \approx 1.8394 \times 10^{12}$ comparisons. Can we reduce the complexity of the computation?

It is easy to find some necessary conditions for detecting non-isomorphic designs:

- two *isomorphic designs* have the same generalized word-length pattern.
- two isomorphic designs have the same *letter pattern* (cf. Draper and Mitchell 1968 for the details);
- two isomorphic designs have the same distribution of Hamming distances between any two distinct rows.

By Definition 1.4.1, the generalized word-length patterns for the designs $L12-5.1$ and $L12-5.2$ are

$$(0, 0, 10/9, 5/9, 0) \text{ and } (0, 0, 10/9, 5/9, 4/9),$$

respectively. Both the designs have the resolution III, but $L12-5.1$ has less generalized aberration than $L12-5.2$; hence, they are not isomorphic.

However, two designs having the same word-length pattern may be non-isomorphic. Draper and Mitchell (1968) gave two $L_{512}(2^{12})$ orthogonal designs which have identical word-length patterns, but are not isomorphic.

Draper and Mitchell (1970) gave a more sensitive criterion for isomorphism, called "letter pattern comparison," and tabulated 1024-run designs of resolution 6. Let a_{ij} be the number of words of length j in which letter i appears in a regular design $\boldsymbol{U}$ and $\boldsymbol{U}^a = (a_{ij})$ be the letter pattern matrix of $\boldsymbol{U}$. They conjectured that two designs $\boldsymbol{U}_1$ and $\boldsymbol{U}_2$ are isomorphic if and only if $\boldsymbol{U}_1^a = \boldsymbol{P}\boldsymbol{U}_2^a$, where $\boldsymbol{P}$ is a permutation matrix. Obviously, two designs having identical letter pattern matrices necessarily have identical word-length patterns. Chen and Lin (1991) gave two non-isomorphic designs 2^{31-15} with identical letter pattern matrices and thus showed that the criterion "letter pattern matrix" is not sufficient for design isomorphism. Note that both the generalized word-length pattern and letter pattern matrix are not easy to calculate and can be applied only to factorial designs with smaller sizes.

Clark and Dean (2001) gave a sufficient and necessary condition for isomorphism of designs. Let $\boldsymbol{H} = (d_{ij})$ be the *Hamming distance matrix* of a design $\boldsymbol{U}$, where d_{ij} is the *Hamming distance* of the ith and jth runs of $\boldsymbol{U}$. Their method is based on the following fact:

Lemma 6.1.1 *Let $\boldsymbol{U}_1$ and $\boldsymbol{U}_2$ be two $D(n; q^s)$ designs. Then $\boldsymbol{U}_1$ and $\boldsymbol{U}_2$ are isomorphic if and only if there exist an $n \times n$ permutation $\boldsymbol{R}$ and a permutation $\{c_1, \ldots, c_s\}$ of $\{1, \ldots, s\}$ such that for $p = 1, \ldots, s$*

$$\boldsymbol{H}_{\boldsymbol{U}_1}^{\{1,\ldots,p\}} = \boldsymbol{R}\boldsymbol{H}_{\boldsymbol{U}_2}^{\{c_1,\ldots,c_p\}}\boldsymbol{R}^T,$$

where $\boldsymbol{H}_{\boldsymbol{U}}^{\{c_1,\ldots,c_p\}}$ is the Hamming distance matrix of the design formed by columns $\{c_1, \ldots, c_p\}$ of design $\boldsymbol{U}$.

This clever method is invariant under the permutations of levels, but the complexity here makes the calculation intractable.

It can be easily found that two isomorphic U-type designs with two levels have the same CD-value; two isomorphic U-type designs with two/three levels have the same WD-value; and two U-type designs with q levels have the same discrete discrepancy. For the two above subdesigns $L12-5.1$ and $L12-5.2$, they have CD-value of 0.166541 and 0.166527, respectively. It is clear that they are non-isomorphic. However, sometimes two designs have the same discrepancy value, but they are non-isomorphic. In this case, the projection discrepancies of the subdesigns are very useful.

Definition 6.1.2 For a given $D(n; q^s)$ design $\boldsymbol{U}$ and k $(1 \leqslant k \leqslant s)$, there are $\binom{s}{k}$ $D(n; q^k)$ subdesigns. Let D be a discrepancy for measuring uniformity. The D-values of these subdesigns form a distribution, denoted by $F_D^k(\boldsymbol{U})$, that is called the *k-marginal D-value distribution* of $\boldsymbol{U}$.

The notations $F_{WD}^k(\boldsymbol{U})$, $F_{CD}^k(\boldsymbol{U})$, and $F_{MD}^k(\boldsymbol{U})$ are taken when D is WD, CD, or MD, respectively. Ma et al. (2001) proposed the following *uniformity criterion for isomorphism* (UCI):

Lemma 6.1.2 *The necessary conditions for two $D(n; 2^s)$ designs $\boldsymbol{U}_1$ and $\boldsymbol{U}_2$ to be isomorphic are*

(a) *they have the same CD_2-value.*

(b) *they have the same distribution $F^k_{CD}(\boldsymbol{U}_1) = F^k_{CD}(\boldsymbol{U}_2)$ for $1 \leqslant k < s$.*

Based on this lemma, they proposed the following algorithm, called *NIU algorithm*, for detecting non-isomorphic $D(n; 2^s)$ designs. Let $\boldsymbol{U}_1$ and $\boldsymbol{U}_2$ be two such designs.

Algorithm 6.1.1 (*NIU Algorithm*)

Step 1. Compare $CD(\boldsymbol{U}_1)$ and $CD(\boldsymbol{U}_2)$, and if $CD(\boldsymbol{U}_1) \neq CD(\boldsymbol{U}_2)$, we conclude $\boldsymbol{U}_1$ and $\boldsymbol{U}_2$ are not isomorphic and terminate the process, otherwise go to step 2.

Step 2. For $k = 1, s-1, 2, s-2, \ldots, \lfloor s/2 \rfloor, s - \lfloor s/2 \rfloor$ where $\lfloor x \rfloor$ denotes the largest integer that is smaller than x, compare $F^k_{CD}(\boldsymbol{U}_1)$ and $F^k_{CD}(\boldsymbol{U}_2)$, and if $F^k_{CD}(\boldsymbol{U}_1) \neq F^k_{CD}(\boldsymbol{U}_2)$, we conclude $\boldsymbol{U}_1$ and $\boldsymbol{U}_2$ are not isomorphic and terminate the process; otherwise, this step goes to the next k-value.

For example, we apply this algorithm to two $L_{32768}(2^{31})$ designs studied by Chen and Lin (1991). The process indicates that the two designs have the same $CD = 4.279$; all the k-dimensional subdesigns have the same CD-value for $k = 1, 30, 2, 29$, but $F^{28}_{CD}(\boldsymbol{U}_1) \neq F^{28}_{CD}(\boldsymbol{U}_2)$. It turns out that two designs are not isomorphic by implementing only a few steps of the algorithm. It shows that the NIU algorithm is powerful in detecting non-isomorphic designs.

The above idea and algorithm can be extended to detect factorial designs with higher levels (see, Ma et al. 2001) and to investigate the design projection properties (see, Lin and Draper 1992).

Let us consider the problem of detecting non-isomorphic $D(n; q^s)$ with $q > 2$ designs. For a $D(n; q^s)$ design $\boldsymbol{U}$, let

$$E_j(\boldsymbol{U}) = \frac{1}{n}\text{card}\{(i,k)|d^H(\boldsymbol{u}_i, \boldsymbol{u}_k) = j\}, \tag{6.1.1}$$

where card($\mathcal{A}$) is the cardinality of the set $\mathcal{A}$, $d^H(\boldsymbol{u}_i, \boldsymbol{u}_k)$ is the Hamming distance between two runs $\boldsymbol{u}_i$ and $\boldsymbol{u}_k$. The vector $(E_0(\boldsymbol{U}), \ldots, E_s(\boldsymbol{U}))$ is called the *distance distribution* of $\boldsymbol{U}$. Denote

$$B_a(\boldsymbol{U}) = \sum_{i=1}^{s} E_i(\boldsymbol{U}) a^i$$

as the *distance enumerator* of $\boldsymbol{U}$ (Roman 1992, p. 226) and a is a positive number. For a two-level design D, Ma et al. (2001) showed that

$$(CD(D))^2 = \frac{1}{n}\left(\frac{5}{4}\right)^s B_{4/5}(D) - 2\left(\frac{35}{32}\right)^s + \left(\frac{13}{12}\right)^s$$

which provides a link between the distance enumerator and uniformity. In fact, the UCI (refer to Lemma 6.1.2) is equivalent to the measure $B_a(\boldsymbol{U})$ with $a = 4/5$ for two-level designs. This measure can naturally be used for high-level factorial designs. Given k ($1 \leqslant k \leqslant s$), the distribution of B_a-values over all k-dimensional projection subdesigns is denoted by $F_{B_{a,k}}(\boldsymbol{U})$. We now can have an NIU version for the high-level designs. As the parameter a is a predetermined value, we omit a from the notation for simplicity.

Algorithm 6.1.2 (*NIU Algorithm for High-level Designs*)

Step 1. Comparing $B(\boldsymbol{U}_1)$ and $B(\boldsymbol{U}_2)$, if $B(\boldsymbol{U}_1) \neq B_2(\boldsymbol{U}_2)$, we conclude $\boldsymbol{U}_1$ and $\boldsymbol{U}_2$ are not isomorphic and terminate the process. Otherwise, go to step 2.

Step 2. For $k = 1, s-1, 2, s-2, \ldots, \lfloor s/2 \rfloor, s - \lfloor s/2 \rfloor$, compare $F_{B_k}(\boldsymbol{U}_1)$ and $F_{B_k}(\boldsymbol{U}_2)$. If $F_{B_k}(\boldsymbol{U}_1) \neq F_{B_k}(\boldsymbol{U}_2)$, we conclude $\boldsymbol{U}_1$ and $\boldsymbol{U}_2$ are not isomorphic and terminate the process; otherwise, this step goes to the next k-value.

For a simple illustration, consider the four $L_{18}(3^7)$ in Table 6.2 from the literature. Taking $a = 4/5$, for example, the four designs have the same distance enumerator 6.685248. However, the distributions of distance enumerator of all six-dimensional projection designs are different as indicated in Table 6.3. Therefore, we conclude that Designs (a), (c), and (d) are non-isomorphic. Note that an exhaustive comparison indicates that Designs (a) and (b) are indeed isomorphic. From our experience, we prefer to choose an irrational number as the parameter a.

Table 6.2 Four $L_{18}(3^7)$ designs

No	(a)	(b)	(c)	(d)
1	1 1 1 1 1 1 1	1 1 1 1 1 1 1	1 2 1 1 3 2 1	3 1 2 1 3 1 2
2	1 2 2 2 2 2 2	1 1 2 3 2 3 1	1 3 1 1 2 1 2	1 1 2 3 1 3 1
3	1 3 3 3 3 3 3	1 2 1 3 3 2 2	1 1 2 3 2 2 3	2 2 2 3 2 1 3
4	2 1 1 2 2 3 3	1 2 3 2 1 3 3	1 3 2 3 1 3 1	1 2 3 1 1 2 2
5	2 2 2 3 3 1 1	1 3 2 2 3 1 3	1 1 3 2 3 3 2	3 1 3 1 2 3 3
6	2 3 3 1 1 2 2	1 3 3 1 2 2 2	1 2 3 2 1 1 3	1 2 1 1 2 1 1
7	3 1 2 1 3 2 3	2 1 1 2 3 3 2	2 1 1 3 1 1 2	1 3 2 2 2 3 2
8	3 2 3 2 1 3 1	2 1 3 3 1 2 3	2 2 1 3 3 3 3	2 3 1 1 3 3 1
9	3 3 1 3 2 1 2	2 2 2 2 2 2 1	2 2 2 2 2 2 2	2 1 3 2 2 2 1
10	1 1 3 3 2 2 1	2 2 3 1 3 1 1	2 3 2 2 3 1 1	3 3 1 3 2 2 2
11	1 2 1 1 3 3 2	2 3 1 3 2 1 3	2 1 3 1 1 2 1	3 2 2 2 3 2 1
12	1 3 2 2 1 1 3	2 3 2 1 1 3 2	2 3 3 1 2 3 3	1 3 3 2 3 1 3
13	2 1 2 3 1 3 2	3 1 2 1 3 2 3	3 1 1 2 2 3 1	2 2 3 3 3 3 2
14	2 2 3 1 2 1 3	3 1 3 2 2 1 2	3 3 1 2 1 2 3	3 3 3 3 1 1 1
15	2 3 1 2 3 2 1	3 2 1 1 2 3 3	3 1 2 1 3 1 3	1 1 1 3 3 2 3
16	3 1 3 2 3 1 2	3 2 2 3 1 1 2	3 2 2 1 1 3 2	2 1 1 2 1 1 2
17	3 2 1 3 1 2 3	3 3 1 2 1 2 1	3 2 3 3 2 1 1	3 2 1 2 1 3 3
18	3 3 2 1 2 3 1	3 3 3 3 3 3 1	3 3 3 3 3 2 2	2 3 2 1 1 2 3

Table 6.3 F_{B_6}'s for four $L_{18}(3^7)$ designs

$B_{(a)}$	freq.	$B_{(b)}$	freq.	$B_{(c)}$	freq.	$B_{(d)}$	freq.
7.6683	1	7.6683	1	7.6719	1	7.6737	3
7.6765	6	7.6765	6	7.6747	2	7.6765	4
				7.6765	4		

There is a close relationship between the two-level orthogonal designs and Hadamard matrices. A Hadamard matrix $\boldsymbol{H}$ of side n is an $n \times n$ matrix with every entry either 1 or -1, which satisfies that $\boldsymbol{H}\boldsymbol{H}^T = n\boldsymbol{I}_n$. Two Hadamard matrices are called *equivalent* if one can be obtained from the other by some sequence of row & column permutations & negations. The Hadamard matrix has played an important role in construction of experimental designs, code theory, and others. To identify the equivalence of two Hadamard matrices by a complete search is an NP-hard problem, when n increases. There is a unique equivalence class of Hadamard matrices of each order 1, 2, 4, 8, and 12.

Fang and Ge (2004) proposed a powerful algorithm that can easily detect inequivalent Hadamard matrices based on the sequence of symmetric Hamming distances. They found that there are at least 382 inequivalent classes for Hadamard matrices of order 36.

6.2 Uniformity and Orthogonality

Orthogonality has played an important role in experimental design and may have a different meaning. In this section "orthogonality" is under the sense of the orthogonal array. For example, the orthogonal array of strength two requires a good balance between levels of each factor and between level-combinations of any two factors; in other words, it requires one- and two-dimensional projection uniformity. The uniform design usually concerns with one-dimensional projection and s-dimensional uniformity. These two kinds of designs should have some relationships, as will be mentioned in this section and in Sect. 6.4.

By a numerical search, Fang and Winker (1998) and Fang et al. (2000) found that many uniform designs with a small number of runs, such as $U_4(2^3)$, $U_8(2^7)$, $U_{12}(2^{11})$, $U_{16}(2^{15})$, $U_9(3^4)$, $U_{12}(3 \times 2^3)$, $U_{16}(4^5)$, $U_{16}(4 \times 2^{12})$, $U_{18}(2 \times 3^7)$, and $U_{25}(5^6)$ are also orthogonal designs under CD. This fact shows that many existing orthogonal designs are also uniform designs under CD and can be found by a numerical search. They conjectured that *any orthogonal design is a uniform design under a certain discrepancy*. Under CD, Ma et al. (2003) gave a study on this conjecture. Consider a set of lattice points $\mathcal{P} = \{\boldsymbol{x}_k, k = 1, \ldots, n\} \in \mathcal{D}(n; q^s)$. Denote by $n(i_1, \ldots, i_s)$ the number of runs at the level-combination $(i_1, \ldots, i_s)$ in $\mathcal{P}$. Section 4.3.1 expressed the squared $CD(\mathcal{P})$ as a quadratic form of $\boldsymbol{y}_{\mathcal{P}}$ in (4.3.3). When $q_1 = \cdots = q_s$, the formula (4.3.3) reduces to

$$[CD(\mathcal{P})]^2 = \left(\frac{13}{12}\right)^s - \frac{2}{n}\boldsymbol{c}^T\boldsymbol{y} + \frac{1}{n^2}\boldsymbol{y}^T\boldsymbol{C}\boldsymbol{y}, \qquad (6.2.1)$$

where $\boldsymbol{y}(\mathcal{P})$ (or $\boldsymbol{y}$ for short) is a q^s-vector with elements $n(i_1, \ldots, i_s)$ arranged lexicographically, $\boldsymbol{c}_0 = (c_1, \ldots, c_q)'$, $\boldsymbol{C}_0 = (c_{ij}, i, j = 1, \ldots, q)$, $\boldsymbol{c} = \otimes^s \boldsymbol{c}_0$, $\boldsymbol{C} = \otimes^s \boldsymbol{C}_0$, $\otimes$ is the Kronecker product,

$$c_i = 1 + \left|\frac{2i-1-q}{4q}\right| - \frac{(2i-1-q)^2}{8q^2}, \text{ and}$$

$$c_{ij} = 1 + \left|\frac{2i-1-q}{4q}\right| + \left|\frac{2j-1-q}{4q}\right| - \left|\frac{i-j}{2q}\right|.$$

This lemma has been useful in theoretical studies. A design is called *complete* (or full) if all the level-combinations of the factors appear equally often. Any complete design is an orthogonal array, and the corresponding vector of integers $\boldsymbol{y}(\mathcal{P})$ is a multiple of $\boldsymbol{1}$. For any factorial design $\mathcal{P} \in \mathcal{D}(n; q^s)$, $y(\mathcal{P})/n$ can be regarded as a measure over q^s level-combinations. Therefore, we can extend $\boldsymbol{y}$ to be a q^s-vector with positive values and constraint $\boldsymbol{y}^T\boldsymbol{1} = n$. Under the above notation, Ma et al. (2003) obtained the following results.

Theorem 6.2.1 *Let $\mathcal{P} \in \mathcal{D}(n; q^s)$ be a set of n lattice points. Then,*

(1) when $q = 2$ or q is odd, $\mathcal{P}$ minimizes $CD(\mathcal{P})$ over $\mathcal{D}(n; q^s)$ if and only if $\boldsymbol{y}(\mathcal{P}) = (n/q^s)\boldsymbol{1}$;

(2) when q is even (but not 2), $\mathcal{P}$ minimizes $CD(\mathcal{P})$ over $\mathcal{D}(n; q^s)$ if and only if

$$\boldsymbol{y}(\mathcal{P}) = \frac{n}{q^s} \otimes^s \begin{pmatrix} \boldsymbol{1}_{q/2-1} \\ 1 - \frac{1}{4(4q+1)} \\ 1 - \frac{1}{4(4q+1)} \\ \boldsymbol{1}_{q/2-1} \end{pmatrix} + n\frac{1-(1-\frac{1}{2q(4q+1)})^s}{2^s} \otimes^s \begin{pmatrix} \boldsymbol{0}_{q/2-1} \\ 1 \\ 1 \\ \boldsymbol{0}_{q/2-1} \end{pmatrix},$$

where $\otimes$ is the Kronecker product;

(3) when q is even (but not 2) and $\mathcal{P} \in \mathcal{U}(n; q^s)$, $\mathcal{P}$ minimizes $CD(\mathcal{P})$ over $\mathcal{U}(n; q^s)$ if and only if

$$\boldsymbol{y}(\mathcal{P}) = n\boldsymbol{C}^{-1}\boldsymbol{c} - n\left[\frac{\boldsymbol{1}^T\boldsymbol{C}^{-1}\boldsymbol{c}}{\boldsymbol{1}^T\boldsymbol{C}^{-1}\boldsymbol{1}}\left(1 - \frac{s(8q^2+2q)}{8q^2+2q-1}\right) + \frac{s-1}{\boldsymbol{1}^T\boldsymbol{C}^{-1}\boldsymbol{1}}\right]\boldsymbol{C}^{-1}\boldsymbol{1}$$
$$-\frac{n}{q}\frac{(\boldsymbol{1}^T\boldsymbol{C}_0^{-1}\boldsymbol{c}_0)^{s-1}-1}{(\boldsymbol{1}^T\boldsymbol{C}_0^{-1}\boldsymbol{1})^{s-1}}\sum_{i=1}^{s}(\boldsymbol{C}_0^{-1}\boldsymbol{1})^{i-1} \otimes \boldsymbol{1} \otimes (\boldsymbol{C}_0^{-1}\boldsymbol{1})^{i-1}.$$

Especially, for $s = 1, 2$, we have $\boldsymbol{y}(\mathcal{P}) = (n/q^s)\boldsymbol{1}$.

For detailed proof, one can refer to Ma et al. (2003). Theorem 6.2.1 shows that the complete design is the uniform design under CD, and the conjecture is true when the design $\mathcal{P}$ is complete. They also showed that the conjecture is true if $q = 2$ and

$n = 2^{s-1}$; the latter is not a complete design. The conjecture in the case of even q is more complicated. For example, when $q = 4$ and $s = 2$, the $n(i, j)$'s of the uniform design are given by

$$N \equiv (n(i, j)) = \frac{n}{16}\begin{pmatrix} 1 & \frac{67}{68} & \frac{67}{68} & 1 \\ \frac{67}{68} & \frac{70}{68} & \frac{70}{68} & \frac{67}{68} \\ \frac{67}{68} & \frac{70}{68} & \frac{70}{68} & \frac{67}{68} \\ 1 & \frac{67}{68} & \frac{67}{68} & 1 \end{pmatrix}.$$

All the elements of N are close to $\frac{n}{16}$. As the objective function in (6.2.1) is a continuous function of y, the optimum y-value is a design in $\mathcal{D}(n; q^s)$ that has the smallest distance to N. Such a solution is $L_n(4^5)$, when n is a divisor of 16 and is not too large. This fact indicates that the conjecture is not always true when q is even and CD is employed as the uniformity measure.

When a lattice design $\mathcal{P}$ is not complete, does the conjecture still hold? Note that $[CD(\mathcal{P})]^2 = \sum_{u \neq \emptyset} I_u(\mathcal{P})^2$, where $I_u(\mathcal{P}) = I_{|u|}(\mathcal{P}^u)$, $\mathcal{P}^u$ is the projection of $\mathcal{P}$ onto C^u, and

$$I_u(\mathcal{P})^2 = \int_{C^u} \left(\frac{N(\mathcal{P}^u, J_{x^u})}{n} - \mathrm{Vol}(J_{x^u}) \right)^2 \mathrm{d}x^u.$$

Define a new measure of uniformity as

$$[CL_{2,t}(\mathcal{P})]^2 = \sum_{0<|u|\leqslant t} I_u(\mathcal{P})^2,$$

Ma et al. (2003) proved the following theorem.

Theorem 6.2.2 *A uniform design $U_n(q^s)$ under $CL_{2,t}$, where $t < s$, n is a multiple of q^t and q equals 2 or q is odd, is an orthogonal array $OA(n; q^s; t)$, if the latter exists.*

Liu (2002) gave a study on this conjecture by the use of the discrete discrepancy (Sect. 2.5.1) as the measure of uniformity. Liu (2002) and Fang et al. (2003) showed that symmetrical saturated orthogonal arrays are uniform designs in terms of the discrete discrepancies. While, from the results of Li et al. (2004) and Liu et al. (2006), we know that asymmetrical saturated orthogonal arrays are also the most uniform ones according to the two-dimensional projection discrepancy defined by (6.5.3) for $j = 2$. Tang (2005) in his Ph.D. thesis gave a further discussion on this conjecture.

From a different angle to the above study, Fang et al. (2002) proposed another measure for non-orthogonality. For a design $\mathcal{P} \in \mathcal{D}(n; q^s)$, $y(\mathcal{P})$ is the same as before, a q^s-vector. Let $v = q^s$ in the following discussion. Let e and I denote, respectively, the $q \times 1$ vector with all elements unity and the $q \times q$ identity matrix. Obvious $J = ee^T$. Let Ω be the set of binary s-tuples with entries 0 or 1. Denote

$$\boldsymbol{L}(0) = q^{-1}\boldsymbol{J},\ \boldsymbol{L}(1) = \boldsymbol{I} - q^{-1}\boldsymbol{J},\ \boldsymbol{G}(0) = \boldsymbol{e}^T,\ \boldsymbol{G}(1) = \boldsymbol{I}. \tag{6.2.2}$$

The t-fold Kronecker products of $\boldsymbol{e}$, $\boldsymbol{I}$, and $\boldsymbol{J}$ will be denoted by $\boldsymbol{e}^{(t)}$, $\boldsymbol{I}^{(t)}$, and $\boldsymbol{J}^{(t)}$, respectively. For any $\boldsymbol{x} = x_1 \cdots x_s \in \Omega$, define the matrices

$$\boldsymbol{W}(\boldsymbol{x}) = \boldsymbol{L}(x_1) \otimes \cdots \otimes \boldsymbol{L}(x_s),\ \boldsymbol{H}(\boldsymbol{x}) = \boldsymbol{G}(x_1) \otimes \cdots \otimes \boldsymbol{G}(x_s), \tag{6.2.3}$$

which are of orders $v \times v$ and $q^{\sum x_i} \times v$, respectively. Here $\otimes$ is the Kronecker product. For $0 \leqslant i \leqslant s$, let Ω_i be the subset of Ω consisting of those binary s-tuples which have exactly i elements unity. Also, let $\Omega^* = \Omega - \Omega_0$ be the set of nonnull members of Ω. By (6.2.2) and (6.2.3), the matrices $W(\boldsymbol{x})$, for $\boldsymbol{x} \in \Omega^*$, add up to $\boldsymbol{I}^{(s)} - v^{-1}\boldsymbol{J}^{(s)}$. Hence

$$\boldsymbol{y}(\mathcal{P})^T\{\boldsymbol{I}^{(s)} - v^{-1}\boldsymbol{J}^{(s)}\}\boldsymbol{y}(\mathcal{P}) = \sum_{\boldsymbol{x}\in\Omega^*} \boldsymbol{y}(\mathcal{P})^T\boldsymbol{W}(\boldsymbol{x})\boldsymbol{y}(\mathcal{P}). \tag{6.2.4}$$

It is easy to see that the matrices $\boldsymbol{W}(\boldsymbol{x})$ are idempotent and $\boldsymbol{W}(\boldsymbol{x})\boldsymbol{W}(\boldsymbol{y}) = \boldsymbol{O}$ for each $\boldsymbol{x} \neq \boldsymbol{y}$. Thus, the right-hand side of (6.2.4) gives an orthogonal partitioning of the left-hand side which can be further interpreted as follows. Define the s factors by $F_1, \ldots, F_s$. For any $\boldsymbol{x} = x_1 \cdots x_s \in \Omega^*$, let $F(\boldsymbol{x})$ represent the interaction $F^{x_1} \cdots F^{x_s}$; as usual, a one-factor interaction is a main effect. Consider now a hypothetical full factorial where each of the $v = q^s$ level-combinations appears exactly once. If the $v \times 1$ vector of lexicographically arranged observations arising from such a hypothetical full factorial equals $\boldsymbol{y}(\mathcal{P})$, then by (6.2.2) and (6.2.3), the sum of squares due to interaction $F(\boldsymbol{x})$ is given by

$$SS(\boldsymbol{x}) = \boldsymbol{y}(\mathcal{P})^T\boldsymbol{W}(\boldsymbol{x})\boldsymbol{y}(\mathcal{P}). \tag{6.2.5}$$

Thus, (6.2.4) is an analysis of variance decomposition. From (6.2.5), it follows that the design $\mathcal{P}$ is represented by an orthogonal array of strength $t(1 \leqslant t \leqslant s)$ if and only if $SS(\boldsymbol{x})$ $(= \boldsymbol{y}(\mathcal{P})^T\boldsymbol{W}(\boldsymbol{x})\boldsymbol{y}(\mathcal{P}))$ vanishes for every $\boldsymbol{x} \in \Omega_1 \bigcup \cdots \bigcup \Omega_t$. Hence, writing

$$B_i(\mathcal{P}) = \sum_{\boldsymbol{x}\in\Omega_i} \boldsymbol{y}(\mathcal{P})^T\boldsymbol{W}(\boldsymbol{x})\boldsymbol{y}(\mathcal{P}),\ 1 \leqslant i \leqslant s, \tag{6.2.6}$$

the departure of $\mathcal{P}$ from being represented by an orthogonal array of strength t can be measured by $\sum_{i=1}^t B_i(\mathcal{P})$. Thus, $B_1(d)$ measures the departure of $\mathcal{P}$ from an orthogonal array of strength unity, $B_2(d)$ is the additional amount needed to measure the departure of $\mathcal{P}$ from an orthogonal array of strength two, and so on. In other words, the quantities $B_1(\mathcal{P}), \ldots, B_s(\mathcal{P})$ capture the departure of $\mathcal{P}$ from orthogonal array of progressively higher strengths. Therefore, in order to ensure proximity to orthogonal array of successively higher strengths, one should choose $\mathcal{P}$ so as to minimize $B_1(\mathcal{P})$, $B_2(\mathcal{P})$, ... sequentially. This resembles what one does under the criterion of minimum aberration for regular fractions. Fang et al. (2002) showed that for a regular fraction $\mathcal{P}$ with $n = q^{s-p}$ runs

$$B_i(\mathcal{P}) = \frac{n^2(q-1)}{q^s} A_i(\mathcal{P}), \; 1 \leqslant i \leqslant s, \tag{6.2.7}$$

where $(A_1(\mathcal{P}), \ldots, A_s(\mathcal{P}))$ is the word-length pattern of $\mathcal{P}$. This result exhibits a connection between aberration and orthogonality for regular fractions. In Sect. 6.4, we will introduced some connections between aberration and uniformity. Then, by (6.2.7), we can establish some connections between orthogonality and uniformity. For details, one can refer to Fang et al. (2002).

6.3 Uniformity and Confounding

Two isomorphic factorial designs have been considered to be equivalent in the sense that they have the same statistical performance in ANOVA model. However, two isomorphic designs may have different uniformity. For example, the two $L_9(3^4)$ designs in Table 6.4, denoted by $\boldsymbol{U}_1$ and $\boldsymbol{U}_2$, are isomorphic to each other, but their CD-values are 0.050059 and 0.0493645, respectively. $\boldsymbol{U}_1$ can be easily found in the literature, while the second one was obtained by minimizing CD over the class of $\mathcal{U}(9; 3^4)$ (see Fang and Winker 1998). Suppose that there are three factors A, B, and C each having three levels in an experiment. We can choose any three columns from $\boldsymbol{U}_1$ or $\boldsymbol{U}_2$ for the factors. Denote the designs formed by the first three columns of $\boldsymbol{U}_1$ and $\boldsymbol{U}_2$ by $\mathcal{P}_1$ and $\mathcal{P}_2$, respectively. Denote the linear and quadratic main effects of A by A_l and A_q, respectively (similarly, for the notations B_l, B_q, C_l and C_q). The interaction $A \times B$ between A and B, if it exists, can be split into four terms A_lB_l, A_lB_q, A_qB_l, and A_qB_q (Fang and Ma 2000; Box and Draper 1987, pp. 236–239). When there are interactions $A \times B$, $A \times C$ and $B \times C$ in the experiment, it is

Table 6.4 Two $L_9(3^4)$ designs

No	$\boldsymbol{U}_1$				$\boldsymbol{U}_2$			
1	1	1	1	1	1	1	1	2
2	1	2	2	2	1	2	3	1
3	1	3	3	3	1	3	2	3
4	2	1	2	3	2	1	3	3
5	2	2	3	1	2	2	2	2
6	2	3	1	2	2	3	1	1
7	3	1	3	2	3	1	2	1
8	3	2	1	3	3	2	1	3
9	3	3	2	1	3	3	3	2

Table 6.5 A uniform minimum aberration design $UL_{27}(3^{13})$ under MD

No	1	2	3	4	5	6	7	8	9	10	11	12	13
1	2	2	2	2	3	2	1	2	3	1	3	3	1
2	2	2	3	2	3	3	2	3	1	3	2	2	3
3	2	2	1	2	3	1	3	1	2	2	1	1	2
4	2	3	2	3	1	2	3	3	2	1	2	2	2
5	2	3	3	3	1	3	2	2	3	2	1	1	1
6	2	3	1	3	1	1	1	1	1	3	3	3	3
7	2	1	2	1	2	2	3	2	1	3	1	1	3
8	2	1	3	1	2	3	1	3	2	2	3	3	2
9	2	1	1	1	2	1	2	1	3	1	2	2	1
10	3	2	2	1	1	3	3	1	1	2	3	2	1
11	3	2	3	1	1	1	1	2	2	1	2	1	3
12	3	2	1	1	1	2	2	3	3	3	1	3	2
13	3	3	2	2	2	3	2	1	2	1	1	3	3
14	3	3	3	2	2	1	3	2	3	3	3	2	2
15	3	3	1	2	2	2	1	3	1	2	2	1	1
16	3	1	2	3	3	3	1	1	3	3	2	1	2
17	3	1	3	3	3	1	3	3	1	1	1	3	1
18	3	1	1	3	3	2	2	2	2	2	3	2	3
19	1	2	2	3	2	1	1	3	3	2	1	2	3
20	1	2	3	3	2	2	2	1	1	1	3	1	2
21	1	2	1	3	2	3	3	2	2	3	2	3	1
22	1	3	2	1	3	1	2	2	1	2	2	3	2
23	1	3	3	1	3	2	1	1	2	3	1	2	1
24	1	3	1	1	3	3	3	3	3	1	3	1	3
25	1	1	2	2	1	1	2	3	2	3	3	1	1
26	1	1	3	2	1	2	3	1	3	2	2	3	3
27	1	1	1	2	1	3	1	2	1	1	1	2	2

impossible to separate the true interactions from the main effects. For the use of $\mathcal{P}_2$, the confounding situations are given by the alias statements:

$$
\begin{aligned}
A_l &= 0.5B_lC_q + 0.5B_qC_l, \\
A_q &= 1.5B_lC_l - 0.5B_qC_q, \\
B_l &= 0.5A_lC_q + 0.5A_qC_l, \\
B_q &= 1.5A_lC_l - 0.5A_qC_q, \\
C_l &= 0.5A_lB_q + 0.5A_qB_l, \\
C_q &= 1.5A_lB_l - 0.5A_qB_q.
\end{aligned}
$$

On the other hand, with the use of $\mathcal{P}_1$, the alias statements are

$$\begin{aligned}
A_l &= -0.75B_lC_l - 0.25B_lC_q + 0.25B_qC_l - 0.25B_qC_q,\\
A_q &= 0.75B_lC_l - 0.75B_lC_q + 0.75B_qC_l + 0.25B_qC_q,\\
B_l &= -0.75A_lC_l - 0.25A_lC_q + 0.25A_qC_l - 0.25A_qC_q,\\
B_q &= 0.75A_lC_l - 0.75A_lC_q + 0.75A_qC_l + 0.25A_qC_q,\\
C_l &= -0.75A_lB_l + 0.25A_lB_q + 0.25A_qB_l + 0.25A_qB_q,\\
C_q &= -0.75A_lB_l - 0.75A_lB_q - 0.75A_qB_l + 0.25A_qB_q.
\end{aligned}$$

If the higher-order interactions $A_lB_q, A_qB_l, A_qB_q, \ldots, B_qC_q$ can be ignored, the alias statements for $\mathcal{P}_2$ become

$$\mathcal{P}_2: \begin{cases} A_q = 1.5B_lC_l,\\ B_q = 1.5A_lC_l,\\ C_q = 1.5A_lB_l.\end{cases}$$

In this case, we can estimate all the linear effects A_l, B_l, and C_l without any confounding. While, the alias statements for $\mathcal{P}_1$ become

$$\mathcal{P}_1: \begin{cases} A_l = -0.75B_lC_l, & A_q = 0.75B_lC_l,\\ B_l = -0.75A_lC_l, & B_q = 0.75A_lC_l,\\ C_l = -0.75A_lB_l, & C_q = -0.75A_lB_l.\end{cases}$$

In this case, the main effects are all confounded with the interactions. Obviously, design $\mathcal{P}_2$ is better than $\mathcal{P}_1$ in the sense of confounding. There are four choices of choosing three columns from $\boldsymbol{U}_1$ or $\boldsymbol{U}_2$. It can be shown that there is only one choice from $\boldsymbol{U}_1$ that has the same confounding situation to $\mathcal{P}_2$ and the rest three choices have the same confounding situation to $\mathcal{P}_1$. On the other hand, all the four choices from $\boldsymbol{U}_2$ have the same confounding situation to $\mathcal{P}_2$. We thus conclude that $\boldsymbol{U}_2$ is better than $\boldsymbol{U}_1$ in the sense of confounding. From this example, Fang and Ma (2000) proposed the following concept.

Definition 6.3.1 For given (n, q, s), an orthogonal design $L_n(q^s)$ is called an *uniformly orthogonal design* and is denoted by $UL_n(q^s)$ if it has the smallest CD-value over all such orthogonal designs.

Obviously, one might choose other measures of uniformity to replace the CD in Definition 6.3.1. Several uniformly orthogonal designs for $q > 2$ are obtained in Fang and Winker (1998). Properties of these designs are yet to be studied. Hickernell and Liu (2002) used the reproducing kernel approach and showed that the uniform designs limit the effects of aliasing to yield reasonable efficiency and robustness together. Recently, Tang et al. (2012) proposed the level permutation technique (see Sect. 3.5) and the concept of the uniform minimum aberration (UMA) design. They obtained a UMA $U_{27}(3^{13})$ under CD. Later, Fang et al. (2016) found a UMA $UL_{27}(3^{13})$

under MD with $MD(UL_{27}) = 62.8011$, that is presented in Table 6.5. This design is an orthogonal array of strength two. The first m columns of UL_{27} are minimum aberration designs, where $m = 4, 5, 6, 10, 11, 12, 13$.

6.4 Uniformity and Aberration

Two important criteria, namely resolution and minimum aberration are based on the word-length pattern (see Sect. 1.4. for their definitions). These criteria were apparently unrelated with the uniformity criterion until Fang and Mukerjee (2000) found a connection between the uniformity and the aberration for regular fractions of two-level factorials. The main result they obtained is presented as follows:

Theorem 6.4.1 *Let* $\boldsymbol{U}$ *be a regular fraction of a* 2^s *factorial involving* $n = 2^{s-p}$ *runs. Then,*

$$\mathrm{CD}^2(\boldsymbol{U}) = \left(\frac{13}{12}\right)^s - 2\left(\frac{35}{32}\right)^s + \left(\frac{9}{8}\right)^s \left\{1 + \sum_{i=1}^{s} \frac{A_i(\boldsymbol{U})}{9^i}\right\}, \tag{6.4.1}$$

where $(A_1(\boldsymbol{U}), \ldots, A_s(\boldsymbol{U}))$ *is the word-length pattern of* $\boldsymbol{U}$.

From this relation, we can see that the minimum aberration and the uniformity measured by CD are almost equivalent for the regular factorial 2^{s-p}. For comparing two designs $\boldsymbol{U}_1$ and $\boldsymbol{U}_2$ via their CD-values, it is equivalent to compare $\sum_{i=1}^{s} \frac{A_i(\boldsymbol{U}_1)}{9^i}$ and $\sum_{i=1}^{s} \frac{A_i(\boldsymbol{U}_2)}{9^i}$. If design $\boldsymbol{U}_1$ has a resolution (say t) that is higher than $\boldsymbol{U}_2$ has (say t'), we have

$$\sum_{i=1}^{s} \frac{A_i(\boldsymbol{U}_1)}{9^i} = \sum_{i=t}^{s} \frac{A_i(\boldsymbol{U}_1)}{9^i}, \text{ and}$$

$$\sum_{i=1}^{s} \frac{A_i(\boldsymbol{U}_2)}{9^i} = \sum_{i=t'}^{s} \frac{A_i(\boldsymbol{U}_2)}{9^i} = \sum_{i=t'}^{t-1} \frac{A_i(\boldsymbol{U}_2)}{9^i} + \sum_{i=t}^{s} \frac{A_i(\boldsymbol{U}_2)}{9^i}.$$

It is easy to see that $\boldsymbol{U}_1$ is more likely to have a smaller CD-value than $\boldsymbol{U}_2$ since the coefficient of $A_i(\cdot)$ in $A_i(\cdot)/9^i$ decreases exponentially with i. Dr. C. X. Ma has checked all two-level regular designs in the catalogue given by Chen et al. (1993) and found that both CD and minimum aberration recommend the same designs.

The connection (6.4.1) in Theorem 6.4.1 holds for the regular two-level case. There have been some extensions of this result in the past years. Ma and Fang (2001) extended this connection to general two-level fractional designs under CD, WD, and symmetric L_2-discrepancy, and to general three-level designs under WD. Fang and Ma (2002) extended this connection to regular fraction 3^{s-1} designs and found that there existed essential difficulties to find more general results under CD. With the help of indicator function, Ye (2003) and Sun et al. (2011) showed that (6.4.1) holds for all two-level factorial designs, regular or non-regular, with or without replicates, and Sun et al. (2011) further extended Theorem 6.4.1 to both general two-level and three-level factorial designs under WD. The following theorem shows their results.

Theorem 6.4.2 *For any $\mathcal{P} \in \mathcal{D}(n; q^s)(q = 2$ or $3)$,*

$$\mathrm{WD}^2(\mathcal{P}) = \begin{cases} -\left(\frac{4}{3}\right)^s + \left(\frac{11}{8}\right)^s \left[1 + \sum_{r=1}^{s} \frac{A_r(\mathcal{P})}{11^r}\right], & \text{if } q = 2, \\ -\left(\frac{4}{3}\right)^s + \left(\frac{73}{54}\right)^s \left[1 + \sum_{r=1}^{s} \left(\frac{4}{73}\right)^r A_r(\mathcal{P})\right], & \text{if } q = 3, \end{cases} \tag{6.4.2}$$

where $(A_1(\mathcal{P}), \ldots, A_s(\mathcal{P}))$ is the generalized word-length pattern of $\mathcal{P}$ defined in (1.4.9).

The proof of Theorem 6.4.2 is based on the quadratic form of $\mathrm{WD}^2(\mathcal{P})$ in (4.3.2) for $q_1 = \cdots = q_s$, the generalized word-length pattern redefined by Cheng and Ye (2004) and some matrix computations. The details are omitted here; interested readers please refer to Sun et al. (2011).

Note that these connections under WD are consistent with that of Ma and Fang (2001) in terms of the generalized word-length pattern $(A_1^g(\mathcal{P}), \ldots, A_s^g(\mathcal{P}))$, but they only proved the result of $q = 2$ for the regular case, not for the general case. Several examples discussed by Ma and Fang (2001) show that the connections can significantly reduce the complexity of the computation for comparing factorial designs and also provide a way for searching minimum aberration designs by uniformity.

From Theorems 6.4.1 and 6.4.2, we notice that the coefficient of $A_r(\mathcal{P})$ in $[WD(\mathcal{P})])^2$ or $[CD(\mathcal{P})]^2$ decreases exponentially with r, so the design with less aberration tends to have smaller $[WD(\mathcal{P})]^2$ or $[CD(\mathcal{P})]^2$. Uniform designs under $[WD(\mathcal{P})]^2$ or $[CD(\mathcal{P})]^2$ and GMA designs are strongly related to each other. In fact, Sun et al. (2011) obtained the following theorem, which shows some conditions under which $[WD(\mathcal{P})]^2$ or $[CD(\mathcal{P})]^2$ agrees with GMA.

Theorem 6.4.3 (i) *Suppose $\mathcal{P}_1, \mathcal{P}_2 \in \mathcal{D}(n; q^s)$, k is some constant and $[kA_r(\mathcal{P}_i)]$ are all integers for $r = 1, \ldots, s$, $i = 1, 2$, then*

(1) *if $q = 2$ and $\max\{kA_r(\mathcal{P}_i), r = 1, \ldots, s, i = 1, 2\} \leqslant 8$, then $[CD(\mathcal{P}_1)]^2 < [CD(\mathcal{P}_2)]^2$ is equivalent to $\mathcal{P}_1$ having less aberration than $\mathcal{P}_2$;*
(2) *if $q = 2$ and $\max\{kA_r(\mathcal{P}_i), r = 1, \ldots, s, i = 1, 2\} \leqslant 10$, then $[WD(\mathcal{P}_1)]^2 < [WD(\mathcal{P}_2)]^2$ is equivalent to $\mathcal{P}_1$ having less aberration than $\mathcal{P}_2$;*
(3) *if $q = 3$ and $\max\{kA_r(\mathcal{P}_i), r = 1, \ldots, s, i = 1, 2\} \leqslant 69/4$, then $[WD(\mathcal{P}_1)]^2 < [WD(\mathcal{P}_2)]^2$ is equivalent to $\mathcal{P}_1$ having less aberration than $\mathcal{P}_2$.*

(ii) *Suppose $\mathcal{P}_1, \mathcal{P}_2 \in \mathcal{D}(n; q^s)$ and there exists a positive integer t such that $A_r(\mathcal{P}_1) = A_r(\mathcal{P}_2)$ for $r < t$ and $A_t(\mathcal{P}_1) \leqslant A_t(\mathcal{P}_2) - 1$, then*

(1) *if $q = 2$ and $\max\{A_r(\mathcal{P}_i), r = 1, \ldots, s, i = 1, 2\} \leqslant 8$, then $[CD(\mathcal{P}_1)]^2 < [CD(\mathcal{P}_2)]^2$;*

(2) *if* $q = 2$ *and* $\max\{A_r(\mathcal{P}_i), r = 1, \ldots, s, i = 1, 2\} \leqslant 10$, *then* $[WD(\mathcal{P}_1)]^2 < [WD(\mathcal{P}_2)]^2$;
(3) *if* $q = 3$ *and* $\max\{A_r(\mathcal{P}_i), r = 1, \ldots, s, i = 1, 2\} \leqslant 69/4$, *then* $[WD(\mathcal{P}_1)]^2 < [WD(\mathcal{P}_2)]^2$.

The proof of Theorem 6.4.3 is similar to that of Theorem 6.4.5, i.e., it follows directly from (6.4.1), (6.4.2) and the following lemma.

Lemma 6.4.1 *Suppose* a_i *and* b_i *are all nonnegative numbers and* $a_i, b_i \leqslant m - 1$, *for* $i = 0, \ldots, k$.

(i) *If* $a_k \leqslant b_k - 1$, *then* $\sum_{i=0}^k a_i m^i < \sum_{i=0}^k b_i m^i$.
(ii) *If* a_i *and* b_i *are integers with* $a_k \neq b_k$, *then* $\sum_{i=0}^k a_i m^i < \sum_{i=0}^k b_i m^i$ *if and only if* $a_k < b_k$.

proof (i) can be proved from

$$\begin{aligned}\sum_{i=0}^{k} a_i m^i &= a_k m^k + \sum_{i=0}^{k-1} a_i m^i \leqslant a_k m^k + (m-1)\sum_{i=0}^{k-1} m^i \\ &= a_k m^k + m^k - 1 < b_k m^k \leqslant \sum_{i=0}^{k} b_i m^i.\end{aligned}$$

And (ii) follows from (i) directly. Thus, the conclusion is true.

If the uniformity is measured by the *discrete discrepancy* defined in (2.5.7), Qin and Fang (2004) obtained the following connection between uniformity and GMA.

Theorem 6.4.4 *For any* $\mathcal{P} \in \mathcal{D}(n; q^s)$,

$$[DD(\mathcal{P})]^2 = \left(\frac{a + (q-1)b}{q}\right)^s \sum_{j=1}^{s} \left(\frac{a - b}{a + (q-1)b}\right)^j A_j(\mathcal{P}),$$

where $DD(\mathcal{P})$ *and* $A_j(\mathcal{P})$ *are defined in* (2.5.7) *and* (1.4.9), *respectively.*

From this connection, we can see that the coefficient of $A_j(\mathcal{P})$ in $DD(\mathcal{P})$ decreases exponentially with j, and we anticipate that designs which keep $A_j(\mathcal{P})$ small for small values of j, that are those having less aberration, should behave well in terms of uniformity in the sense of keeping $DD(\mathcal{P})$ small. This shows that uniform designs under the discrete discrepancy and GMA designs are strongly related to each other and provides a justification for the criterion of GMA by consideration of uniformity measured by the discrete discrepancy. Theorem 6.4.4 also shows us that the minimum discrete discrepancy (2.5.7) does not completely agree with the GMA criterion. When are those two criteria equivalent to each other? Sun et al. (2011) obtained Theorem 6.4.4 in a bit more intuitive way and provided a condition for their equivalency in the following theorem.

Theorem 6.4.5 *Suppose $\mathcal{P}_1$ and $\mathcal{P}_2$ are two designs from $\mathcal{D}(n; q^s)$, both of which have no replicates. If*

$$\frac{a+(q-1)b}{a-b} - 1 = \frac{qb}{a-b} \geqslant n(q^s - n),$$

then the fact that $[DD(\mathcal{P}_1)]^2 < [DD(\mathcal{P}_2)]^2$ is equivalent to $\mathcal{P}_1$ having less aberration than $\mathcal{P}_2$.

proof From the proof of Theorem 2 in Hickernell and Liu (2002), we have

$$A_r(\mathcal{P}) = \frac{1}{n^2} \sum_{|\boldsymbol{u}|=r} \sum_{i,k=1}^{n} \prod_{l \in \boldsymbol{u}} (-1 + q\delta_{a_{il}a_{kl}}), \tag{6.4.3}$$

where $\boldsymbol{u}$ is a subset of $\{1, \ldots, s\}$, $|\boldsymbol{u}|$ denotes the cardinality of $\boldsymbol{u}$, and $\mathcal{P} = (a_{ij})$; thus, $n^2 A_r(\mathcal{P})$ is an integer. In addition, from Theorem 4.1 of Cheng and Ye (2004), we know that for any $\mathcal{P} \in \mathcal{D}(n; q^s)$ without replicates,

$$\sum_{r=1}^{s} A_r(\mathcal{P}) = \frac{q^s}{n} - 1.$$

So for any r, $n^2 A_r(\mathcal{P}) \leqslant n(q^s - n)$. Thus from Theorem 6.4.4 and Lemma 6.4.1, the conclusion can be reached easily.

From this theorem, we know that if orthogonal designs $L(n; q^s)$ without replicates exist and $qb/(a-b) \geqslant n(q^s - n)$, then the uniform design under $\left([DD(\mathcal{P})]^2\right)$ is an orthogonal design and has GMA among all designs in $\mathcal{D}(n; q^s)$ without replicates. From Lemma 6.4.1, Sun et al. (2011) further provided some conditions under which the minimum discrete discrepancy agrees with GMA:

Theorem 6.4.6 (i) *Suppose $\mathcal{P}_1, \mathcal{P}_2 \in \mathcal{D}(n; q^s)$ and $[kA_r(\mathcal{P}_i)]$ are all integers for $r = 1, \ldots, s$, and $i = 1, 2$, where k is some constant. If $qb/(a-b) \geqslant \max\{kA_r(\mathcal{P}_i), r = 1, \ldots, s, i = 1, 2\}$, then $\left([DD(\mathcal{P}_1)]^2 < [DD(\mathcal{P}_2)]^2\right)$ is equivalent to $\mathcal{P}_1$ having less aberration than $\mathcal{P}_2$.*

(ii) *Suppose $\mathcal{P}_1$ and $\mathcal{P}_2$ are two regular designs from $\mathcal{D}(n; q^s)$. If $qb/(a-b) \geqslant \max\{A_r(\mathcal{P}_i), r = 1, \ldots, s, i = 1, 2\}$, then $\left([DD(\mathcal{P}_1)]^2 < [DD(\mathcal{P}_2)]^2\right)$ is equivalent to $\mathcal{P}_1$ having less aberration than $\mathcal{P}_2$.*

(iii) *Suppose $\mathcal{P}_1, \mathcal{P}_2 \in \mathcal{D}(n; q^s)$. If $qb/(a-b) \geqslant \max\{A_r(\mathcal{P}_i), r = 1, \ldots, s, i = 1, 2\}$ and there exists a positive integer t such that $A_r(\mathcal{P}_1) = A_r(\mathcal{P}_2)$ for $r < t$ and $A_t(\mathcal{P}_1) \leqslant A_t(\mathcal{P}_2) - 1$, then $\left([DD(\mathcal{P}_1)]^2 < [DD(\mathcal{P}_2)]^2\right)$.*

Besides, Hickernell and Liu (2002) also showed that GMA designs and uniform designs are equivalent in a certain limit, see Theorem 6.5.2 in Sect. 6.5.1 for this result. The connection between projection uniformity and aberration was also studied by Hickernell and Liu (2002) and Fang and Qin (2004), please refer to Sect. 6.5 for the details.

Next, consider the relationship between any uniformity criterion and the GMA criterion for any level design. Let (n, q^s)-design be a design of n runs, s factors and q levels. For an $(n, q^s)-$design $\mathcal{P}$, when considering all $q!$ possible level permutations for every factor, we obtain $(q!)^s$ combinatorially isomorphic designs. Denote the set of these designs as $\mathcal{H}(\mathcal{P})$. Because reordering the rows or columns does not change the geometrical structure and statistical properties of a design, there is no need to consider row or column permutations. For an n-point design $\mathcal{P} = \{\boldsymbol{x}_1, \ldots, \boldsymbol{x}_n\}$ and a nonnegative function $F(\boldsymbol{x}_i, \boldsymbol{x}_j) \geqslant 0$, define

$$\phi(\mathcal{P}) = \frac{1}{n^2} \sum_{i,j=1}^{n} F(\boldsymbol{x}_i, \boldsymbol{x}_j). \tag{6.4.4}$$

Call $\phi(\mathcal{P})$ as a space-filling measure of $\mathcal{P}$ with respect to F. All designs in $\mathcal{H}(\mathcal{P})$ share the same GWP but may have different $\phi(\mathcal{P})$. We can compute $\phi(\mathcal{P})$ for each design, as well as the average value, denoted by $\bar{\phi}(\mathcal{P})$, of all designs in $\mathcal{H}(\mathcal{P})$. More precisely, define

$$\bar{\phi}(\mathcal{P}) = \frac{1}{(q!)^s} \sum_{\mathcal{P}' \in \mathcal{H}(\mathcal{P})} \phi(\mathcal{P}'). \tag{6.4.5}$$

The following result shows that the average value $\bar{\phi}(\mathcal{P})$ in (6.4.5) can be expressed as a linear combination of GWP for a wide class of space-filling measures.

Lemma 6.4.2 *Suppose* $F(\boldsymbol{x}_i, \boldsymbol{x}_j) = \prod_{k=1}^{s} f(x_{ik}, x_{jk})$ *and* $f(\cdot, \cdot)$ *satisfies*

$$\begin{cases} f(x, x) + f(y, y) > f(x, y) + f(y, x), \\ f(x, y) \geqslant 0, \quad \text{for any } x \neq y, \; x, y \in [0, 1]. \end{cases} \tag{6.4.6}$$

For an $(n, q^s)-$*design* $\mathcal{P}$,

$$\bar{\phi}(\mathcal{P}) = \left(\frac{c_1(c_2 + q - 1)}{q^2(q-1)} \right)^s \sum_{i=0}^{s} \left(\frac{c_2 - 1}{c_2 + q - 1} \right)^i A_i(\mathcal{P}), \tag{6.4.7}$$

where $c_1 = \sum_{k=0}^{q-1} \sum_{l \neq k} f(k, l)$ *and* $c_2 = (q-1) \sum_{k=0}^{q-1} f(k, k)/c_1$.

The conditions of $F(\cdot, \cdot)$ in Lemma 6.4.2 are nonrestrictive and satisfied by many commonly used discrepancies and other measures. The requirement (6.4.6) makes $c_2 > 1$ so that the coefficient of $A_i(\mathcal{P})$ in (6.4.7) decreases geometrically as i increases. As a result, when all level permutations are considered, $\bar{\phi}(\mathcal{P})$ tends to agree with the GMA criterion. Tang et al. (2012) and Tang and Xu (2013) showed that the average CD-value is a linear function of the GWP, when all level permutations are considered. Here we generalize their results for any discrepancy defined by a reproducing kernel.

Note that we can express the discrepancy in (2.4.6) as a space-filling measure defined in (6.4.4) with

$$F(\boldsymbol{x}_i, \boldsymbol{x}_j) = \mathcal{K}(\boldsymbol{x}_i, \boldsymbol{x}_j) - \mathcal{K}_1(\boldsymbol{x}_i) - \mathcal{K}_1(\boldsymbol{x}_j) + \mathcal{K}_2, \tag{6.4.8}$$

where $\mathcal{K}_1(\boldsymbol{x}) = \int_{\mathcal{X}} \mathcal{K}(\boldsymbol{x}, \boldsymbol{y}) dF_{\mathrm{u}}(\boldsymbol{y})$ and $\mathcal{K}_2 = \int_{\mathcal{X}^2} \mathcal{K}(\boldsymbol{x}, \boldsymbol{y}) dF_{\mathrm{u}}(\boldsymbol{x}) dF_{\mathrm{u}}(\boldsymbol{y})$ is a constant. In other words, any discrepancy defined by the reproducing kernel method is a special space-filling measure.

Commonly used reproducing kernels for discrepancies in the literature are defined on $\mathcal{X} = [0, 1]^s$ and have a multiplicative form

$$\mathcal{K}(\boldsymbol{x}, \boldsymbol{y}) = \prod_{k=1}^{s} f(x_k, y_k), \tag{6.4.9}$$

where $f(x, y)$ is defined on $[0, 1]^2$. Then, the corresponding discrepancy in (2.4.6) can be expressed by

$$D^2(\mathcal{P}, \mathcal{K}) = \mathcal{K}_2 - \frac{2}{n} \sum_{i=1}^{n} \prod_{k=1}^{s} f_1(x_{ik}) + \frac{1}{n^2} \sum_{i,j=1}^{n} \prod_{k=1}^{s} f(x_{ik}, x_{jk}), \tag{6.4.10}$$

where $f_1(x) = \int_0^1 f(x, y) dy$. The various kernel functions are as follows:

(i) for CD, $f(x, y) = 1 + (|x - 0.5| + |y - 0.5| - |x - y|)/2$;

(ii) for WD, $f(x, y) = 1.5 - |x - y| + |x - y|^2$;

(iii) for MD, $f(x, y) = 15/8 - |x - 0.5|/4 - |y - 0.5|/4 - 3|x - y|/4 + |x - y|^2/2$;

(iv) for Lee discrepancy, $f(x, y) = 1 - \min\{|x - y|,\ 1 - |x - y|\}$.

Now we consider level permutations of any given fractional factorial design $\mathcal{P}$ and calculate the average discrepancy of all permuted designs, denoted by $\overline{D}(\mathcal{P}, \mathcal{K})$. For any row $\boldsymbol{x}_i$ of $\mathcal{P}$, when one considers all level permutations, each $s-$tuple in Z_q^s occurs $((q-1)!)^s$ times. Then,

$$\begin{aligned} \sum_{\mathcal{P}' \in \mathcal{H}(\mathcal{P})} \sum_{i=1}^{n} \prod_{k=1}^{s} f_1(x_{ik}) &= \sum_{i=1}^{n} \sum_{\mathcal{P}' \in \mathcal{H}(\mathcal{P})} \prod_{k=1}^{s} f_1(x_{ik}) \\ &= n \left((q-1)! \sum_{k=0}^{q-1} f_1(k) \right)^s, \end{aligned} \tag{6.4.11}$$

which is a constant. From Lemma 6.4.2, (6.4.10) and (6.4.11), Zhou and Xu (2014) obtained the following result.

Theorem 6.4.7 *Suppose that* $\mathcal{K}(\boldsymbol{x}, \boldsymbol{y}) = \prod_{k=1}^{s} f(x_k, y_k)$ *and* $f(\cdot, \cdot)$ *satisfies (6.4.6). For an* $(n, q^s)-$*design* $\mathcal{P}$*, when all level permutations of* $\mathcal{P}$ *are considered,*

$$\overline{D}(\mathcal{P}, \mathcal{K}) = \mathcal{K}_0 + \left(\frac{c_1(c_2+q-1)}{q^2(q-1)}\right)^s \sum_{i=0}^{s} \left(\frac{c_2-1}{c_2+q-1}\right)^i A_i(\mathcal{P}), \quad (6.4.12)$$

where $\mathcal{K}_0 = \mathcal{K}_2 - 2\left(\sum_{k=0}^{q-1} f_1((k+0.5)/q)/q\right)^s$ *is a constant,* $\mathcal{K}_2$ *and* $f_1(\cdot)$ *are, respectively, defined in (6.4.8) and (6.4.10),* $c_1 = \sum_{k=0}^{q-1}\sum_{l\neq k} f((k+0.5)/q, (l+0.5)/q)$*, and* $c_2 = (q-1)\sum_{k=0}^{q-1} f((k+0.5)/q, (k+0.5)/q)/c_1$.

From Theorem 6.4.7, for any discrepancy defined by a reproducing kernel satisfying (6.4.6), the average discrepancy is a linear combination of GWP under all level permutations, and the commonly used discrepancies such as WD and CD satisfy the condition (6.4.6). For example, from the kernel of WD, we have $\mathcal{K}_2 = (4/3)^s$, $\sum_{k=0}^{q-1} f_1((k+0.5)/q)/q = 4/3$, $c_1 = \sum_{k=0}^{q-1}\sum_{l\neq k}\left(1.5 - |k-l|/q + |k-l|^2/q^2\right)$ $= (q-1)(8q-1)/6$, and $c_2 = 9q/(8q-1) > 1$. Then, for the average WD-value, we have

$$\overline{WD}(\mathcal{P}) = -\left(\frac{4}{3}\right)^s + \left(\frac{8q^2+1}{6q^2}\right)^s \sum_{i=0}^{s} \left(\frac{q+1}{8q^2+1}\right)^i A_i(\mathcal{P}). \quad (6.4.13)$$

Especially, when $q = 2$, all level permuted designs have the same WD value since WD is invariant under coordinate rotation; therefore, Eq. (6.4.13) shows the exact relationship between GWP and WD for two-level designs. When $q = 3$, Tang et al. (2012) showed that we need only to consider linear level permutations when computing $\bar{\phi}(\mathcal{P})$, then based on the expression of WD in (2.3.6), any linear level permutation does not change the WD-value, which means that Eq. (6.4.13) also shows the exact relationship between GWP and WD for three-level designs. In other words, Eq. (6.4.13) includes the result of WD for two- and three-level designs in Ma and Fang (2001). Similarly, applying Theorem 6.4.7 to CD, we obtain the relationship between average CD and GWP, which was reported by Fang and Mukerjee (2000), Ma and Fang (2001), Tang et al. (2012), and Tang and Xu (2013) for two-, three-, and multi-level designs, respectively. Moreover, Theorem 6.4.7 includes the results on Lee discrepancy for two- and three-level designs in Zhou et al. (2008) and the result on MD for two-level designs in Zhou at al. (2013). In summary, Theorem 6.4.7 gives a unified result for any type of discrepancy defined by a reproducing kernel.

Before ending this section, we should note that Sun et al. (2011) investigated the close relationships in Theorems 6.4.2 and 6.4.4 by expressing WD in the quadratic form of (4.3.2) and the discrete discrepancy in the quadratic form of

$$[DD(\mathcal{P})]^2 = -\left[\frac{a+(q-1)b}{q}\right]^s + \frac{b^s}{n^2}\boldsymbol{y}^T\boldsymbol{D}_s\boldsymbol{y},$$

where $\boldsymbol{y}$ is the same as in (6.2.1), $\boldsymbol{D}_s = \otimes^s \boldsymbol{D}_0$, $\boldsymbol{D}_0 = (d_{ij})$, and $d_{ij} = a/b$, if $i = j$ and 1 otherwise, $i, j = 0, \ldots, q-1$. Those expressions of the discrepancies in the quadratic forms of the indicator functions are useful for us to find optimal designs under each of the criteria. In fact, in order to find a design minimizing the quadratic term like $\boldsymbol{y}^T \mathsf{B} \boldsymbol{y}$, we need to only solve

$$\begin{cases} \min\limits_{y} \boldsymbol{y}^T \mathsf{B} \boldsymbol{y}, \\ \text{s.t. } \boldsymbol{y}^T \mathbf{1} = n, \ \ y_i = 0, \ldots, n. \end{cases} \tag{6.4.14}$$

One such approach is provided in Sun et al. (2009), and many optimal designs under GMA as well as a uniformity criterion are tabulated there. For more approaches for solving the problem like (6.4.14), please refer to Sect. 4.3.

6.5 Projection Uniformity and Related Criteria

Although many criteria were proposed for comparing U-type designs, none of these criteria can directly distinguish non-isomorphic saturated designs. Definition and related discussion on the isomorphism have been given in Sect. 6.1. Let $\mathcal{P}$ be a design of n runs and s factors. There are many subdesigns, s subdesigns for one-factor experiments, $\binom{s}{2}$ subdesigns for two-factor experiments, and so on. A specific criterion can measure all these subdesigns, and the related values are called its projection pattern. We can use the distribution or the vector of these projection values as a tool to distinguish the underlying designs. Hickernell and Liu (2002) proposed the *projection discrepancy pattern* and related criteria, Fang and Qin (2004) suggested a different uniformity pattern and related criteria for two-level factorials, and Fang and Zhang (2004) suggested the minimum aberration majorization based on the majorization theory and projection aberration. This section will give a brief introduction to these approaches.

6.5.1 Projection Discrepancy Pattern and Related Criteria

This subsection introduces *projection discrepancy pattern* and related criteria proposed by Hickernell and Liu (2002).

A discrepancy can be defined as a norm with a specific kernel (see Definition 2.4.4). To define the *projection discrepancy pattern*, we restrict ourselves to the case where the *experimental domain*, $\mathcal{X}$, is a Cartesian product of one-dimensional domains, i.e., $\mathcal{X} = \mathcal{X}_1 \times \cdots \times \mathcal{X}_s$. Suppose the *reproducing kernel* $\mathcal{K}$ is a product of one-dimensional kernels, i.e., $\mathcal{K} = \prod_{j=1}^{s} \mathcal{K}_j$ with reproducing kernel $\mathcal{K}_j = 1 + \hat{\mathcal{K}}_j$. Let $\mathcal{S}$ denote the set $\{1, \ldots, s\}$, and u be any subset of $\mathcal{S}$. Let $\boldsymbol{x}_u$ denote the elements of the vector $\boldsymbol{x}$ indexed by the elements of u, and $\mathcal{X}_u$ denote the Cartesian product of $\mathcal{X}_j$ with $j \in u$. Let $\mathcal{P} = \{\boldsymbol{x}_1, \ldots, \boldsymbol{x}_n\}$ denote a design with n points on the domain $\mathcal{X}$, and $\mathcal{P}_u$ denote the projection of $\mathcal{P}$ into the domain $\mathcal{X}_u$. One would normally desire that for small $|u|$, the projections $\mathcal{P}_u$ would be good designs on the

$\mathcal{X}_u$. This is the motivation behind the definitions of resolution (p. 385, Box et al. 1978) and aberration (Fries and Hunter 1980). The discrepancy, as defined above, does not necessarily guarantee this, but one may defined an aberration in terms of the pieces of the squared discrepancy. Now let us recall the corresponding definitions due to Hickernell and Liu (2002). They showed that the reproducing kernel $\mathcal{K}$ may be written as

$$\mathcal{K}(\boldsymbol{x}, \boldsymbol{w}) = \prod_{j=1}^{s} \mathcal{K}_j(x_j, w_j) = \sum_{\emptyset \subseteq u \subseteq \mathcal{S}} \hat{\mathcal{K}}_u(\boldsymbol{x}_u, \boldsymbol{w}_u), \tag{6.5.1a}$$

$$\hat{\mathcal{K}}_u(\boldsymbol{x}_u, \boldsymbol{w}_u) = \prod_{j \in u} \hat{\mathcal{K}}_j(x_j, w_j). \tag{6.5.1b}$$

Note that $\hat{\mathcal{K}}_\emptyset = 1$ by convention.

Example 6.5.1 Suppose that the jth factor has q_j levels, i.e., $\mathcal{X}_j = \{0, 1, \ldots, q_j - 1\}$, for $j = 1, \ldots, s$, and the reproducing kernel is defined by

$$\mathcal{K}_j(x, w) = 1 + \gamma(-1 + q_j \delta_{xw}).$$

Then, one can identify

$$\hat{\mathcal{K}}_u(\boldsymbol{x}_u, \boldsymbol{w}_u) = \gamma^{|u|} \prod_{j \in u} (-1 + q_j \delta_{x_j w_j}).$$

For kernels of the form (6.5.1), one may write

$$D^2(\mathcal{P}; \mathcal{K}) = \sum_{\emptyset \subset u \subseteq \mathcal{S}} D^2(\mathcal{P}_u; \hat{\mathcal{K}}_u) = \sum_{j=1}^{s} D^2_{(j)}(\mathcal{P}; \mathcal{K}), \quad \text{where} \tag{6.5.2a}$$

$$D^2_{(j)}(\mathcal{P}; \mathcal{K}) = \sum_{|u|=j} D^2(\mathcal{P}_u; \hat{\mathcal{K}}_u). \tag{6.5.2b}$$

Since $\hat{\mathcal{K}}_\emptyset = 1$, it follows that $D(\mathcal{P}; \hat{\mathcal{K}}_\emptyset) = 0$.

Definition 6.5.1 Suppose that $\mathcal{X} = \mathcal{X}_1 \times \cdots \times \mathcal{X}_s$, and that the reproducing kernel, $\mathcal{K}$, is of the form (6.5.1). The *projection discrepancy pattern* (PDP) is defined as the s-vector

$$\mathrm{PD}(\mathcal{P}; \mathcal{K}) = (D_{(1)}(\mathcal{P}; \mathcal{K}), \ldots, D_{(s)}(\mathcal{P}; \mathcal{K})).$$

The dictionary ordering can be used for comparing any two designs $\mathcal{P}, \tilde{\mathcal{P}} \subseteq \mathcal{X}$, one says that $\mathcal{P}$ has *better projection uniformity* than $\tilde{\mathcal{P}}$, or equivalently, $\mathrm{PD}(\mathcal{P}; \mathcal{K}) < \mathrm{PD}(\tilde{\mathcal{P}}; \mathcal{K})$, if and only if the first (from the left) nonzero component of $\mathrm{PD}(\mathcal{P}; \mathcal{K}) - \mathrm{PD}(\tilde{\mathcal{P}}; \mathcal{K})$ is negative. If $t = \min\{j : D_{(j)}(\mathcal{P}; \mathcal{K}) > 0\}$, then $\mathcal{P}$ is said to have *uniformity resolution* t. A design is said to have *minimum projection uniformity* (MPU) if no other design has better projection uniformity than it.

For Example 6.5.1, the $D^2_{(j)}(\mathcal{P};\mathcal{K})$ is given by

$$D^2_{(j)}(\mathcal{P};\mathcal{K}) = \frac{\gamma^j}{n^2}\sum_{|u|=j}\sum_{i,k=1}^{n}\prod_{l\in u}(-1+q_l\delta_{x_{il}x_{kl}}). \tag{6.5.3}$$

Definition 6.5.1 does not assume that the design is a regular fractional factorial design (FFD, for short) or even that each factor has a finite number of levels. It only assumes that the experimental domain is a Cartesian product of one-factor domains, and that the reproducing kernel is product of one-dimensional kernels.

For the special case of (6.5.3), Definition 6.5.1 reduces to the GMA proposed by Xu and Wu (2001), as shown below by Hickernell and Liu (2002).

Theorem 6.5.1 *For the case where the components of PDP are given by* (6.5.3), *the resulting MPU as given in Definition 6.5.1 is equivalent to the GMA defined by Xu and Wu (2001), i.e.,*

$$\gamma^{-j}D^2_{(j)}(\mathcal{P};\mathcal{K}) = A_j(\mathcal{P}), \ \textit{for}\ j=1,\ldots,s.$$

For the case of two-level designs, the MPU defined here is equivalent to the minimum G_2-aberration of Tang and Deng (1999).

Some comments are in order regarding the parameter γ that enters into the definition of the discrepancy. This parameter has no effect when comparing the PDPs of different designs. Thus, as far as the PDP is concerned, one might as well set $\gamma = 1$. By doing so, Liu et al. (2006) investigated the connections among MPU, GMA, *minimum moment aberration* (Xu 2003), and *minimum χ^2 criterion* (Liu et al. 2006). The connections provide strong statistical justification for each of them. Some general optimality results are developed, which not only unify several results, but also are useful for constructing asymmetrical supersaturated designs.

However, the value of γ does affect the comparison of the discrepancies of different designs. Recall from (6.5.2) that the squared discrepancy is a sum of the $D^2_{(j)}(\mathcal{P};\mathcal{K})$, and note that if each $\hat{\mathcal{K}}_j$ has a leading factor of γ, then $D^2_{(j)}(\mathcal{P};\mathcal{K})$ has a leading factor of γ^j. A larger value of γ gives a relatively heavier weight to the $D^2_{(j)}(\mathcal{P};\mathcal{K})$ with large j and implies a preference for better uniformity in the high-dimensional projections of $\mathcal{P}$, whereas a small value of γ implies a preference for better uniformity in the low-dimensional projections of $\mathcal{P}$. Thus, comparing the aberration of two designs is equivalent to comparing their discrepancies for vanishing γ, as is explained by Hickernell and Liu (2002) in the following theorem whose proof is straightforward.

Theorem 6.5.2 *Suppose that the reproducing kernel is of the form* (6.5.1), *and that $\hat{\mathcal{K}}_j$ has a leading coefficient γ. For a fixed number of experiments, n, let $\mathcal{P}_a$ be a GMA design, and t be its resolution. For any $\gamma > 0$, let $\mathcal{P}_\gamma$ denote a minimum discrepancy design. Then, if one assumes that $D(\mathcal{P}_a;\mathcal{K}) > 0$, it follows that*

$$\lim_{\gamma\downarrow 0} D(\mathcal{P}_\gamma;\mathcal{K})/D(\mathcal{P}_a;\mathcal{K}) = 1, \quad \lim_{\gamma\downarrow 0} D_{(j)}(\mathcal{P}_\gamma;\mathcal{K}) = D_{(j)}(\mathcal{P}_a;\mathcal{K})\ (j=1,\ldots,t).$$

6.5.2 *Uniformity Pattern and Related Criteria*

For U-type designs with two levels, Fang and Qin (2004) suggested the uniformity pattern and related criteria. Let $\mathcal{P}$ be a $U(n, 2^s)$. For convenience, we consider only CD and its uniformity pattern. Let u be any subset of $\mathcal{S}$ and the subdesign of $\mathcal{P}$ is denoted by $\mathcal{P}_u$. Its CD is written as $CD_{2,u}(\mathcal{P})$. It is easy to find

$$[CD_{2,u}(\mathcal{P})]^2 = \left(\frac{1}{12}\right)^{|u|} - \frac{3^{|u|}}{2^{5|u|-1}} + \frac{1}{n2^{2|u|}} E_0(\mathcal{P}_u), \tag{6.5.4}$$

where $E_0(\mathcal{P}_u)$ is the first component of the distance distribution of the subdesign $\mathcal{P}_u$ (see (6.1.1)). Let

$$[I_i(\mathcal{P})]^2 = \sum_{|u|=i} [CL_{2,u}(\mathcal{P})]^2,$$

which measures the overall uniformity of $\mathcal{P}$ on i-subdimension. Obviously, $CD(\mathcal{P})$ has the decomposition $[CL_2(\mathcal{P})]^2 = \sum_{i=1}^{s} [I_i(\mathcal{P})]^2$. From (6.5.4), it is evident that $\mathcal{P}$ has strength t if and only if for $1 \leqslant j \leqslant t$

$$[I_j(\mathcal{P})]^2 = \binom{s}{j} \left[\left(\frac{1}{12}\right)^j - \frac{3^j}{2^{5j-1}} + \frac{1}{8^j} \right],$$

and

$$[I_{t+1}(\mathcal{P})]^2 > \binom{s}{t+1} \left[\left(\frac{1}{12}\right)^{t+1} - \frac{3^{t+1}}{2^{5(t+1)-1}} + \frac{1}{8^{(t+1)}} \right].$$

Definition 6.5.2 Define

$$U_i(\mathcal{P}) = [I_i(\mathcal{P})]^2 - \binom{s}{i} \left[\left(\frac{1}{12}\right)^i - \frac{3^i}{2^{5i-1}} + \frac{1}{8^i} \right], \; 1 \leqslant i \leqslant s. \tag{6.5.5}$$

The vector $(U_1(\mathcal{P}), \ldots, U_s(\mathcal{P}))$ is called a *uniformity pattern* (UP, for short) of design $\mathcal{P}$.

Then, the MPU criterion based on this pattern can be defined similarly as in Definition 6.5.1.

Fang and Qin (2004) gave the following theorems, where the second one builds an analytic relationship between $\{U_i(\mathcal{P})\}$ and $\{A_i(\mathcal{P})\}$.

Theorem 6.5.3 *Suppose* $\mathcal{P} \in \mathcal{U}(n; 2^s)$, *then* $\mathcal{P}$ *with strength* t *if and only if* $U_j(\mathcal{P}) = 0, 1 \leqslant j \leqslant t$ *and* $U_{t+1}(\mathcal{P}) \neq 0$.

Theorem 6.5.4 *Suppose $\mathcal{P} \in \mathcal{P}(n; 2^s)$. For any $1 \leqslant i \leqslant s$, the U_j's and A_j's are linearly related through the following equations*

$$U_i(\mathcal{P}) = \frac{1}{8^i} \sum_{v=1}^{i} \binom{s-v}{s-i} A_v(\mathcal{P}), \tag{6.5.6}$$

and

$$A_i(\mathcal{P}) = \sum_{v=1}^{i} (-1)^{i-v} \binom{s-v}{i-v} 8^v U_v(\mathcal{P}). \tag{6.5.7}$$

Note that the leading coefficient $\frac{1}{8^i}\binom{s-v}{s-i}$ in (6.5.6) is positive. It is clear that sequentially minimizing $U_i(\mathcal{P})$ for $i = 1, \ldots, s$ is equivalent to sequentially minimizing $A_i(\mathcal{P})$ for $i = 1, \ldots, s$. Therefore, for two-level designs, the MPU criterion coincides with the GMA criterion. In particular, a two-level design has GMA if and only if it has MPU.

There is much more work on the connection between MPU and other criteria. For two-level designs, Zhang and Qin (2006) showed that MPU criterion, V-criterion proposed by Tang (2001) and nearest balance criterion proposed by Fang et al. (2003) are mutually equivalent, and the popular $E(s^2)$-optimality is a special case of MPU. Qin et al. (2011) also found that MPU criterion is a good surrogate for the design efficiency criterion proposed by Cheng et al. (2002). For multi-level designs, Qin et al. (2012) discussed the issue of MPU based on the discrete discrepancy proposed in Qin and Fang (2004) and provided connection between MPU and other optimality criteria.

6.6 Majorization Framework

Let $\boldsymbol{U}$ be an $n \times s$ design in $\mathcal{U}$. A criterion for assessing designs on $\mathcal{U}$ can use a function $\phi(\boldsymbol{U})$ or a vector $(\phi_1(\boldsymbol{U}), \ldots, \phi_m(\boldsymbol{U}))$. The world-length pattern is an example of using a vector. How to compare two such vectors? Section 1.4.1 introduces two ordering ways: the dictionary ordering and majorization ordering. This section concerns with two different vector functions: pairwise coincidence and generalized word-length pattern under the majorization ordering.

6.6.1 Based on Pairwise Coincidence Vector

Let $\delta_{ik}(\boldsymbol{U})$ be the *pairwise coincidence* (PC) between runs $\boldsymbol{u}_i$ and $\boldsymbol{u}_k$ in the design matrix $\boldsymbol{U} = (u_{ij})$. Obviously, the coincidence between rows $\boldsymbol{u}_i$ and $\boldsymbol{u}_k$ (see (2.5.2) has $\delta_{ik}(U) = s - d_{ik}^H(U)$, where $d_{ik}^H(U)$ is the Hamming distance between runs $\boldsymbol{u}_i$ and $\boldsymbol{u}_k$. Let $\boldsymbol{\delta}(U) = (\delta_1, \delta_2, \ldots, \delta_m)'$ be the PC vector of $\delta_{ik}(U)$ for $1 \leqslant i < k \leqslant n$ such that δ_{ik} corresponds to the element $\delta_{n(i-1)+k-i(i+1)/2}$, where $m = n(n-1)/2$, i.e.,

$$\boldsymbol{\delta}(U) = (\delta_{12}(U), \ldots, \delta_{1n}(U), \delta_{23}(U), \ldots, \delta_{2n}(U), \ldots, \delta_{(n-1)n}(U)). \qquad (6.6.1)$$

We call two designs PC-*different* if their PC-vectors cannot be exchanged by permutation. The PC-sum $\sum_{r=1}^m \delta_r = \sum_{i<k} \sum_{j=1}^s \delta_{u_{ij}u_{kj}}$ keeps invariant in any U-type design, by observing that $1 + \sum_{k \neq i} \delta_{u_{ij}u_{kj}} = n/q$ for any i, j.

Lemma 6.6.1 *For any* $U \in \mathcal{U}(n, q^s)$*, its PC-sum is* $\frac{ns}{2}(\frac{n}{q} - 1)$.

Based on the decision theory, majorization theory, and Lemma 6.6.1, Zhang et al. (2005) proposed a general framework for U-type designs via their PC vectors.

If we chooses some suitable Schur-convex function $\Psi(\boldsymbol{\delta}(U))$ ($\Psi(\boldsymbol{x}) = \sum_{r=1}^m \psi(x_r)$, where ψ is convex on R_+, please refer to Sect. 1.4.1), the function $\Psi(\boldsymbol{\delta}(U))$ can be used as a criterion for comparing two U-type designs in $\mathcal{U}(n, q^s)$.

Definition 6.6.1 In the design space $\mathcal{U} = \mathcal{U}(n, q^s)$ define

1. *Admissibility:* we say design U_1 is better than U_2 if $\boldsymbol{\delta}(U_1) \prec \boldsymbol{\delta}(U_2)$. A design U is inadmissible if there exists a design U^* such that $\boldsymbol{\delta}(U^*) \prec \boldsymbol{\delta}(U)$. A design which is not inadmissible is called admissible.
2. *Majorant:* If there exists a design U such that

$$\boldsymbol{\delta}(U) \preceq \boldsymbol{\delta}(U^*), \ \text{for any } U^* \in \mathcal{U},$$

we call U a majorum design in the space $\mathcal{U}$.
3. *Schur-convex optimality:* For a predefined Schur-convex kernel function $\Psi(\cdot)$ on R_+^m, a design U is called Schur-Ψ-optimal with respect to $\Psi(\cdot)$

$$\Psi(\boldsymbol{\delta}(U)) \leqslant \Psi(\boldsymbol{\delta}(U^*)), \ \text{for any } U^* \in \mathcal{U}.$$

The three parts in the above definition can be divided hierarchically into two stages of investigation, namely *stringent majorization check* and *flexible Schur-convex comparison*. At the first stage, for competing designs in $\mathcal{U}$, compute their PC vectors with elements sorted in increasing order. Compare the cumulative summations in the sense of majorization ordering (1.4.1). By Definition 6.6.1, any inadmissible design should be prohibited for experimentation; the majorant design(s) if it exists is the winner and absolutely recommended; otherwise, we go to the second stage for comparing admissible designs. The first stage is stringent since majorization requires strong conditions between PC vectors. At the second stage, specify a convex kernel and compute the Schur-Ψ value for each admissible design. Since the above Schur-Ψ criterion is single-valued, all the designs are pairwise comparable and able to be rank-ordered. For different specific purposes, it is very flexible to predefine kernels, as long as they are convex functions. The above two stages are based on the following theory.

Theorem 6.6.1 *For two designs* $\boldsymbol{U}_1, \boldsymbol{U}_2 \in \mathcal{U}(n, q^s)$, $\boldsymbol{\delta}(\boldsymbol{U}_1) \preceq \boldsymbol{\delta}(\boldsymbol{U}_2)$ *if and only if* $\Psi(\boldsymbol{\delta}(\boldsymbol{U}_1)) \leqslant \Psi(\boldsymbol{\delta}(\boldsymbol{U}_2))$ *for every Schur-convex kernel* $\Psi(\cdot)$.

Zhang et al. (2005) illustrated the ideas with an example and some toy convex kernels. Their example demonstrates both stringency and flexibility of majorization framework for assessing designs. Formally, they have the following main theorem to characterize the necessary and sufficient conditions between majorant designs and Schur-Ψ optimum designs.

Zhang et al. (2005) found that many existing criteria, including the discrete discrepancy, WD for the case of $q = 2, 3$ and CD for the case of $q = 2$, can be expressed as a separable convex function, i.e., a Schur-Ψ function with the form (1.4.2). This fact can be extended to many criteria, such as $E(s^2)$, $Ave(\chi^2)$, and Ave(f^2) used in supersaturated design (refer to Chap. 7). Therefore, Zhang et al. (2005) proposed a united approach to find a tight lower bound for a separable convex function.

Consider the PC-mean of any U-type design $\boldsymbol{U} \in \mathcal{U}(n; q^s)$, which is a constant $\bar{\delta} = \frac{s(n-q)}{q(n-1)}$ by Lemma 6.6.1. Now we can apply Lemmas 1.4.1 and 1.4.2 into the vector set of $\boldsymbol{\delta}(\boldsymbol{U})$ on $\mathcal{U}$. For integer-valued $\boldsymbol{\delta}(\boldsymbol{U})$ with length m, let

$$\overline{\boldsymbol{\delta}} \equiv (\underbrace{\bar{\delta}, \ldots, \bar{\delta}}_{m})' \quad \text{and} \quad \widetilde{\boldsymbol{\delta}} \equiv (\underbrace{\theta, \ldots, \theta}_{m(1-f)}, \underbrace{\theta + 1, \ldots, \theta + 1}_{mf})',$$

where θ and f are the integral part and fractional part of $\bar{\delta}$, respectively. It is clear that $\overline{\boldsymbol{\delta}} \preceq \widetilde{\boldsymbol{\delta}} \preceq \boldsymbol{\delta}$, where $\widetilde{\boldsymbol{\delta}}$ reduces to $\overline{\boldsymbol{\delta}}$ when $f = 0$.

Theorem 6.6.2 *A U-type design is majorant if and only if it is Schur-optimum w.r.t every convex kernel. For any well-defined Schur-*ψ *criterion, it has a lower bound* $m(1 - f)\psi(\theta) + mf\psi(\theta + 1)$.

In the proof of the above theorem, Lemma 5.2.1 in Dey and Mukerjee (1999) plays an important role. Zhang et al. (2005) also discussed how to choose suitable kernels for investigating the orthogonality, aberration, and uniformity properties of designs. Some new criteria are also proposed.

6.6.2 *Minimum Aberration Majorization*

In the previous subsection, the majorization framework based on the PC-vector shows its perfect performance for comparing U-type designs. However, any good criterion may meet difficulty for some cases. For example, the Hamming distance of any two rows of a saturated orthogonal design $L_n(q^s)$ is n/q (see Mukerjee and Wu 1995 and Definition 1.3.4). Therefore, PC-vector and the majorization framework cannot distinguish non-isomorphism saturated orthogonal designs, because they have the same PC-vector. Two designs are called *isomorphic* if one can be obtained from the other by relabeling factors, reordering the runs or switching the levels of the factors. A famous example is about $L_{16}(2^{15})$. There are five non-isomorphic $L_{16}(2^{15})$ designs, denoted by HM16.1 to HM16.5 according to the order given by Hall (1961). We list the first two HM16.1 and HM16.2 below, where two levels are 0 and 1. Several authors were interested in how to distinguish them, as well as their subdesigns. Sun and Wu (1993) defined the word-length pattern through the *incomplete* defining contrast subgroup and studied algebraic structure of these five designs. Lin and Draper (1995) considered the pure geometric projection patterns of these five designs up to the lower dimension 5. Fang and Zhang (2004) proposed a so-called *minimum aberration majorization* criterion for this purpose. For each subdesign of $L_{16}(2^p)$, $p \leqslant 15$ of a given $L_{16}(2^{15})$, we can calculate its generalized word-length pattern (GWP) (cf. Sect. 1.4). For a given p, there are $N_p = \binom{s}{p}$ subdesigns. A natural idea is to consider the average projection GWP of all N_p subdesigns. Fang and Zhang (2004) found that the average projection GWPs for non-isomorphic saturated designs are coincident and that the average projection GWP can be expressed in terms of the overall GWP, i.e., all designs in $\boldsymbol{L}_n(q^s)$ (either non-saturated or saturated) have the same average projection GWP, provided that they have the same overall GWP.

$$
HM16.1 = \begin{pmatrix}
1 & 1 & 1 & 1 & 1 & 1 & 1 & 1 & 1 & 1 & 1 & 1 & 1 & 1 & 1 \\
0 & 1 & 0 & 1 & 0 & 1 & 0 & 1 & 0 & 1 & 0 & 1 & 0 & 1 & 0 \\
1 & 0 & 0 & 1 & 1 & 0 & 0 & 1 & 1 & 0 & 0 & 1 & 1 & 0 & 0 \\
0 & 0 & 1 & 1 & 0 & 0 & 1 & 1 & 0 & 0 & 1 & 1 & 0 & 0 & 1 \\
1 & 1 & 1 & 0 & 0 & 0 & 0 & 1 & 1 & 1 & 1 & 0 & 0 & 0 & 0 \\
0 & 1 & 0 & 0 & 1 & 0 & 1 & 1 & 0 & 1 & 0 & 0 & 1 & 0 & 1 \\
1 & 0 & 0 & 0 & 0 & 1 & 1 & 1 & 1 & 0 & 0 & 0 & 0 & 1 & 1 \\
0 & 0 & 1 & 0 & 1 & 1 & 0 & 1 & 0 & 0 & 1 & 0 & 1 & 1 & 0 \\
1 & 1 & 1 & 1 & 1 & 1 & 1 & 0 & 0 & 0 & 0 & 0 & 0 & 0 & 0 \\
0 & 1 & 0 & 1 & 0 & 1 & 0 & 0 & 1 & 0 & 1 & 0 & 1 & 0 & 1 \\
1 & 0 & 0 & 1 & 1 & 0 & 0 & 0 & 0 & 1 & 1 & 0 & 0 & 1 & 1 \\
0 & 0 & 1 & 1 & 0 & 0 & 1 & 0 & 1 & 1 & 0 & 0 & 1 & 1 & 0 \\
1 & 1 & 1 & 0 & 0 & 0 & 0 & 0 & 0 & 0 & 0 & 1 & 1 & 1 & 1 \\
0 & 1 & 0 & 0 & 1 & 0 & 1 & 0 & 1 & 0 & 1 & 1 & 0 & 1 & 0 \\
1 & 0 & 0 & 0 & 0 & 1 & 1 & 0 & 0 & 1 & 1 & 1 & 1 & 0 & 0 \\
0 & 0 & 1 & 0 & 1 & 1 & 0 & 0 & 1 & 1 & 0 & 1 & 0 & 0 & 1
\end{pmatrix}
$$

and

$$
HM16.2 = \begin{pmatrix}
1 & 1 & 1 & 1 & 1 & 1 & 1 & 1 & 1 & 1 & 1 & 1 & 1 & 1 & 1 \\
0 & 1 & 0 & 1 & 0 & 1 & 0 & 1 & 0 & 1 & 0 & 1 & 0 & 1 & 0 \\
1 & 0 & 0 & 1 & 1 & 0 & 0 & 1 & 1 & 0 & 0 & 1 & 1 & 0 & 0 \\
0 & 0 & 1 & 1 & 0 & 0 & 1 & 1 & 0 & 0 & 1 & 1 & 0 & 0 & 1 \\
1 & 1 & 1 & 0 & 0 & 0 & 0 & 1 & 1 & 1 & 1 & 0 & 0 & 0 & 0 \\
0 & 1 & 0 & 0 & 1 & 0 & 1 & 1 & 0 & 1 & 0 & 0 & 1 & 0 & 1 \\
1 & 0 & 0 & 0 & 0 & 1 & 1 & 1 & 1 & 0 & 0 & 0 & 0 & 1 & 1 \\
0 & 0 & 1 & 0 & 1 & 1 & 0 & 1 & 0 & 0 & 1 & 0 & 1 & 1 & 0 \\
1 & 1 & 1 & 1 & 1 & 1 & 1 & 0 & 0 & 0 & 0 & 0 & 0 & 0 & 0 \\
0 & 1 & 0 & 1 & 0 & 0 & 1 & 0 & 1 & 0 & 1 & 0 & 1 & 1 & 0 \\
1 & 0 & 0 & 1 & 1 & 0 & 0 & 0 & 0 & 1 & 1 & 0 & 0 & 1 & 1 \\
0 & 0 & 1 & 1 & 0 & 1 & 0 & 0 & 1 & 1 & 0 & 0 & 1 & 0 & 1 \\
1 & 1 & 1 & 0 & 0 & 0 & 0 & 0 & 0 & 0 & 0 & 1 & 1 & 1 & 1 \\
0 & 1 & 0 & 0 & 1 & 1 & 0 & 0 & 1 & 0 & 1 & 1 & 0 & 0 & 1 \\
1 & 0 & 0 & 0 & 0 & 1 & 1 & 0 & 0 & 1 & 1 & 1 & 1 & 0 & 0 \\
0 & 0 & 1 & 0 & 1 & 0 & 1 & 0 & 1 & 1 & 0 & 1 & 0 & 1 & 0
\end{pmatrix},
$$

Let us take an example of HM16.2 in $\boldsymbol{L}_{16}(2^{15})$ and $p = 5$ for illustrating the distribution of projection GWP. There are $3003 = \binom{15}{5}$ subdesigns with nine different GWPs: 1056 subdesigns with GWP $(0, 0, 1, 0, 0)$, 384 subdesigns with GWP $(0, 0, 0, 1, 0)$, etc. In fact, the average projection GWP of HM16.2 coincides with that of HM16.i, $i = 1, 3, 4, 5$. By writing GWP together with the corresponding frequency, we sort 3003 five-dimension projection GWPs of HM16.2 by the dictionary ordering $\models$ in Sect. 1.4.1 as follows:

$$
\begin{pmatrix}
0 & 0 & 1 & 0 & 0 & 1056 \\
0 & 0 & 0 & 1 & 0 & 384 \\
0 & 0 & \frac{1}{2} & \frac{1}{2} & 0 & 576 \\
0 & 0 & \frac{3}{2} & \frac{1}{2} & 0 & 288 \\
0 & 0 & 0 & 0 & 1 & 72 \\
0 & 0 & \frac{1}{4} & \frac{1}{2} & \frac{1}{4} & 192 \\
0 & 0 & 2 & 1 & 0 & 99 \\
0 & 0 & 1 & 1 & 0 & 144 \\
0 & 0 & \frac{3}{4} & 0 & \frac{1}{4} & 192
\end{pmatrix}
\Rightarrow
\begin{pmatrix}
0 & 0 & 0 & 0 & 1 & 72 \\
0 & 0 & 0 & 1 & 0 & 384 \\
0 & 0 & \frac{1}{4} & \frac{1}{2} & \frac{1}{4} & 192 \\
0 & 0 & \frac{1}{2} & \frac{1}{2} & 0 & 576 \\
0 & 0 & \frac{3}{4} & 0 & \frac{1}{4} & 192 \\
0 & 0 & 1 & 0 & 0 & 1056 \\
0 & 0 & 1 & 1 & 0 & 144 \\
0 & 0 & \frac{3}{2} & \frac{1}{2} & 0 & 288 \\
0 & 0 & 2 & 1 & 0 & 99
\end{pmatrix}
$$

Now we can define the new concept "projection GWP distribution" as follows:

Definition 6.6.2 For a saturated design $L_n(q^s)$, its p-dimension $(1 \leqslant p \leqslant s)$ projection GWP distribution is defined by the distribution of generalized word-length patterns of its $N_p = \binom{s}{p}$ projection designs. Symbolically, we write it as an ordered statistic $\Psi_p = \left\{ W_{[i]} \right\}_{i=1}^{N_p}$, where $W_{[i]} \models W_{[j]}$ whenever $i \leqslant j$.

Let $\mathcal{P}_1$ and $\mathcal{P}_2$ be two non-isomorphic saturated $L_n(q^s)$ designs. Fang and Zhang (2004) showed that

$$\sum_{i=1}^{N_p} W_{[i]}(\mathcal{P}_1) = \sum_{i=1}^{N_p} W_{[i]}(\mathcal{P}_2). \tag{6.6.2}$$

This fact gives us a possibility to apply the majorization theory to compare non-isomorphic saturated designs, and we can similarly define the majorization relation between $\Psi_p(\mathcal{P}_1)$ and $\Psi_p(\mathcal{P}_2)$.

Definition 6.6.3 (*Minimum Aberration Majorization*) For saturated designs $\mathcal{P}_1$ and $\mathcal{P}_2 \in \boldsymbol{L}_n(q^s)$ and p $(1 \leqslant p \leqslant s)$, we say $\mathcal{P}_1$ has less aberration majorization than $\mathcal{P}_2$ and write

$$\Psi_p(\mathcal{P}_1) \overset{\alpha}{\succeq} \Psi_p(\mathcal{P}_2) \quad (\text{or } \Psi_p(\mathcal{P}_1) \overset{\alpha}{\succ} \Psi_p(\mathcal{P}_2) \text{ if } \Psi_p(\mathcal{P}_1) \neq \Psi_p(\mathcal{P}_2)) \tag{6.6.3}$$

if and only if

$$\sum_{i=1}^{t} W_{[i]}(\mathcal{P}_1) \models \sum_{i=1}^{t} W_{[i]}(\mathcal{P}_2), \ t = 1, 2, \ldots, N_p - 1. \tag{6.6.4}$$

A saturated design $\mathcal{P}_*$ is said to be of minimum aberration majorization (MAM) if no designs have less aberration majorization than it. Furthermore, if $\Psi_p(\mathcal{P}_1) \overset{\alpha}{\succeq} \Psi_p(\mathcal{P}_2)$ for all $p = 1, \ldots, s$ with at least one strict majorization relationship $\overset{\alpha}{\succ}$, we say that $\mathcal{P}_1$ dominates $\mathcal{P}_2$ globally under the MAM criterion and write $\mathcal{P}_1 \overset{\alpha}{\gg} \mathcal{P}_2$.

The MAM criterion concerns the capacity of less aberration in all subdesigns. If $\mathcal{P}_1 \overset{\alpha}{\gg} \mathcal{P}_2$, for any m p-dimensional subdesigns, where $1 \leqslant m \leqslant N_p$, $p = 1, \ldots, s$, we can find m subdesigns of $\mathcal{P}_1$ such that the average GWP of these m subdesigns has less generalized aberration than the average GWP of any m subdesigns of $\mathcal{P}_2$. The MAM criterion can be used for detecting non-isomorphism of the designs. For two designs $\mathcal{P}_1$ and $\mathcal{P}_2$ in $\boldsymbol{L}_n(q^s)$, they have the same Ψ_p for all $p = 1, \ldots, s$ if they are isomorphic. If there is p such that $\Psi_p(\mathcal{P}_1) \overset{\alpha}{\succ} \Psi_p(\mathcal{P}_2)$, it indicates that they are non-isomorphic. By applying the MAM criterion to the five $L_{16}(2^{15})$ non-isomorphic designs, Fang and Zhang (2004) found

$$\text{HM16.1} \overset{\alpha}{\gg} \text{HM16.2} \overset{\alpha}{\gg} \text{HM16.3} \overset{\alpha}{\gg} \begin{cases} \text{HM16.4} \\ \text{HM16.5} \end{cases}.$$

and found that HM16.4 and HM16.5 have the same projection GWP distributions at low dimensions $p = 1, \ldots, 5$ and high dimensions $p = 8, \ldots, 15$, while different at $p = 6, 7, 8$ and 9. There is no clear majorization relationship between HM16.4 and HM16.5.

Exercises

6.1

Prove that the following two designs, D_1 and D_2, are isomorphic

$$D_1 = \begin{bmatrix} 1 & 1 & 1 \\ 1 & -1 & 1 \\ 1 & 1 & -1 \\ -1 & -1 & -1 \\ -1 & 1 & -1 \\ -1 & -1 & 1 \end{bmatrix} \quad D_2 = \begin{bmatrix} -1 & -1 & -1 \\ 1 & -1 & -1 \\ -1 & 1 & -1 \\ 1 & 1 & 1 \\ -1 & 1 & 1 \\ 1 & -1 & 1 \end{bmatrix}.$$

Indicate that these designs belong to:
(i) U-type design; (ii) orthogonal design; (iii) fractional factorial design.

6.2

Denote by $\mathcal{U}_9(3^4)$ all the possible orthogonal designs $L_9(3^4)$ with levels 1, 2 and 3. Answer the following questions:

(1) Show that all these designs in $\mathcal{U}_9(3^4)$ form only one isomorphic group.

(2) Calculate WD, CD and MD for all designs in $\mathcal{U}_9(3^4)$. Give your conclusion.

(3) Table 6.4 list two designs $L_9(3^4)$, where $\boldsymbol{U}_2$ was obtained by minimizing CD on $\mathcal{U}_9(3^4)$. The $\boldsymbol{U}_2$ is called uniformly orthogonal design under CD. Find uniformly orthogonal design under MD/WD.

(4) Calculate the discrete discrepancy for all designs in $\mathcal{U}_9(3^4)$. Give your conclusion.

(5) Give a discussion on two concepts: the uniformly orthogonal design and uniform minimum aberration design.

(6) Calculate the projection discrepancy pattern defined in Definition 6.5.1.

6.3

Calculate the uniform pattern (refer to Definition 6.5.2) for all subdesigns $L_8(2^5)$ of $L_8(2^7)$ in Table 1.3.2.

6.4

Table 6.5 gives a uniform minimum aberration design $UL_{27}(3^{13})$ under MD. This design involves several uniform minimum aberration subdesigns $UL_{27}(3^s)$ for $s < 13$ under MD. Give two such subdesigns for $s = 6$ and $s = 10$ with detailed calculation, respectively.

6.5

There are five non-isomorphic $L_{16}(2^{15})$ designs. This chapter lists two of them. Give other three from the literature.

6.6

There are two non-isomorphic $L_{27}(3^{13})$ designs (refer to Fang and Zhang 2004). Apply Algorithm 6.1.2 to detect their non-isomorphism.

6.7

Table 6.2 list four orthogonal designs $L_{18}(3^7)$. Ma and Fang (2001) pointed out that there are at least three non-isomorphic $L_{18}(3^7)$ designs. Furthermore, Evangelaras et al. (2007) confirmed that there exist exactly three non-isomorphic three-level orthogonal arrays with 18 runs and 7 columns. They obtained several minimum aberration subdesigns $L_{18}(3^s)$, $s \leqslant 7$. List these minimum aberration subdesigns as many as possible.

References

Box, G.E.P., Draper, N.R.: Empirical Model-Building and Respose Surfaces. Wiley, New York (1987)

Box, G.E.P., Hunter, W.G., Hunter, J.S.: Statistics for Experimenters, An Introduction to Design, Data Analysis, and Model Building. Wiley, New York (1978)

Clark, J.B., Dean, A.M.: Equivalence of fractional factorial designs. Stat. Sin. **11**, 537–547 (2001)

Chen, J., Lin, D.K.J.: On the identity relationship of 2^{k-p} designs. J. Stat. Plan. Inference **28**, 95–98 (1991)

Chen, J., Sun, D.X., Wu, C.F.J.: A catalogue of two-level and three-level fractional factorial designs with small runs. Int. Stat. Rev. **61**, 131–145 (1993)

Cheng, C.S., Deng, L.W., Tang, B.: Generalized minimum aberration and design efficiency for non-regular fractional factorial designs. Stat. Sin. **12**, 991–1000 (2002)

Cheng, S.W., Ye, K.Q.: Geometric isomorphism and minimum aberration for factorial designs with quantitative factors. Ann. Stat. **32**, 2168–2185 (2004)

Dey, A., Mukerjee, R.: Fractional Factorial Plans. Wiley, New York (1999)

Draper, N.R., Mitchell, T.J.: Construction of the set of 256-run designs of resolution $\geqslant 5$ and set of even 512-run designs of resolution $\geqslant 6$ with special reference to the unique saturated designs. Ann. Math. Stat. **39**, 246–255 (1968)

Draper, N.R., Mitchell, T.J.: Construction of a set of 512-run designs of resolution $\geqslant 5$ and a set of even 1024-run designs of resolution $\geqslant 6$. Ann. Math. Stat. **41**, 876–887 (1970)

Evangelaras, H., Koukouvinos, C., Lappas, E.: 18-run nonisomorphic three level orthogonal arrays. Metrika **66**, 31–37 (2007)

Fang, K.T., Ma, C.X.: The usefulness of uniformity in experimental design. In: Kollo, T., Tiit, E.-M., Srivastava, M. (eds.) New Trends in Probability and Statistics, pp. 51–59. De Gruyter, The Netherlands (2000)

Fang, K.T., Ma, C.X.: Relationship between uniformity, aberration and correlation in regular fractions 3^{s-1}. In: Fang, K.T., Hickernell, F.J., Niederreiter, H. (eds.) Monte Carlo and Quasi-Monte Carlo Methods 2000, pp. 213–231. Springer, Berlin (2002)

Fang, K.T., Mukerjee, R.: A connection between uniformity and aberration in regular fractions of two-level factorials. Biometrika **87**, 1993–198 (2000)

Fang, K.T., Qin, H.: Uniformity pattern and related criteria for two-level factorials. Sci. China Ser. A. **47**, 1–12 (2004)

Fang, K.T., Winker, P.: Uniformity and orthogonality. Technical Report MATH-175, Hong Kong Baptist University (1998)

Fang, K.T., Zhang, A.: Minimum aberration majorization in non-isomorphic saturated designs. J. Stat. Plan. Inference **126**, 337–346 (2004)
Fang, K.T., Lin, D.K.J., Winker, P., Zhang, Y.: Uniform design: theory and applications. Technometrics **42**, 237–248 (2000)
Fang, K.T., Ma, C.X., Mukerjee, R.: Uniformity in fractional factorials. In: Fang, K.T., Hickernell, F.J., Niederreiter, H. (eds.) Monte Carlo and Quasi-Monte Carlo Methods 2000, pp. 232–241. Springer, Berlin (2002)
Fang, K.T., Lin, D.K.J., Liu, M.Q.: Optimal mixed-level supersaturated design. Metrika **58**, 279–291 (2003)
Fang, K.T., Lu, X., Winker, P.: Lower bounds for centered and wrap-around L_2-discrepancies and construction of uniform. J. Complex. **20**, 268–272 (2003)
Fang, K.T., Ge, G.N.: An efficient algorithm for the classification of hadamard matrices. Math. Comput. **73**, 843–851 (2004)
Fang, K.T., Ke, X. Elsawah, A.M.: Construction of a new 27-run uniform orthogonal design. J. Complex. (2016). (submitted)
Fries, A., Hunter, W.G.: Minimum aberration 2^{k-p} designs. Technometrics **22**, 601–608 (1980)
Hall, J.M.: Hadamard matrix of order 16. Jet Propulsion Laboratory Res Summery **1**, 21–26 (1961)
Hickernell, F.J., Liu, M.Q.: Uniform designs limit aliasing. Biometrika **89**, 893–904 (2002)
Liu, M.Q.: Using discrepancy to evaluate fractional factorial designs. In: Fang, K.T., Hickernell, F.J., Niederreiter, H. (eds.) Monte Carlo and Quasi-Monte Carlo Methods 2000, pp. 357–368. Springer, Berlin (2002)
Lin, D.K.J., Draper, N.R.: Projection properties of plackett and burman designs. Technometrics **34**, 423–428 (1992)
Lin, D., Draper, N.R.: Screening properties of certain two-level designs. Metrika **42**, 99–118 (1995)
Liu, M.Q., Fang, K.T., Hickernell, F.J.: Connections among different criteria for asymmetrical fractional factorial designs. Stat. Sin. **16**, 1285–1297 (2006)
Li, P.F., Liu, M.Q., Zhang, R.C.: Some theory and the construction of mixed-level supersaturated designs. Stat. Probab. Lett. **69**, 105–116 (2004)
Ma, C.X., Fang, K.T.: A note on generalized aberration in factorial designs. Metrika **53**, 85–93 (2001)
Ma, C.X., Fang, K.T., Lin, D.K.J.: On isomorphism of factorial designs. J. Complex. **17**, 86–97 (2001)
Ma, C.X., Fang, K.T., Lin, D.K.J.: A note on uniformity and orthogonality. J. Stat. Plan. Inference **113**, 323–334 (2003)
Mukerjee, R., Wu, C.F.J.: On the existence of saturated and nearly saturated asymmetrical orthogonal arrays. Ann. Stat. **23**, 2102–2115 (1995)
Qin, H., Fang, K.T.: Discrete discrepancy in factorial designs. Metrika **60**, 59–72H (2004)
Qin, H., Zou, N., Zhang, S.L.: Design efficiency for minimum projection uniform designs with two-level. J. Syst. Sci. Complex. **24**, 761–768 (2011)
Qin, H., Wang, Z.H., Chatterjee, K.: Uniformity pattern and related criteria for q-level factorials. J. Stat. Plan. Inference **142**, 1170–1177 (2012)
Roman, S.: Coding and Information Theory. Wiley, New York (1992)
Sun, F.S., Liu, M.Q., Hao, W.R.: An algorithmic approach to finding factorial designs with generalized minimum aberration. J. Complex. **25**, 75–84 (2009)
Sun, F.S., Chen, J., Liu, M.Q.: Connections between uniformity and aberration in general multi-level factorials. Metrika **73**, 305–315 (2011)
Sun, D.X., Wu, C.F.J.: Statistical properties of hadamard matrices of order 16. In: Quality Through Engineering Design (1993)
Tang, B.: Theory of J-characteristics for fractional factorial designs and projection justification of minimum G_2-aberration. Biometrika **88**, 401–407 (2001)
Tang, Y.: Combinatorial properties of uniform designs and their applications in the constructions of low-discrepancy designs. Ph.D. thesis, Hong Kong Baptist University (2005)

Tang, B., Deng, L.Y.: Minimum G_2-aberration for nonregular fractional designs. Ann. Stat. **27**, 1914–1926 (1999)

Tang, Y., Xu, H.: An effective construction method for multilevel uniform designs. J. Stat. Plan. Inference **143**, 1583–1589 (2013)

Tang, Y., Xu, H., Lin, D.K.J.: Uniform fractional factorial designs. Ann. Stat. **40**, 891–907 (2012)

Xu, H.: Minimum moment aberration for nonregular designs and supersaturated designs. Stat. Sin. **13**, 691–708 (2003)

Xu, H.Q., Wu, C.F.J.: Generalized minimum aberration for asymmetrical fractional factorial designs. Ann. Stat. **29**, 1066–1077 (2001)

Ye, K.Q.: Indicator function and its application in two-level factorial design. Ann. Stat. **31**, 984–994 (2003)

Zhang, A., Fang, K.T., Li, R., Sudjianto, A.: Majorization framework for balanced lattice designs. Ann. Stat. **33**, 2837–2853 (2005)

Zhang, S.L., Qin, H.: Minimum projection uniformity criterion and its application. Stat. Probab. Lett. **76**, 634–640 (2006)

Zhou, Y.D., Xu, H.: Space-filling fractional factorial designs. J. Am. Stat. Assoc. **109**, 1134–1144 (2014)

Zhou, Y.D., Ning, J.H., Song, X.B.: Lee discrepancy and its applications in experimental designs. Stat. Probab. Lett. **78**, 1933–1942 (2008)

Zhou, Y.D., Fang, K.T., Ning, J.H.: Mixture discrepancy for quasi-random point sets. J. Complex. **29**, 283–301 (2013)

Chapter 7
Applications of Uniformity in Other Design Types

From the previous chapter, we see that there exist close relationships between uniformity and several other design criteria. In this chapter, we will show that the uniformity is also a useful criterion in some classical designs, such as block design, supersaturated design, Latin hypercube design and so on.

7.1 Uniformity in Block Designs

Block design is an important kind of experimental design. Its basic ideas come from agricultural and biological experiments. But now the applications of these ideas are found in many areas of sciences and engineering. The most widely used one is the *balanced incomplete block design* (BIBD, see Definition 3.6.2). Another important one is the resolvable incomplete block design (RIBD, for short). See Sect. 3.6.1 for some basic knowledge of block designs. For a thorough discussion of block designs, refer to Dey (1986).

As we know, the definitions in block designs reflect some "*balance*" among the treatments, the block, or the parallel class. This kind of balance is easy to be accepted intuitively. While in existing works on block designs, the criterion of balance was introduced from the estimation point of view. In fact, the balance criterion can be regarded as a kind of *uniformity*. Liu and Chan (2004) and Liu and Fang (2005) studied the uniformity of block designs and obtained some satisfactory results.

7.1.1 Uniformity in BIBDs

A BIBD can be characterized by the balanced arrangement of its design points. Liu and Chan (2004) investigated incomplete block designs from the perspective

K.-T. Fang et al., *Theory and Application of Uniform Experimental Designs*, Lecture Notes in Statistics 221,
https://doi.org/10.1007/978-981-13-2041-5_7

of uniformity. They used the discrete discrepancy defined in Sect. 2.5.1 to prove theoretically that BIBDs are the most uniform ones among all *equireplicate*, *proper*, and *binary*.

Suppose n treatments are arranged in s blocks, such that the jth block contains t_j experimental units and the ith treatment appears r_i times in the entire design, $i = 1, \ldots, n$; $j = 1, \ldots, s$. Let $\mathbf{Z} = (z_i^j)_{n \times s}$ be the *incidence matrix* of the design, where z_i^j is the number of times that the ith treatment appears in the jth block. Following Definition 3.6.1 for block design, we denote an equireplicate (i.e., $r_i = r$), proper (i.e., $t_j = t$), and binary (i.e., $z_i^j = 1$ or 0) incomplete block design by IB(n, s, r, t). Regard the s blocks as s-factors each having two levels, 0 and 1, and regard the allocation of each treatment to these s blocks as a point with elements 0 and 1, where 1 means that this treatment appears in the corresponding block and 0 means it does not. Then, the n points of an IB(n, s, r, t) just correspond to the n rows of the incidence matrix $\mathbf{Z}$. For the kernel defined by (2.5.3) and (2.5.4), Liu and Chan (2004) showed that the squared discrete discrepancy of $\mathbf{Z}$ is

$$\mathrm{DD}^2(\mathbf{Z}) = -\left(\frac{a+b}{2}\right)^s + \frac{a^s}{n} + \frac{a^s}{n^2}\left(\frac{b}{a}\right)^{2r} \sum_{i,j=1, j\neq i}^{n} \left(\frac{a}{b}\right)^{2n_{ij}^{11}}, \qquad (7.1.1)$$

where n_{ij}^{11} is the number of blocks in which the pair of treatments i and j appear together. Based on expression (7.1.1), Liu and Chan (2004) further obtained the following theorem.

Theorem 7.1.1 *Let* $\mathbf{Z}$ *be the incidence matrix of an IB*(n, s, r, t). *Then,*

$$\mathrm{DD}^2(\mathbf{Z}) \geqslant -\left(\frac{a+b}{2}\right)^s + \frac{a^s}{n} + \frac{n-1}{n} a^s \left(\frac{b}{a}\right)^{2r-2\lambda}, \qquad (7.1.2)$$

where $\lambda = r(t-1)/(n-1)$, *and the lower bound of* $DD^2(\mathbf{Z})$ *on the right-hand side of* (7.1.2) *can be achieved if and only if* λ *is a positive integer and every pair of treatments appears in altogether* λ *blocks, i.e., the design is a BIBD*(n, s, r, t, λ).

Theorem 7.1.1 shows that for a BIBD(n, s, r, t, λ), the lower bound of $\mathrm{DD}^2(\mathbf{Z})$ is attained, and thus BIBD(n, s, r, t, λ)'s are the most uniform ones among all IB(n, s, r, t)'s. This is an important characteristic of BIB designs in terms of uniformity.

7.1.2 Uniformity in PRIBDs

Furthermore, Liu and Fang (2005) considered equireplicate and binary incomplete block designs and did not care whether they are proper or not. If a block design is *resolvable* (see Definition 3.6.3), it is obvious that the design is also equireplicate and binary. Let $\boldsymbol{t} = (t_1, t_2, \ldots, t_s)'$, and RIB$(n, s, r, \boldsymbol{t}, \mathbf{Z})$ denote a resolvable incomplete block design (RIBD, for simplicity). In particular, for an RIBD$(n, s, r, \boldsymbol{t}, \boldsymbol{Z})$, if each parallel class is proper, i.e., in the ith ($1 \leqslant i \leqslant r$) parallel class, there are q_i blocks each of size n/q_i, then the design is called a PRIBD, denoted by PRIBD$(n, s, r, \boldsymbol{t}, \boldsymbol{Z})$. Here, $q_1, \ldots, q_r$ are positive divisors of n. Liu and Fang (2005) obtained a sufficient and necessary condition under which a PRIBD is the most uniform one in the sense of the discrete discrepancy in (2.5.8).

Theorem 7.1.2 *For a PRIBD$(n, s, r, \boldsymbol{t}, \boldsymbol{Z})$, we have*

$$\mathrm{DD}^2(\boldsymbol{Z}) \geqslant -\left(\frac{a+b}{2}\right)^s + \frac{a^s}{n} + \frac{n-1}{n}a^s\left(\frac{b}{a}\right)^{2r-2\lambda}, \tag{7.1.3}$$

where $\lambda = (\sum_{k=1}^{r} n/q_k - r)/(n-1)$, and the lower bound of $\mathrm{DD}^2(\boldsymbol{Z})$ *on the right-hand side of* (7.1.3) *can be achieved if and only if λ is a positive integer and every pair of treatments appears altogether in λ blocks.*

Based on this theorem, we call a PRIBD$(n, s, r, \boldsymbol{t}, \boldsymbol{Z})$ a *uniform PRIBD* if its $\mathrm{DD}^2(\boldsymbol{Z})$ achieves the lower bound in (7.1.3). In such a uniform PRIBD, every pair of treatments appears altogether in the same number of blocks, which means that there exists some "balance" among the treatments. In fact, if a uniform PRIBD is also proper, i.e., $q_i = q$ (say), for all $1 \leqslant i \leqslant r$, then we can easily have the following conclusion.

Corollary 7.1.1 *If a uniform PRIBD$(n, s, r, \boldsymbol{t}, \lambda)$ is proper, then it is a resolvable BIBD$(n, rq, r, n/q, \lambda)$, where $t_i = n/q$ for all $1 \leqslant i \leqslant s$, and $\lambda = r(n/q - 1)/(n - 1)$.*

From this corollary, the criterion "*balance*" can be regarded as a kind of *uniformity*.

Liu and Fang (2005) further showed that for a uniform PRIBD, all elementary treatment contrasts are estimable, so are those among block effects. They also proposed a construction approach for uniform PRIBDs via a kind of U-type designs. This approach sets up a strong link between U-type designs and PRIBDs.

All these results confirm our judgment that the "*balance*" criterion can be regarded as a kind of *uniformity*. Note that these results are obtained in the sense of the discrete discrepancy in (2.5.8), but they also hold for any of the reflection-invariant L_2-discrepancies proposed by Hickernell (1998a, b), any of the four discrepancies due to Hickernell (1998a, b), i.e., the centered L_2-discrepancy, wrap-around L_2-discrepancy, symmetric L_2-discrepancy, and unanchored L_2-discrepancy; see Sect. 2.3 for the definitions and expressions of the former two.

7.1.3 Uniformity in POTBs

Main effects plans (MEPs) occupy an important position in many industrial experiments when interest lies only in the main effects, assuming that all interactions

Table 7.1 Two $3^3 2^3$ POTBs

A dummy block factor	D_1			D_2		
	F_1	F_2	F_3	F_1	F_2	F_3
Batch 1	0	0	0	0	0	1
	0	1	1	0	1	0
	1	0	1	2	0	0
	1	1	0	2	1	1
Batch 2	1	1	1	2	1	0
	1	2	2	2	2	2
	2	1	2	1	1	2
	2	2	1	1	2	0
Batch 3	2	2	2	1	2	2
	2	0	0	1	0	1
	0	2	0	0	2	1
	0	0	2	0	0	2

between factors are negligible. An orthogonal MEP permits the estimation of all main effects of a factorial arrangement without correlation. The only problem with an orthogonal MEP is that the plan often requires a large number of runs. Thus, many alternative approaches can be found in the literature such as nearly orthogonal arrays (e.g., Wang and Wu 1992; Ma et al. 2000), MEPs with blocks (e.g., Mukerjee et al. 2002), MEPs in which the treatment factors are pairwise orthogonal through the block factor (Bose and Bagchi 2007; Bagchi 2010). Among them, Bagchi (2010) obtained saturated plans orthogonal through the block factor (POTBs) for an $s^{3m}2^{3m}$ experiment. Recently, Chen et al. (2015) presented direct as well as recursive constructions for asymmetrical saturated POTBs.

A $3^3 2^3$ POTB with three-level columns, denoted by D_1, is listed in Table 7.1. Design D_2 in Table 7.1 is obtained by level permutations of factor F_1 (map $(0, 1, 2)$ to $(0, 2, 1)$) and factor F_3 (map $(0, 1, 2)$ to $(1, 0, 2)$). Note that the run size of a POTB is much smaller than the one of an orthogonal MEP. In such plans, the s-level factors are non-orthogonal to the block factor but are pairwise orthogonal through the block factor.

Consider a plan D for an experiment with factors $F_1, F_2, \ldots$ (possibly including a block factor L) at s-level on N runs. Let an $N \times s$ matrix $\mathbf{X}_{F_1}$ denote the incidence matrix of factor F_1, in which the (u, i)th entry is 1 if in the uth run factor F_1 is set at level i and 0 otherwise. For two factors F_1 and F_2, the incidence matrix is denoted by $\mathbf{M}_{F_1F_2}$, where the (i, j)th entry is the number of runs in which F_1 is set at level i and F_2 is set at level j. Factors F_1 and F_2 are said to be *orthogonal through the block factor* if the following condition satisfies,

$$\mathbf{M}_{F_1L}\mathbf{M}_{LF_2} = k\mathbf{M}_{F_1F_2}, \tag{7.1.4}$$

where k is the block size. Hence, design D is said to be a *plan orthogonal through the block factor* (POTB) if every pair of factors of D is orthogonal through the block factor.

Chen et al. (2015) investigated the relationship between level permutation and POTB and obtained the following theorem.

Theorem 7.1.3 *If D_1 is a POTB and D_2 is obtained by any level permutations of factors of D_1, then D_2 is also a POTB.*

Chen et al. (2015) further used a modified optimization algorithm below to search uniform or nearly uniform $s^{3m}2^{3m}$ POTBs under level permutation, where CD is used as the uniformity measure. Noted that searching for all $(s!)^m$ level permuted designs (denote the set of all these designs as $\mathcal{P}(D)$) becomes an NP-hard problem as s or m increases. In Chen et al. (2015), the implementation (with the number of iterations being $\tau = 10{,}000$) was done in R program on a personal computer with Intel Core(TM) i5-3230M Duo CPU 2.6GHz and 4 GB memory. And the most complex case of constructing uniform POTBs took only a few minutes to complete.

Algorithm. Pseudo code for prototype local search heuristic

1. Initialize τ (number of iterations)
2. Initialize δ (impairment threshold) and $c_0 \in (0.1, 0.25)$
3. Generate an initial $s^{3m}2^{3m}$ POTB D by Algorithm 1 in (Chen et al., 2015) and let $D^{min} = D^c = D$
4. **for** $i = 1$ to τ **do**
5. Generate a new design $D^{new} \in \mathcal{P}(D)$ based on D^c
6. Compute $\nabla = CD^2(D^{new}) - CD^2(D^c)$ and generate c ($c \sim U(0, 1)$)
7. **if** ($\nabla < 0$) or ($\nabla < \delta$ and $c < c_0$) **then** let $D^c = D^{new}$
8. **if** $CD^2(D^c) < CD^2(D^{min})$ **then** let $D^{min} = D^c$
9. **end for**

Remark 7.1 In the above algorithm, there are two major differences between the algorithm in Fang et al. (2006) and the current one. The first difference is the input starting design D^c. Instead of randomly generating a U-type design as the starting design, the current algorithm chooses a POTB first. The second difference is the definition of neighbor $\mathcal{P}(D)$ to current solution. A normal TA algorithm exchanges two distinct elements within the same columns, while Chen et al. (2015) exchanged all elements of two distinct levels within the same column as done in Tang et al. (2012).

7.2 Uniformity in Supersaturated Designs

The supersaturated design (SSD, for short) is a fractional factorial design whose run size is insufficient for estimating all the main effects represented by the columns of the design matrix. In industrial and scientific experiments, especially in their preliminary stages, very often there are a large number of factors to be studied and the run size is limited because of expensive costs. However, in many situations only a few factors are believed to have significant effects. Under this assumption of *effect sparsity*, SSDs can be effectively used to identify the dominant factors.

7.2.1 Uniformity in Two-Level SSDs

Most studies on SSDs have focused on the two-level case. Booth and Cox (1962), in the first systematic construction of SSDs, proposed the $E(s^2)$ criterion, which is a measure of *non-orthogonality* under the assumption that *only two out of the* m factors are active. After Booth and Cox (1962), there was not much work on the subject of SSDs until Lin (1993). Other early work focusing on constructions of $E(s^2)$-optimal SSDs includes, e.g., Nguyen (1996); Cheng et al. (1997); Tang and Wu (1997); Liu and Zhang (2000); Butler et al. (2001); Liu and Dean (2004); Bulutoglu and Cheng (2004), and more references on $E(s^2)$-optimal SSDs can be found in the review paper Georgiou (2014).

Let $\boldsymbol{U} \in \mathcal{U}(n; 2^m)$; when $n < m+1$, the design is a two-level SSD. The commonly used $E(s^2)$ criterion for comparing SSD is

$$E(s^2) = \frac{2}{m(m-1)} \sum_{1 \leqslant i < j \leqslant m} s_{ij}^2,$$

where s_{ij} is the (i, j)th entry of $\boldsymbol{X}^T\boldsymbol{X}$. Liu and Hickernell (2002) studied the connection between $E(s^2)$-optimality and minimum discrepancy in two-level SSDs. They defined the discrete discrepancy by taking $a = 1 + \beta$ and $b = 1 + \beta\rho$ ($\beta > 0, -1 \leqslant \rho < 1$) in (2.5.4) for $q = 2$ and showed that for two-level factorial designs both the $E(s^2)$ and the discrete discrepancy can be expressed in terms of the *Hamming distances* (or the *coincidence numbers*) between any two runs of the design. These expressions in terms of Hamming distances lead to lower bounds for $E(s^2)$ and the discrete discrepancy. It is interesting to note that if a design $\boldsymbol{U}$ can attain one of these lower bounds, then it attains both of them. In other words, an $E(s^2)$-optimal design is also uniform (minimal discrepancy) in terms of this discrete discrepancy. They further showed that in what cases these lower bounds can be achieved, even though the discrete discrepancy is *not* equivalent to the $E(s^2)$ criterion.

Theorem 7.2.1 *Let* $\boldsymbol{U}$ *be a two-level design with* n *runs and* m *factors, where each column has the same number of* ± 1 *elements. Suppose that* $\rho\beta > -1$ *and that* $m = c(n-1)+e$ *for* $e = -1, 0$ *or* 1. *Also, suppose that either (a)* n *is a multiple of* 4 *and there exists an* $n \times n$ *Hadamard matrix, or (b)* c *is even and there exists a* $2n \times 2n$ *Hadamard matrix. Then, the lower bounds of* $E(s^2)$ *and the discrete discrepancy can be attained.*

Moreover, the discrete discrepancy is a more general and thus more flexible criterion than $E(s^2)$. For example, $E(s^2)$ ignores possible interactions of more than one factor. However, the discrete discrepancy includes interactions of all possible orders, and their importance may be increased or decreased by changing the value of β in the definition of the discrete discrepancy.

7.2.2 Uniformity in Mixed-Level SSDs

Two-level SSDs can be used for screening the factors in simple linear models. When the relationship between a set of factors and a response is nonlinear or approximated by a polynomial response surface model, designs with multi-level and mixed-level are often required, e.g., to exploring nonlinear effects of the factors. In the past two decades, the study of SSDs with multi-level and mixed-level has also raised great attention; see, for example, Sun et al. (2011); Liu and Liu (2011, 2012, 2013); Chen et al. (2013); Georgiou (2014) and the references therein. It is clear that all SSDs cannot be completely orthogonal among columns of the design. The block orthogonality (meaning that columns of the design are grouped as blocks and columns in each block are orthogonal) has been considered by many authors. Fang et al. (2000) proposed a way that collapses a uniform design to an orthogonal array for constructing multi-level SSDs.

There are many criteria, such as aveχ^2 (Yamada and Lin 1999), $\chi^2(D)$ (Yamada and Matsui 2002), $E(f_{NOD})$ (Fang et al. 2003), minimum moment aberration (Xu 2003), and minimum χ^2 (Liu et al. 2006) for comparing multi-level and mixed-level SSDs in the literature. In particular, for a design $\boldsymbol{U} \in \mathcal{U}(n; q_1 \times \cdots \times q_m)$, the $E(f_{NOD})$ criterion is defined as minimizing

$$E(f_{NOD}) = \sum_{1 \leqslant i < j \leqslant m} f_{NOD}^{ij} \Big/ \binom{m}{2}, \tag{7.2.1}$$

where

$$f_{NOD}^{ij} = \sum_{u=1}^{q_i} \sum_{v=1}^{q_j} \left(n_{uv}^{(ij)} - \frac{n}{q_i q_j} \right)^2,$$

$n_{uv}^{(ij)}$ is the number of (u, v)-pairs in the ith and j columns, and $n/(q_i q_j)$ stands for the average frequency of level-combinations in this pair of columns. Here, the subscript NOD stands for *non-orthogonality of the design*. It is obvious that $E(f_{NOD}) = 0$ for an orthogonal array. Fang et al. (2003) proved that the $E(s^2)$ and ave χ^2 criteria are in fact special cases of the $E(f_{NOD})$ criterion, and obtained a lower bound for $E(f_{NOD})$ which can serve as a benchmark of design optimality. They also studied the connection between the discrete discrepancy (2.5.6) and $E(f_{NOD})$. Fang et al. (2004a) provided the following lower bound and the sufficient and necessary condition to achieve it for $E(f_{NOD})$, which includes the bound and condition of Fang et al. (2003) as a special case.

Theorem 7.2.2 *Let $\boldsymbol{U} \in \mathcal{U}(n; q_1 \times \cdots \times q_m)$, then*

$$E(f_{NOD}) \geqslant \frac{n(n-1)}{m(m-1)}\left[(\gamma+1-\lambda)(\lambda-\gamma)+\lambda^2\right]+C(n, q_1, \ldots, q_m), \quad (7.2.2)$$

where $C(n, q_1, \ldots, q_m) = \frac{nm}{m-1} - \frac{1}{m(m-1)}\left(\sum_{i=1}^{m}\frac{n^2}{q_i} + \sum_{1\leqslant i\neq j\leqslant m}\frac{n^2}{q_iq_j}\right)$, λ, γ and the sufficient and necessary conditions for the lower bound to be achieved are the same as those of Theorem 2.6.20.

Based on these two papers' results, we have

Theorem 7.2.3 *Let $\boldsymbol{U}$ be a U-type design, and if the difference among all the Hamming distances between any two different rows of $\boldsymbol{U}$ does not exceed one, then $\boldsymbol{U}$ is both a uniform design in terms of the discrete discrepancy* (2.5.6) *and an $E(f_{NOD})$-optimal design.*

This theorem generalizes the result obtained by Liu and Hickernell (2002) for two-level case. And it leads to a strong relation between $E(f_{NOD})$ optimality and uniformity measured by the discrete discrepancy (2.5.6) of any SSD. The uniformity of $E(s^2)$- and aveχ^2-optimal SSDs can be obtained directly based on this theorem, as special cases of SSDs with equal-level factors.

Fang et al. (2003) and Fang et al. (2004a) also proposed ways for constructing $E(f_{NOD})$-optimal as well as uniform SSDs with mixed levels. And more uniform SSDs under the discrete discrepancy (2.5.6) have been obtained by the combinatorial approaches. See Sect. 3.6 for the details of these approaches. All these studies show that the uniformity plays an important role in evaluating and constructing SSDs.

7.3 Uniformity in Sliced Latin Hypercube Designs

Recently, computer experiments with both quantitative and qualitative factors have raised increasing interests. To suit such a computer experiment, Qian et al. (2012) first proposed a general method for constructing sliced Latin hypercube designs (SLHDs). An $m \times q$ matrix is called a Latin hypercube design (LHD), denoted by $L(m, q)$, if each of its q columns includes m equally spaced levels, say $1, \ldots, m$. An $n \times q$ matrix S is called an SLHD with s slices, denoted by $SL(n, q, s)$, if S is an $L(n, q)$ and can be partitioned into s slices each of which is an $L(m, q)$ with $m = n/s$ after the n levels are collapsed to m equally spaced levels according to $\lceil i/s \rceil$ for level i, where $\lceil a \rceil$ means the smallest integer greater than or equal to a. SLHDs inherit the good property of LHDs, i.e., they possess maximum stratification in any one dimension as well as their slices. Further studies on SLHDs include some constructions ensuring good projection in more than one dimension or orthogonality between columns, i.e.,

Yang et al. (2013), Huang et al. (2014), Cao and Liu (2015), Yang et al. (2016) and Wang et al. (2017) proposed methods to construct orthogonal and nearly orthogonal SLHDs, Yin et al. (2014) constructed SLHDs with an attractive low-dimensional stratification via orthogonal arrays (OAs), and Yang et al. (2014) obtained SLHDs based on resolvable orthogonal arrays (ROAs) and then highly improved the uniformity of the resulting designs with respect to the centered L_2-discrepancy criterion.

There is a disadvantage of SLHDs that remains to be addressed: SLHDs usually do not possess a good uniformity over the experimental region. Such a disadvantage goes against the space-filling principle and is undesirable for computer experiments. Although Yang et al. (2014) considered such a problem, the existence of their designs depends heavily on the existence of the resolvable orthogonal arrays, whose numbers of runs and factors are constrained.

Note that in applications of computer experiments with both qualitative and quantitative factors, each slice of an SLHD corresponds to one level-combination of the qualitative factors. Thus, the design points under each level-combination of the qualitative factors should spread evenly over the experimental region when the response surfaces at different level-combinations of the qualitative factors are similar (see Huang et al. 2016). However, there is no one-to-one correspondence between the uniformity of a whole SLHD and that of its slices. In order to avoid the possible inconsistency between the uniformity of the whole design and that of its slices, recently Chen et al. (2016) proposed a new optimization criterion by combining the two uniformity measures of the whole design and its slices. The design obtained under such a combined measure, called a uniform sliced Latin hypercube design (USLHD), not only has good uniformity in terms of the whole design but also spreads the points of each slice evenly over the experimental region.

7.3.1 A Combined Uniformity Measure

Chen et al. (2016) used the centered L_2-discrepancy (CD) for evaluating the uniformity of SLHDs. For an SLHD $D = (D_{(1)}^T, \ldots, D_{(s)}^T)^T$, let $\mathrm{CD}(D)$ and $\mathrm{CD}(D_{(i)})$ be the CD-values of D and $D_{(i)}$, respectively, where $D_{(1)}, \ldots, D_{(s)}$ are the s slices of D. A combined uniformity measure can be of the following form:

$$\mathrm{CD}(D, \xi) = \xi \mathrm{CD}(D) + (1 - \xi) \sqrt[s]{\prod_{i=1}^{s} \mathrm{CD}(D_{(i)})}, \tag{7.3.1}$$

where $0 \leqslant \xi \leqslant 1$ is a real weighting parameter and $\sqrt[s]{\prod_{i=1}^{s} \mathrm{CD}(D_{(i)})}$ is the geometric mean of $\mathrm{CD}(D_{(i)})$'s. Note that $\mathrm{CD}(D)$ and $\mathrm{CD}(D_{(i)})$'s may have different magnitudes, which may bring unfair comparison for the two kinds of uniformity and difficulty for determining the proper value of ξ. Instead of directly using the CD-values as in (7.3.1), one needs a measure that not only has the same magnitude order for the whole design and its slices but also can reflect the uniformity of designs. For this purpose, Chen et al. (2016) introduced the uniformity efficiency (U-efficiency for short) of D, which is defined as

$$E_U(D) = \frac{\mathrm{CD}(U)}{\mathrm{CD}(D)}, \tag{7.3.2}$$

where U is a uniform design with the same parameters as D under the CD. Many uniform designs can be found on Web site http://dst.uic.edu.hk/en/isci/uniform-design/uniform-design-tables. It is clear that $0 < E_U(D) \leqslant 1$, and $E_U(D)$ has the same order as $E_U(D_{(i)})$'s, where $E_U(D)$ and $E_U(D_{(i)})$'s are the U-efficiencies of D and $D_{(i)}$ for $i = 1, \ldots, s$, respectively. The proposed combined uniformity measure is then of the following form:

$$E_U(D, \omega) = \omega E_U(D) + (1 - \omega) \sqrt[s]{\prod_{i=1}^{s} E_U(D_{(i)})}, \tag{7.3.3}$$

where $0 \leqslant \omega \leqslant 1$. Since $\mathrm{CD}(U)$ is fixed, so the larger $E_U(D)$ is, the smaller the $\mathrm{CD}(D)$ is. Thus for a fixed ω, the objective is to find a design $D^* \in \mathcal{D}$ such that

$$E_U(D^*, \omega) = \max_{D \in \mathcal{D}} (E_U(D, \omega)). \tag{7.3.4}$$

Here, D^* is called a USLHD. Optimization algorithms for finding a D^* are given in next subsection.

7.3.2 *Optimization Algorithms*

As a stepping stone to USLHDs, the neighbor of an SLHD is an important concept which will be used in the proposed optimization procedure. Let $\mathcal{D}$ be the set consisting of all the $SL(n, q, s)$'s, then a neighbor of an SLHD $D_0 \in \mathcal{D}$ can be constructed by the following algorithm (Chen et al. 2016).

Algorithm 7.3.1

Step 1. Randomly choose one column of D_0, and then from each slice of this column, choose one element such that these s elements, say $a_1, \ldots, a_s$, are "equal" in the sense that $\lceil a_1/s \rceil = \cdots = \lceil a_s/s \rceil$.

Step 2. Randomly choose two elements among $a_1, \ldots, a_s$, say a_{i_1} and a_{i_2}, and exchange their positions in the column.

Step 3. Randomly choose one of a_{i_1} and a_{i_2}, say a_{i_1}, and select an element from the same column in the same slice of a_{i_1}, say b_1 ($b_1 \neq a_{i_1}$), exchange their positions.

The resulting design is one neighbor of D_0. It is obvious that such an exchanging procedure does not change the sliced structure of an SLHD. This is necessary for a design to be a neighbor of an SLHD.

To search USLHDs, Chen et al. (2016) used the threshold-accepting (TA) algorithm (see Sect. 4.2). The step-by-step guidelines for the proposed optimization algorithm are given as follows.

Algorithm 7.3.2

Step 1. Give $n,\ q,\ s,\ \omega$, randomly generate an $SL(n, q, s)$ as the initial design D_0 using the method in Qian et al. (2012), and calculate $E_U(D_0, \omega)$. Set a non-positive sequence of threshold parameters $T_1 < \cdots < T_L = 0$. Denote the iteration number by I under each T_l for $l = 1, \ldots, L$. Set two indexes $l = 1,\ i = 1$.

Step 2. Randomly construct a neighbor of D_0 by Algorithm 7.3.1, denoted by D_c, and calculate $E_U(D_c, \omega)$.

Step 3. If $E_U(D_c, \omega) - E_U(D_0, \omega) \geqslant T_l$, replace D_0 by D_c; else leave D_0 unchanged.

Step 4. Update $i = i + 1$, if $i \leqslant I$, go to Step 2.

Step 5. Update $l = l + 1$, if $l \leqslant L$, reset $i = 1$, and go to Step 2; else output $D_{\text{best}} = D_0$.

Remark 7.2 Setting the sequence of $T_1, \ldots, T_L$ is a critical step. Several candidate sequences can be tried, and the one that can bring a quicker convergence and a more remarkable improvement is adopted.

The design obtained by Algorithm 7.3.2 is called a USLHD in terms of the combined uniformity with weight ω, denoted by $USL(n, q, s, \omega)$. It jointly considers the uniformity of the whole SLHD and that of its slices, so the design points of both the whole design and each slice are distributed evenly over the experimental region.

The resulting design D_{best} may be locally optimal depending on the selection of initial design. Hence, it is strongly recommended to run the algorithm a number of times with different initial designs and then select the best one among the resulting designs. Moreover, for an $SL(n, q, s)$, determining what a particular value ω_0 should be assigned to ω so that the $USL(n, q, s, \omega_0)$ is the most effective one among $SL(n, q, s, \omega)$'s is another important issue. This will be discussed next.

7.3.3 Determination of the Weight ω

From (7.3.3) and (7.3.4), as ω decreases from 1 to 0, $E_U(D^*)$ decreases, while $\sqrt[s]{\prod_{i=1}^{s} E_U(D^*_{(i)})}$ increases, where $D^*_{(1)}, \ldots, D^*_{(s)}$ are the s slices of D^*. For such a trade-off, it is appropriate to avoid a low $E_U(D^*)$. So, Chen et al. (2016) imposed a lower threshold l_u on $E_U(D^*)$, meanwhile maximizing $\sqrt[s]{\prod_{i=1}^{s} E_U(D^*_{(i)})}$, and this leads to the following multi-objective optimization problem

$$\max_{\omega} \sqrt[s]{\prod_{i=1}^{s} E_U(D^*_{(i)})}, \quad \text{subject to } E_U(D^*) \geqslant l_u. \tag{7.3.5}$$

To solve (7.3.5), Chen et al. (2016) introduced a tool called "ω-trace", which plots $E_U(D^*)$ and $\sqrt[s]{\prod_{i=1}^{s} E_U(D^*_{(i)})}$ as functions of ω. For convenience, Chen et al. (2016) took a sequence of values $\{0, 0.05, 0.1, \ldots, 1\}$ for ω. As ω decreases from 1 to 0, $E_U(D^*)$ will decrease to the lower threshold l_u at some ω_0, and at the same time, $\sqrt[s]{\prod_{i=1}^{s} E_U(D^*_{(i)})}$ takes the maximum value among all the ω's at which $E_U(D^*)$ is larger than l_u. That is, ω_0 is just the value of ω we are looking for. The following is an illustrative example due to Chen et al. (2016).

Example 7.3.1 Suppose S_0 is a randomly generated $SL(18, 3, 3)$, and then we carry out Algorithm 7.3.2 with 21 values of ω from 0 to 1 by 0.05 to search for the desired design, i.e., $\omega_i = 0.05(i-1)$ for $i = 1, \ldots, 21$. In the algorithm, set the threshold parameters $T_1, \ldots, T_{11}$ to be $T_i = -10^{-5} + 10^{-6}(i-1)$ for $i = 1, \ldots, 11$. For each ω_i, we obtain a $USL(18, 3, 3, \omega_i)$, denoted by S_i for $i = 1, \ldots, 21$, and compute $E_U(S_i)$ and $\sqrt[3]{\prod_{j=1}^{3} E_U(S_{i(j)})}$, where $S_{i(j)}$ for $j = 1, 2, 3$ are the three slices of S_i. To obtain the U-efficiencies, the CD-values of the corresponding uniform designs $U_{18}(18^3)$ and $U_6(6^3)$, which are 0.0506 and 0.1365, respectively, can be found from the above Web site. Then, $E_U(S_i)$ and $\sqrt[3]{\prod_{j=1}^{3} E_U(S_{i(j)})}$ for $i = 1, \ldots, 21$ can be computed through (7.3.2). Now, we plot the "ω-trace," which is presented in Fig. 7.1. As for the lower threshold of the U-efficiency for the whole design, we take the upper five percent quartile of the U-efficiencies of 10,000 randomly generated $SL(18, 3, 3)$'s, which is 0.7314, i.e., $l_u = 0.7314$. Such a threshold can ensure that the obtained $USL(18, 3, 3, \omega)$ has a CD smaller than about 95% randomly generated $SL(18, 3, 3)$'s when considering the uniformity of the whole design. From Fig. 7.1, we find that as ω decreases to 0.45, the lower threshold has been reached, and thus we can assign $\omega = 0.45$ for this example.

Remark 7.3 Taking the upper five percent quartile of the U-efficiencies of a large number of randomly generated $SL(n, q, s)$'s as the lower threshold of the U-efficiency of the whole design is not the essence, and some other values can also be taken according to the user's need, such as the upper two percent quartile, which can ensure that the obtained USLHD has a CD smaller than about 98% of the randomly generated $SL(n, q, s)$'s when considering the uniformity of the whole design.

Remark 7.4 From Fig. 7.1, we find that the lower threshold of the U-efficiency of the whole design is reached coincidentally at $\omega = \omega_{10} = 0.45$, which is just among the initial values taken for ω, i.e., $\omega_1, \ldots, \omega_{21}$, in Example 7.3.1. In general, Chen et al. (2016) suggested taking ω to be the smallest value among the initial ω_i's that are larger than the one at which the lower bound l_u is reached. To obtain a USLHD for any given size, Chen et al. (2016) suggested 0.5 for ω if we cannot afford so much computational burden to determine the value of ω. In fact, the case of $\omega = 0.5$ gives

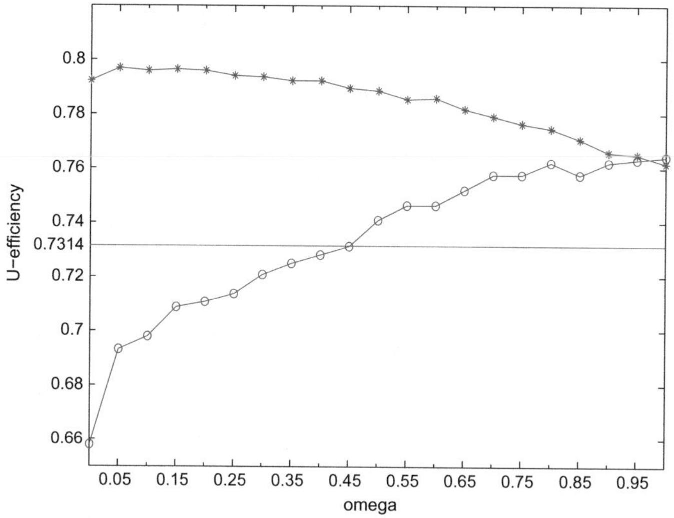

Fig. 7.1 Plot of "ω-trace," where the broken line with "o" corresponds to the U-efficiencies of the whole designs; the broken line with "$*$" corresponds to the geometric means of the U-efficiencies of the slices

approximately equal importance to both the uniformity of the whole design and that of its slices, which appears to be a fair choice.

Chen et al. (2016) further provided two simulated examples to illustrate the performance of USLHDs when used for building Gaussian process models for computer experiments with both quantitative and qualitative factors. The simulations indicated that USLHDs with better uniformity in terms of whole design tend to have better performance when the real response functions have big differences from each other, and USLHDs with better uniformity in terms of slices probably result in smaller root mean square prediction errors (RMSPEs) when the real response functions are similar to each other. Thus, Chen et al. (2016) suggested adopting $\omega = 0.5$ for the sake of robust application since the real response functions are usually unknown.

7.4 Uniformity Under Errors in the Level Values

When the experiments are carried out in practice, the actually performed factor-level values may be accompanied by errors; see, for example, Box (1963) and Draper and Beggs (1971). Consider the design $U(6; 6^5)$ in Table 7.2, obtainable from the Web site http://www.math.hkbu.edu.hk/UniformDesign/ for uniform designs. The WD^2-value of the design is 0.2030. When the factor level values are contaminated with errors, the actually performed factor levels become $u'_{ik} = u_{ik} + \varepsilon_{ik}$, with ε_{ik} being the random error. Consider the bounded support uniform distribution for the errors,

Table 7.2 A $U(6; 6^5)$ design

Run	1	2	3	4	5
1	3	1	6	5	4
2	5	2	2	2	5
3	6	5	3	6	3
4	4	4	5	1	1
5	2	6	4	3	6
6	1	3	1	4	2

Table 7.3 A $U(6; 6^5)$ design under errors of Unif$(-1/3, 1/3)$

Run	1	2	3	4	5
1	2.6644	0.5613	5.8923	4.7964	3.4486
2	4.9332	1.5265	1.7120	1.7475	4.9865
3	5.8811	4.5608	2.5803	5.8632	2.5143
4	3.7633	3.6894	4.8018	0.9712	0.5016
5	1.8786	5.8181	3.6977	2.6817	5.9172
6	0.7735	2.5395	0.6299	3.9201	1.8249

namely $\varepsilon_{ik} \sim$ i.i.d. Unif$(-\tau, \tau)$. Then, the actually performed design with $\tau = 1/3$ may be as in Table 7.3, the WD^2-value of which is 0.2182, as compared to 0.2030 of the original WD^2.

Yang et al. (2010) studied the design uniformity when the factor-level values are contaminated with random errors. They first considered the WD-values for designs with errors in *all* factors and then investigated the cases in which errors only occur in *some* factors. Finally, they applied the results to the construction of uniform designs.

For any $U(n; n^s)$ design $\boldsymbol{U}$, when the factor levels are contaminated with uniformly distributed errors, it becomes $\boldsymbol{Z} = \boldsymbol{U} + \boldsymbol{\varepsilon}$, where $\boldsymbol{\varepsilon} = (\varepsilon_{ik})$ with $\varepsilon_{ik} \sim$ i.i.d. Unif$(-\tau, \tau)$. In this case, the value of WD^2 for the resulting design Z becomes

$$\mathrm{WD}^2(\boldsymbol{Z}) = -\left(\frac{4}{3}\right)^s + \frac{1}{n^2}\sum_{i=1}^{n}\sum_{j=1}^{n}\prod_{k=1}^{s}\left(\frac{3}{2} - |x_{ik} - x_{jk} + \delta_{ik} - \delta_{jk}| + |x_{ik} - x_{jk} + \delta_{ik} - \delta_{jk}|^2\right),$$

where $\delta_{ik}(= \varepsilon_{ik}/n) \sim$ i.i.d. Unif$(-a, a)$ with $a = \tau/n$. Let $B_Z(n, s, a) = \mathrm{E}(\mathrm{WD}^2(\boldsymbol{Z}) - \mathrm{WD}^2(\boldsymbol{U}))$ represent the expected difference between the WD^2-values for design $\boldsymbol{Z}$ and for design $\boldsymbol{U}$. Yang et al. (2010) had the following result.

Theorem 7.4.1 *For a $U(n; n^s)$ design $\boldsymbol{U}$, with $a < 1/(2n)$, we have*

(i) $\frac{n-1}{n}\left(\frac{5}{4}\right)^s \sum_{k=1}^{s}\binom{s}{k}\left(\frac{8}{15}a^2\right)^k < B_Z(n, s, a) < \frac{n-1}{n}\left(\frac{3}{2}\right)^s \sum_{k=1}^{s}\binom{s}{k}\left(\frac{4}{9}a^2\right)^k.$

(ii) *For any fixed n and s, $B_Z(n, s, a)$ is an increasing function of a.*

(iii) *For any fixed n and a, $B_Z(n, s, a)$ is an increasing function of s.*

Theorem 7.4.1 shows how the discrepancies of designs with and without errors in their level values will differ. It is shown that designs with errors are on average less

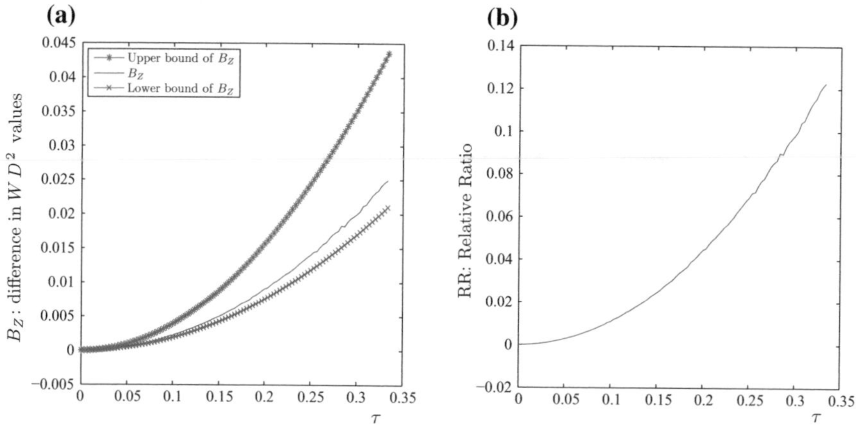

Fig. 7.2 Plots of (**a**) B_Z against τ and (**b**) RR against τ for the design in Table 7.2

uniform than those without any error. Furthermore, the larger the error, the larger the expected WD value for the design will be. For example, Fig. 7.2 displays the result for the uniform design in Table 7.2. The middle curve of Fig. 7.2a shows the relationship between the values of $B_Z(n, s, a)$ and τ. Here, Yang et al. (2010) took the mean value of 100 replications for each τ as $B_Z(n, s, a)$ in the simulation. The upper and lower curves are the upper and lower bounds for $B_Z(n, s, a)$, respectively. To illustrate the magnitude of $B_Z(n, s, a)$, Yang et al. (2010) defined the relative ratio (RR) of this expected difference to $\mathrm{WD}^2(\boldsymbol{U})$ as $\mathrm{RR} = B_Z(n, s, a)/\mathrm{WD}^2(\boldsymbol{U})$. A larger RR value implies that the uniformity discrepancy changes more significantly. Figure 7.2b shows how RR varies, as τ increases. It is shown that the uniformity discrepancy is robust if the errors are relatively small (e.g., $\tau < 0.2$). Here, robustness refers to a situation when the expected change of discrepancies between the original design (without error) and the contaminated design (with errors) is rather insignificant.

Yang et al. (2010) further generalized the results to the case when errors only occur in some but not all factors. Without loss of generality, they assumed the first s_1-factors have errors in their levels and the true design matrix is $\boldsymbol{Y} = \boldsymbol{U} + \boldsymbol{\varepsilon}$, where $\boldsymbol{\varepsilon} = (\varepsilon_{ik})$ with $\varepsilon_{ik} \sim$ i.i.d. Unif$(-\tau, \tau)$ for $k \leqslant s_1$ and $\varepsilon_{ik} = 0$ for $k > s_1$. The value of WD^2 for design $\boldsymbol{Y}$ then becomes

$$\mathrm{WD}^2(\boldsymbol{Y}) = -\left(\frac{4}{3}\right)^s + \frac{1}{n^2}\sum_{i=1}^{n}\sum_{j=1}^{n}\left[\prod_{k=s_1+1}^{s}\left(\frac{3}{2} - |x_{ik} - x_{jk}| + |x_{ik} - x_{jk}|^2\right)\right]$$
$$\cdot\left[\prod_{k=1}^{s_1}\left(\frac{3}{2} - |x_{ik} - x_{jk} + \delta_{ik} - \delta_{jk}| + |x_{ik} - x_{jk} + \delta_{ik} - \delta_{jk}|^2\right)\right],$$

where $\delta_{ik} = \varepsilon_{ik}/q_k$ for $k = 1, \ldots, s_1$. And they had the following theorem.

Theorem 7.4.2 *For a $U(n; n^s)$ design $\boldsymbol{U}$, let $B_Y(n, s, a, s_1) = E(\mathrm{WD}^2(\boldsymbol{Y}) - \mathrm{WD}^2(\boldsymbol{U}))$, then*

(i) $$\frac{n-1}{n}\left(\frac{5}{4}\right)^s \sum_{k=1}^{s_1}\binom{s_1}{k}\left(\frac{8}{15}a^2\right)^k < B_Y(n, s, a, s_1) < \frac{n-1}{n}\left(\frac{3}{2}\right)^s \sum_{k=1}^{s_1}\binom{s_1}{k}\left(\frac{4}{9}a^2\right)^k;$$

(ii) *For any fixed n, s and s_1, $B_Y(n, s, a, s_1)$ will increase as the value a increases.*

(iii) *For any fixed n, s and a, $B_Y(n, s, a, s_1)$ will increase as s_1, the number of factors with errors, increases.*

The theorem shows that the fewer factors with errors and/or the smaller the errors are, the better the WD uniformity is. The results in Theorems 7.4.1 and 7.4.2 indicate that if the errors are relatively small, the traditional uniform designs are rather robust. Also, the results can be used in the construction of uniform designs. There are two conventional frameworks for construction of uniform designs on $[0, 1]^s$. The traditional one, called lattice sampling (Patterson 1954), is to select experimental points from the centers of grids, namely $x_{ik} = (u_{ik} - 1/2)/n$, for $1 \leqslant i \leqslant n$ and $1 \leqslant k \leqslant s$. The other one, called Latin hypercube sampling (McKay et al. 1979), is to select experimental points randomly within the grids, namely $\tilde{x}_{ik} = (u_{ik} - \epsilon_{ik})/n$, where $\epsilon_{ik} \sim$ i.i.d. Unif(0, 1) for $1 \leqslant i \leqslant n$ and $1 \leqslant k \leqslant s$. These two frameworks have been generalized to orthogonal arrays for space fillings by Owen (1992b). Fang et al. (2002) proved that lattice sampling minimizes the WD discrepancy on [0, 1] for one dimension ($s = 1$). Theorem 7.4.1(ii) extends their result to indicate that this is also true in expectation for higher dimensions ($s \geqslant 2$).

Note that, the above results are based on the WD as the measure of uniformity. In fact, a similar study has been applied to other discrepancies, including the centered L_2-discrepancy. The results are similar to Theorems 7.4.1 and 7.4.2.

Though only uniform random errors are considered, it can be shown that the main results of Theorems 7.4.1 and 7.4.2 also hold for other random error structures, such as normal and beta distributions. This can be theoretically derived for the normal distribution and via simulation for the beta distribution. For the normal distribution, because of its unbounded support, it is more appropriate to adopt the truncated normal random errors for which the truncation points are chosen in a way to ensure that all design points will stay within the unit cube.

Exercises

7.1

Consider the resolvable incomplete block design given in Example 3.6.2.

(1) Write down its incidence matrix $\mathbf{Z} = (z_i^j)_{6\times 11}$.

(2) From this $\mathbf{Z}$, what relationships among the z_i^j, r_i, and t_j for $i = 1, \ldots, 6$, $j = 1, \ldots, 11$ can you find?

(3) Let n_{ij}^{lh} be the number of (l, h)-pairs in the two rows z_i and z_j of $\boldsymbol{Z}$, then what relationships among n_{ij}^{lh}'s can you find?
(4) Generalize the findings in (2) and (3) to a general RIB$(n, s, r, \boldsymbol{t}, \boldsymbol{Z})$.

7.2

Show that the expression in (7.1.1) also holds for an RIB$(n, s, r, \boldsymbol{t}, \boldsymbol{Z})$. Then based on this expression, prove that (7.1.3) is true.

7.3

Let δ_{ij} be the number of coincidences between the ith and jth runs of design $\boldsymbol{U} \in \mathcal{U}(n; q_1 \times \cdots \times q_m)$, and express the $E(f_{NOD})$ in (7.2.1) in terms of δ_{ij}'s.

7.4

Suppose A, B, and C are three $SL(12, 2, 2)$'s as shown below:

$$A = \left(\begin{array}{cccccc|cccccc} 11 & 8 & 1 & 6 & 4 & 10 & 7 & 12 & 3 & 2 & 9 & 5 \\ 1 & 3 & 11 & 7 & 9 & 6 & 4 & 2 & 8 & 10 & 12 & 5 \end{array}\right)^T,$$

$$B = \left(\begin{array}{cccccc|cccccc} 9 & 1 & 5 & 8 & 12 & 4 & 2 & 3 & 10 & 11 & 7 & 6 \\ 10 & 4 & 6 & 8 & 11 & 2 & 9 & 12 & 5 & 1 & 7 & 3 \end{array}\right)^T \text{ and}$$

$$C = \left(\begin{array}{cccccc|cccccc} 1 & 7 & 12 & 4 & 9 & 6 & 5 & 10 & 2 & 8 & 3 & 11 \\ 10 & 12 & 7 & 1 & 3 & 6 & 5 & 4 & 9 & 11 & 2 & 8 \end{array}\right)^T.$$

(1) Calculate the CD-values of A, B, and C.
(2) Calculate the combined CD-values of A, B, and C in (7.3.1) by setting $\xi = 0.5$.
(3) Give the scatter plots of A, B, and C.
(4) From (1), (2), and (3), what can you find on the designs' whole uniformity, combined uniformity, and the relationship between the uniformity of a whole SLHD and that of its slices?

7.5

Consider the two designs given in Tables 7.2 and 7.3.

(1) Which design is the better one in terms of the CD-value?
(2) Set $\tau = 0.1, 0.2, 0.3, 0.4$, for each τ, generate 50 $U(6; 6^5)$ designs under errors of Unif$(-\tau, \tau)$, and compute the mean of the CD-values of these 50 replications. What can you find based on these mean values in terms of different τ and the CD-value of the $U(6; 6^5)$ in Table 7.2?

References

Bagchi, S.: Main-effect plans orthogonal through the block factor. Technometrics **52**, 243–249 (2010)

Box, G.: The effect of error in the factor levels and experimental design. Technometrics **5**, 247–262 (1963)

Bose, M., Bagchi, S.: Optimal main effect plans on blocks of small size. Stat. Probab. Lett. **77**, 142–147 (2007)

Booth, K.H.V., Cox, D.R.: Some systematic supersaturated designs. Technometrics **4**, 489–495 (1962)

Bulutoglu, D.A., Cheng, C.S.: Constructionof $E(s^2)$-optimal supersaturated designs. Ann. Stat. **32**, 1662–1678 (2004)

Butler, N.A., Mead, R., Eskridge, K.M., Gilmour, S.G.: A general method of constructing $E(s^2)$-optimal supersaturated designs. J. R. Stat. Soc. Ser. B **63**, 621–632 (2001)

Cao, R.Y., Liu, M.Q.: Construction of second-order orthogonal sliced Latin hypercube designs. J. Complex. **31**, 762–772 (2015)

Cheng, C.S.: E(s^2)-optimal superdaturated designs. Stat. Sin. **7**, 929–939 (1997)

Chen, J., Liu, M.Q., Fang, K.T., Zhang, D.: A cyclic construction of saturated and supersaturated designs. J. Stat. Plan. Inference **143**, 2121–2127 (2013)

Chen, X.P., Lin, J.G., Yang, J.F., Wang, H.X.: Construction of main-effect plans orthogonal through the block factor. Stat. Probab. Lett. **106**, 58–64 (2015)

Chen, X.P., Lin, J.G., Huang, X.F.: Construction of main effects plans orthogonal through the block factor based on level permutation. J. Korean Stat. Soc. **44**, 538–545 (2015)

Chen, H., Huang, H.Z., Lin, D.K.J., Liu, M.Q.: Uniform sliced Latin hypercube designs. Appl. Stoch. Model. Bus. Ind. **32**, 574–584 (2016)

Draper, N.R., Beggs, W.J.: Errors in the factor levels and experimental design. Ann. Math. Stat. **41**, 46–56 (1971)

Dey, A.: Theory of Block Designs. Wiley, New York (1986)

Fang, K.T., Lin, D.K.J., Ma, C.X.: On the construction of multi-level supersaturated designs. J. Stat. Plan. Inference **86**, 239–252 (2000)

Fang, K.T., Ma, C.X., Winker, P.: Centered L_2- discrepancy of random sampling and Latin Hypercube design, and construction of uniform designs. Math. Comput. **71**, 275–296 (2002)

Fang, K.T., Lin, D.K.J., Liu, M.Q.: Optimal mixed-level supersaturated design. Metrika **58**, 279–291 (2003)

Fang, K.T., Ge, G.N., Liu, M.Q., Qin, H.: Combinatorial constructions for optimal supersaturated designs. Discret. Math. **279**, 191–202 (2004a)

Fang, K.T., Ge, G.N., Liu, M.Q., Qin, H.: Construction of uniform designs via super-simple resolvable t-designs. Util. Math. **66**, 15–32 (2004b)

Fang, K.T., Maringer, D., Tang, Y., Winker, P.: Lower bounds and stochastic optimization algorithms for uniform designs with three or four levels. Math. Comput. **75**, 859–878 (2006)

Georgiou, S.D.: Supersaturated designs: a review of their construction and analysis. J. Stat. Plan. Inference **144**, 92–109 (2014)

Hickernell, F.J.: A generalized discrepancy and quadrature error bound. Math. Comput. **67**, 299–322 (1998a)

Hickernell, E.J.: Lattice rules: how well do they measure up? In: Hellekalek, P., Larcher, G. (eds.) Random and Quasi-Random Point Sets, pp. 106–166. Springer, New York (1998b)

Huang, H.Z., Yang, J.F., Liu, M.Q.: Construction of sliced (nearly) orthogonal Latin hypercube designs. J. Complex. **30**, 355–365 (2014)

Huang, H.Z., Lin, D.K.J., Liu, M.Q., Yang, J.F.: Computer experiments with both qualitative and quantitative variables. Technometrics **58**, 495–507 (2016)

Lin, D.K.J.: A new class of supersaturated designs. Technometrics **35**, 28–31 (1993)

Liu, M.Q., Hickernell, F.J.: $E(s^2)$-optimality and minimum discrepancy in 2-level supersaturated designs. Stat. Sin. **12**, 931–939 (2002)

Liu, M.Q., Zhang, R.C.: Construction of $E(s^2)$-optimal supersaturated designs. J. Stat. Plan. Inference **86**, 229–238 (2000)
Liu, M.Q., Chan, L.Y.: Uniformity of incomplete block designs. Int. J. Mater. Prod. Technol. **20**, 143–149 (2004)
Liu, Y., Dean, A.: k-circulant supersaturated designs. Technometrics **46**(1), 32–43 (2004)
Liu, M.Q., Fang, K.T.: Some results on resolvable incomplete block designs. Sci. China Ser. A **48**, 503–512 (2005)
Liu, Y., Liu, M.Q.: Construction of optimal supersaturated design with large number of levels. J. Stat. Plan. Inference **141**, 2035–2043 (2011)
Liu, Y., Liu, M.Q.: Construction of equidistant and weak equidistant supersaturated designs. Metrika **75**, 33–53 (2012)
Liu, Y., Liu, M.Q.: Construction of supersaturated design with large number of factors by the complementary design method. Acta Math. Appl. Sin. (English Ser.) **29**, 253–262 (2013)
Liu, M.Q., Fang, K.T., Hickernell, F.J.: Connections among different criteria for asymmetrical fractional factorial designs. Stat. Sin. **16**, 1285–1297 (2006)
Ma, C.X., Fang, K.T., Liski, E.: A new approach in constructing orthogonal and nearly orthogonal arrays. Metrika **50**, 255–268 (2000)
McKay, M., Beckman, R., Conover, W.: A comparison of three methods for selecting values of input variables in the analysis of output from a computer code. Technometrics **21**, 239–245 (1979)
Mukerjee, R., Dey, A., Chatterjee, K.: Optimal main effect plans with non-orthogonal blocking. Biometrika **89**, 225–229 (2002)
Nguyen, N.K.: An algorithmic approach to constructing supersaturated designs. Technometrics **38**, 69–73 (1996)
Owen, A.B.: Randomly orthogonal arrays for computer experiments, integration and visualization. Stat. Sin. **2**, 439–452 (1992b)
Patterson, H.D.: The errors of lattice sampling. J. R. Stat. Soc. Ser. B **16**, 140–149 (1954)
Qian, P.Z.G.: Sliced Latin hypercube designs. J. Am. Stat. Assoc. **107**, 393–399 (2012)
Sun, F.S., Lin, D.K.J., Liu, M.Q.: On construction of optimal mixed-level supersaturated designs. Ann. Stat. **39**, 1310–1333 (2011)
Tang, B., Wu, C.F.J.: A method for constructing supersaturated designs and its $E(s^2)$ optimality. Can. J. Stat. **25**, 191–201 (1997)
Tang, Y., Xu, H., Lin, D.K.J.: Uniform fractional factorial designs. Ann. Stat. **40**, 891–907 (2012)
Wang, J.C., Wu, C.F.J.: Nearly orthogonal arrays with mixed levels and small runs. Technometrics **34**, 409–422 (1992)
Wang, X.L., Zhao, Y.N., Yang, J.F., Liu, M.Q.: Construction of (nearly) orthogonal sliced Latin hypercube designs. Statist. Probab. Lett. **125**, 174–180 (2017)
Xu, H.: Minimum moment aberration for nonregular designs and supersaturated designs. Stat. Sin. **13**, 691–708 (2003)
Yamada, S., Lin, D.K.J.: 3-level supersaturated designs. Stat. Probab. Lett. **45**, 31–39 (1999)
Yamada, S., Matsui, T.: Optimality of mixed-level supersaturated designs. J. Stat. Plan. Inference **104**(2), 459–468 (2002)
Yang, J.F., Sun, F.S., Lin, D.K.J., Liu, M.Q.: A study on design uniformity under errors in the level values. Stat. Probab. Lett. **80**, 1467–1471 (2010)
Yang, J.F., Lin, C.D., Qian, P.Z.G., Lin, D.K.J.: Construction of sliced orthogonal Latin hypercube designs. Stat. Sin. **23**, 1117–1130 (2013)
Yang, X., Chen, H., Liu, M.Q.: Resolvable orthogonal array-based uniform sliced Latin hypercube designs. Stat. Probab. Lett. **93**, 108–115 (2014)
Yang, J.Y., Chen, H., Lin, D.K.J., Liu, M.Q.: Construction of sliced maximin-orthogonal Latin hypercube designs. Stat. Sin. **26**, 589–603 (2016)
Yin, Y.H., Lin, D.K.J., Liu, M.Q.: Sliced Latin hypercube designs via orthogonal arrays. J. Stat. Plan. Inference **149**, 162–171 (2014)

Chapter 8
Uniform Design for Experiments with Mixtures

This chapter introduces uniform design and modeling for experiments with mixtures and for experiments with restricted mixtures. Firstly, some designs for experiments with mixtures including the Scheffé simplex-lattice, simplex-centroid designs, and axial designs are introduced. Secondly, the uniform design of experiments with mixtures and the corresponding uniformity criteria are introduced. Finally, various modeling techniques for designs with mixtures are given.

8.1 Introduction to Design with Mixture

Many products are formed by mixing several ingredients together, for example, the building construction concrete consists of sand, water, and one or more types of cement. The manufacturer or experimenter who is responsible for mixing the ingredients may be interesting in the hardness or compression strength of the concrete, where the hardness is a function of the percentages of cement, sand, and water in the mix. Designs for deciding how to mix the ingredients are called *experimental designs with mixtures* that have played an important role in various fields such as chemical engineering, rubber industry, material, and pharmaceutical engineering. Here, we rewrite Example 1.1.6 as follows.

Example 8.1.1 There are 11 components in a coffee bread: flour, water, sugar, vegetable shortening, flaked coconut, salt, yeast, emulsifier, calcium propionate, coffee powder, and liquid flavor. Choosing a suitable percentage for each ingredient to let the bread to have good taste needs a design of experiments with mixtures on T^{11} which is defined in (8.1.1).

Assume the number of factors be s, and the ith ingredient be x_i, $i = 1, \cdots, s$. Then, the experimental domain is the simplex

K.-T. Fang et al., *Theory and Application of Uniform Experimental Designs*, Lecture Notes in Statistics 221,
https://doi.org/10.1007/978-981-13-2041-5_8

$$T^s = \{(x_1, \ldots, x_s) : x_j \geqslant 0,\ j = 1, \ldots, s,\ x_1 + \cdots + x_s = 1\}. \tag{8.1.1}$$

A design of n runs for mixtures of s ingredients is a set of n points in the domain T^s. Due to the constraint $x_1 + \cdots + x_s = 1$, to find a design for experiments with mixtures is quite different from the factorial design where there is no constraints on the factors.

However, in most experiments with mixtures, some constraints have to be placed on the ingredients. For example, in making a coffee bread in Example 8.1.1, water and flour should be the major ingredients while sugar, salt, and others have a small percentage. The most popular constraints are $0 \leqslant a_i \leqslant x_i \leqslant b_i \leqslant 1, i = 1, \ldots, s$, or $\mathbf{0} \leqslant \boldsymbol{a} \leqslant \boldsymbol{x} \leqslant \boldsymbol{b} \leqslant \mathbf{1}$ where $\boldsymbol{a} = (a_1, \ldots, a_s), \boldsymbol{x} = (x_1, \ldots, x_s), \boldsymbol{b} = (b_1, \ldots, b_s)$ and $\mathbf{0}$ and $\mathbf{1}$ are vectors of 0's and 1's, respectively. Then, the experimental domain becomes

$$T^s(\boldsymbol{a}, \boldsymbol{b}) = \left\{ \boldsymbol{x} : \mathbf{0} \leqslant \boldsymbol{a} \leqslant \boldsymbol{x} \leqslant \boldsymbol{b} \leqslant \mathbf{1}, \sum_{i=1}^{s} x_i = 1 \right\}, \tag{8.1.2}$$

which is a partial region of the entire simplex factor space. More general (multiple-component) constraints are of the form

$$d_k \leqslant \sum_{i=1}^{s} a_{ki} x_i \leqslant e_k,\ \ k = 1, \ldots, m. \tag{8.1.3}$$

The designs on $T^s(\boldsymbol{a}, \boldsymbol{b})$ are called as *experimental designs with restricted mixtures*, which will be discussed in Sect. 8.2.3.

The following problem demonstrates the need of a space-filling design in the domain $T^s(\boldsymbol{a}, \boldsymbol{b})$ and had been studied by Piepel et al. (1993), Piepel et al. (2002), and Borkowski and Piepel (2009).

Example 8.1.2 The Waste Treatment and Immobilization Plant (WTP) is being constructed on the Hanford site near Richland, Washington. The WTP will produce glass waste forms to immobilize high-level waste (HLW) and low-activity waste (LAW) fractions of nuclear waste currently in large, underground storage tanks. The HLW and LAW glass must meet requirements associated with waste loading, chemical durability, processing, and other properties. The strategy to meet the requirements involves generating an experimental design, making the experimental design glasses in the laboratory, measuring the glass properties of interest, and developing property-composition models to use before and during WTP operations.

Prior to generating an experimental design, glass scientists want to (i) identify the components that may significantly affect glass properties of interest and (ii) define the experimental glass composition region of interest (Table 8.1). A region is selected to include glasses that are acceptable as well as somewhat unacceptable with respect to various requirements. Then, the experimental data collected at design points in the region provide for developing models that can adequately predict whether glass properties will be acceptable or unacceptable.

In this chapter, we will discuss both the designs with mixtures on the entire simplex factor space and the designs of experiments with restricted mixtures. The first design with mixtures suggested by Scheffé (1958) is the *simplex-lattice design* based on a quadratic regression model. Scheffé (1963) proposed the so-called *simplex-centroid design*. A lot of designs have been proposed in the past. Cornell (2002, 2011) and references therein gave a comprehensive review on designs of experiments with mixtures. Chan (2000) gave a review on optimal designs for experiments with mixtures.

8.1.1 *Some Types of Designs with Mixtures*

In this subsection, we introduce some types of designs with mixtures such as the simplex-lattice design, simplex-centroid design, axial design, and Scheffé type design.

(a) Simplex-Lattice Design

To represent the response surface on the entire simplex region, a natural choice for a design is to spread the design points evenly on the whole simplex factor space.

Table 8.1 Components and their lower and upper limits in Example 8.1.2

No.	Component	Lower limit (a_i)	Upper limit (b_i)
1	SiO_2	0.38	0.53
2	B_2O_3	0.05	0.14
3	Na_2O	0.04	0.15
4	Fe_2O_3	0.08	0.14
5	ZrO_2	0.00	0.06
6	MnO	0.00	0.05
7	SrO	0.00	0.05
8	Al_2O_3	0.04	0.085
9	Li_2O	0.02	0.06
10	CdO	0.0005	0.015
11	Spike	0.0015	0.015
12	NiO	0.001	0.01
13	Tl_2O_3	0.0002	0.002
14	Sb_2O_3	0.0002	0.002
15	SeO_2	0.0002	0.002
16	Other	0.042	0.042

Suppose that the experiments with mixtures has s components. Let m be a positive integer and suppose that each component takes $(m+1)$ equally spaced places from 0 to 1, i.e.,

$$x_i = 0, 1/m, 2/m, \ldots, 1, \ \ for\ i = 1, \ldots, s. \tag{8.1.4}$$

A s-ingredient simplex-lattice is denoted by $\{s, m\}$ which has $\binom{s+m-1}{m}$ design points, i.e., it consists of all possible combinations of the components under the constraint $x_1 + \cdots + x_s = 1$, and each component is an element from the set $\{0, 1/m, 2/m, \ldots, 1\}$. For example, when $s = 3$, the simplex-lattice $\{s, m\}$ is as follows:

$m = 1$: 3 design points (1, 0, 0), (0, 1, 0), (0, 0, 1);
$m = 2$: 6 design points (1, 0, 0), (0, 1, 0), (0, 0, 1), (1/2, 1/2, 0), (1/2, 0, 1/2), (0, 1/2, 1/2);
$m = 3$: 10 design points (1, 0, 0), (0, 1, 0), (0, 0, 1), (1/3, 2/3, 0), (1/3, 0, 2/3), (0, 1/3, 2/3), (2/3, 1/3, 0), (2/3, 0, 1/3), (0, 2/3, 1/3) (1/3, 1/3, 1/3).

The {3, 2} and {3, 3} simplex-lattice designs are shown in Fig. 8.1. It can be seen that many points locate at the boundary of the simplex factor space.

(b) Simplex-Centroid Design

In an s-component simplex-centroid design, the design points form as follows:

- s points correspond to the permutations of $(1, 0, 0, \ldots, 0)$,
- $\binom{s}{2}$ points correspond to the permutations of $(1/2, 1/2, 0, \ldots, 0)$,
- $\binom{s}{3}$ points correspond to the permutations of $(1/3, 1/3, 1/3, 0, \ldots, 0)$,
- $\ldots$,
- the centroid point $(1/s, \ldots, 1/s)$.

Then, the total number of design points is $2^s - 1$. For example, when $s = 3$, the design points are (1, 0, 0), (0, 1, 0), (0, 0, 1), (1/2, 1/2, 0), (1/2, 0, 1/2), (0, 1/2, 1/2), and (1/3, 1/3, 1/3). Comparing the {3, 2} simplex-lattice design in Fig. 8.1a, the

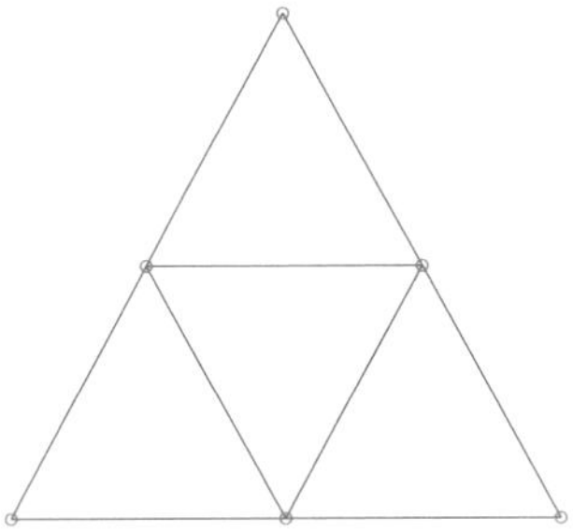

(a) $\{3, 2\}$ design

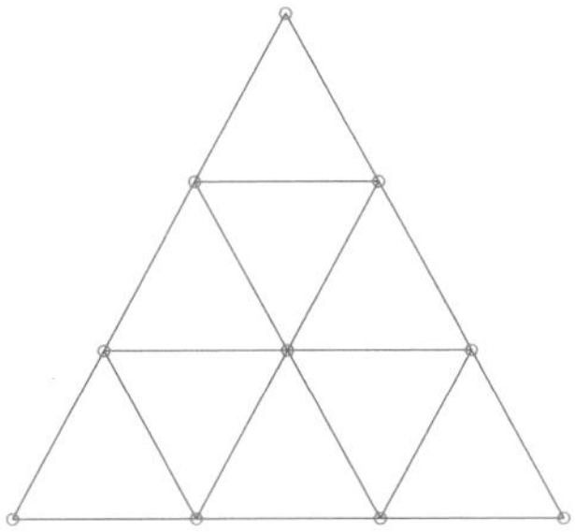

(b) $\{3, 3\}$ design

Fig. 8.1 {3, 2} and {3, 3} simplex-lattice designs

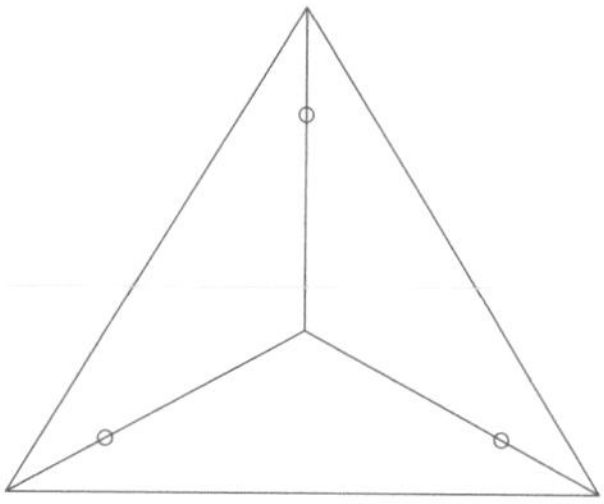

Fig. 8.2 Axial design on T^3

simplex-centroid design has one more design point, i.e., the centroid point. It can be seen that most of the design points of simplex-centroid design are positioned on the boundaries (vertices, edges, faces, etc.) of the simplex factor space.

However, there are some weaknesses of the simplex-lattice design and simplex-centroid design. For example, these designs do not provide many choices of designs for the user and there are so many points at the boundary of T^s. The experiment is often impossible for many chemical experiments if a component has zero value, and the boundary experimental points are meaningless in that case. To overcome such disadvantage, the axial design may be a suitable choice. The axial design is a type of designs with mixtures whose design points are located on the inner region of simplex factor space. Another natural way is to keep the pattern of original design and to contract the boundary points toward the centroid of T^s such as the Scheffé type designs proposed by Fang and Wang (1994).

(c) **Axial Design**

The line segment joining a vertex of the simplex T^s with its centroid $(1/s, \ldots, 1/s)$ is called an *axis*. The distance between the centroid point and each vertex is $\sqrt{(s-1)/s}$. Let d be a positive number such that $0 < d < \sqrt{(s-1)/s}$. The experimental points of the axial design are s points on the s axes such that each point to the centroid has the same distance d. Then each point of axial design locates at the inner region of the simplex factor space. Figure 8.2 shows one axial design with $s = 3$. Different d obtains different axial design, and the optimal d can be determined by some criteria which will be discussed in the next section.

(d) **Scheffé Type Design**

Fang and Wang (1994) proposed the *contraction method* for construction of Scheffé type designs. The key idea is to keep the pattern of original design and to contract the boundary points toward the centroid of T_s.

We now illustrate the method by an example of simplex-lattice design $\{3, 3\}$. Suppose that the original design is shown by Fig. 8.1b. Let a be a number which will be determined later. Then we move the three vertices as follows:

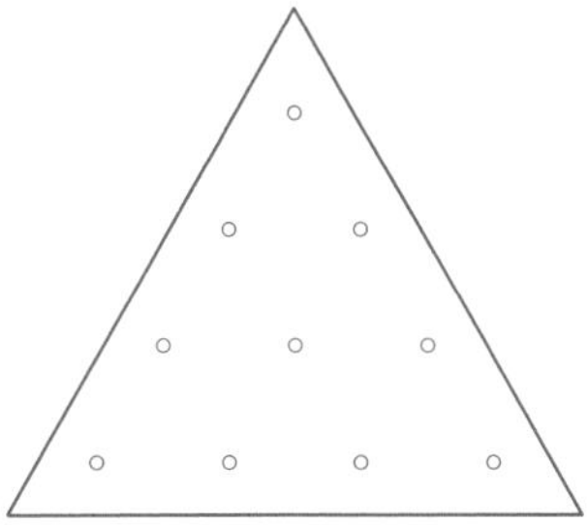

Fig. 8.3 Scheffé type design

$$(1, 0, 0) \Rightarrow \left(1 - \frac{1}{a}, \frac{1}{2a}, \frac{1}{2a}\right),$$
$$(0, 1, 0) \Rightarrow \left(\frac{1}{2a}, 1 - \frac{1}{a}, \frac{1}{2a}\right),$$
$$(0, 0, 1) \Rightarrow \left(\frac{1}{2a}, \frac{1}{2a}, 1 - \frac{1}{a}\right),$$

and the other boundary and interior points can be easily computed; see Fig. 8.3. One wishes to find a suitable number a under a certain criterion.

8.1.2 *Criteria for Designs with Mixtures*

In axial design and Scheffé type design, one needs to determine the optimal parameter. In this subsection, some criteria under distance and uniformity are given.

Fang and Wang (1994) proposed the so-called F-discrepancy, but it is not easy to compute. Therefore, they suggested to use the *mean Square Distance (MSD)* (MSD) criterion for assessing the quality of the designs. Recently, Borkowski and Piepel (2009) considered the *root mean square distance* (RMSD), the *average distance* (AD), and the *maximum Distance (MD)* (MD) as criteria. Let $\mathcal{X}$ be an experimental region. It can be T^s defined in (8.1.1) or $T^s(\boldsymbol{a}, \boldsymbol{b})$ defined in (8.1.2), or others. Let $\mathcal{P} = \{\boldsymbol{x}_1, \ldots, \boldsymbol{x}_n\}$ be a design on $\mathcal{X}$ and the random vector $\boldsymbol{x}$ follow the uniform distribution on $\mathcal{X}$. The above criteria are defined as follows:

Mean square distance and root mean square distance:

$$\mathrm{MSD}(\mathcal{P}) = E[d^2(\boldsymbol{x}, \mathcal{P})], \tag{8.1.5}$$

where $d(\boldsymbol{x}, \mathcal{P}) = \min_{1 \leqslant j \leqslant n} d(\boldsymbol{x}, \boldsymbol{x}_j)$, and $\mathrm{RMSD}(\mathcal{P}) = \sqrt{\mathrm{MSD}(\mathcal{P})}$.

Average distance:

$$\mathrm{AD}(\mathcal{P}) = E[d(\boldsymbol{x}, \mathcal{P})]. \tag{8.1.6}$$

Maximum distance:

$$\mathrm{MD}(\mathcal{P}) = \max_{\boldsymbol{x}} d(\boldsymbol{x}, \mathcal{P}). \tag{8.1.7}$$

Under these criteria, one may minimize the value to find a best design. When these criteria cannot be easily computed, we can use a Monte Carlo simulation to find an approximated value. Let $z_1, \ldots, z_N$ be a random sample of size N from the uniform distribution on $\mathcal{X}$. Then we can use

$$\begin{aligned}
\mathrm{MSD}(\mathcal{P}) &= \frac{1}{N}\sum_{k=1}^{N} d^2(z_k, \mathcal{P}),\\
\mathrm{AD}(\mathcal{P}) &= \frac{1}{N}\sum_{k=1}^{N} d(z_k, \mathcal{P}),\\
\mathrm{MD}(\mathcal{P}) &= \max_{1\leqslant k\leqslant N} d(z_k, \mathcal{P}).
\end{aligned}$$

to estimate the above criteria. It is not easy to generate the random sample on T^s directly. Fang and Wang (1994) applied contraction method to the simplex-lattice $\{3, 3\}$ and the simplex-centroid design with $s = 3$ for the construction of the Scheffé type designs. Under the MSD criterion, the best a-value in the contraction method defined in the Sect. 8.1.1 can be obtained. Their results are given in Table 8.2, which shows that the contraction method can decrease the MSD-value of these designs. Recently, Prescott (2008) gave another way to construct Scheffé type designs by placing lower and upper bounds on some or all of the ingredients, i.e., $a_i \leqslant x_i \leqslant b_i, i = 1, \ldots, s$. We omit the detailed procedure here.

One of the disadvantages of Scheffé and Scheffé type designs is that the numbers of experimental points of the designs are restricted, i.e., the number of experimental points is of the type $\binom{s+m-1}{m}$ in the simplex-lattice design, and of the form $2^s - 1$ in the simplex-centroid design. In most chemical or industrial experiments, we meet the requirement that the number of experiments is considerably flexible, and therefore Wang and Fang (1990) proposed a so-called *uniform design for experiments with mixtures* (UDEM) or simply called as *uniform mixture design* (UMD) by Borkowski and Piepel (2009), to seek experimental points to be uniformly scattered in the domain T^s.

Table 8.2 MSD for Scheffé and Scheffé type designs

Design	MSD of Scheffé design	MSD of Scheffé type design	a-value
Simplex-lattice $\{3, 3\}$	0.03087	0.01568	4.836
Simplex-centroid	0.05553	0.02296	3.761

Wang and Fang (1990) proposed the *transformation method* to generate random sample from the uniform distribution on T^s. Let $\boldsymbol{y} = (y_1, \ldots, y_{s-1})$ follow the uniform distribution on the $(s-1)$-dimensional unit hypercube $[0, 1]^{s-1}$. Let

$$\begin{cases} z_i = \left(1 - y_i^{\frac{1}{s-i}}\right) \prod_{j=1}^{i-1} y_j^{\frac{1}{s-j}}, \ i = 1, \ldots, s-1, \\ z_s = \prod_{j=1}^{s-1} y_j^{\frac{1}{s-j}}. \end{cases} \tag{8.1.8}$$

Then $\boldsymbol{z} = (z_1, \ldots, z_s)$ follows the uniform distribution on T^s. Such transformation method can be used to construct uniform mixture designs on T^s, i.e., one firstly constructs the uniform design on $[0, 1]^{s-1}$ and then obtains the n-point design with mixtures on T^s by (8.1.8). For measuring uniformity of the designs with mixtures, the corresponding discrepancy may be given. The discrepancies discussed in Chap. 2 cannot be used directly for designs with mixtures. Section 8.2 will discuss the uniformity criteria and the construction method of uniform designs of experiments with mixtures.

8.2 Uniform Designs of Experiments with Mixtures

In this section, an introduction to the methodology of the uniform design of experiments with mixtures without and with constraints is given. For simplicity, we may call uniform designs of experiments with mixtures by uniform mixture designs.

8.2.1 *Discrepancy for Designs with Mixtures*

For constructing uniform design of experiments with mixtures, the uniformity criterion should be given first. There are two types of uniformity criteria, indirect and direct methods.

One indirect method for measuring the uniformity of designs with mixtures is to measure the uniformity of the corresponding design on the hypercube C^{s-1} by the transformation (8.1.8). Assume $\mathcal{P} = \{\boldsymbol{x}_1, \ldots, \boldsymbol{x}_n\}$ be a n-point set on the simplex T^s, and $\mathcal{P}_0 = \{\boldsymbol{y}_1, \ldots, \boldsymbol{y}_n\}$ be the corresponding point set on C^{s-1} by the inverse transformation of (8.1.8). Then we define the discrepancy of $\mathcal{P}$ be equal to $D(\mathcal{P}_0)$, where $D(\cdot)$ can be chosen as any type of discrepancy in Chap. 2. Then, one can choose a uniform design $\mathcal{P}_0$ on C^{s-1} and obtain the design $\mathcal{P}$ by the transformation (8.1.8). However, the indirect method may not measure the uniformity of designs with mixtures accurately, i.e., the design $\mathcal{P}$ on T^s may be not a uniform design when $\mathcal{P}_0$ is the uniform design on C^{s-1}.

Ning et al. (2011b) proposed another uniformity criterion, *DM_2-discrepancy*, for measuring the uniformity of designs with mixtures. The DM_2-discrepancy is defined

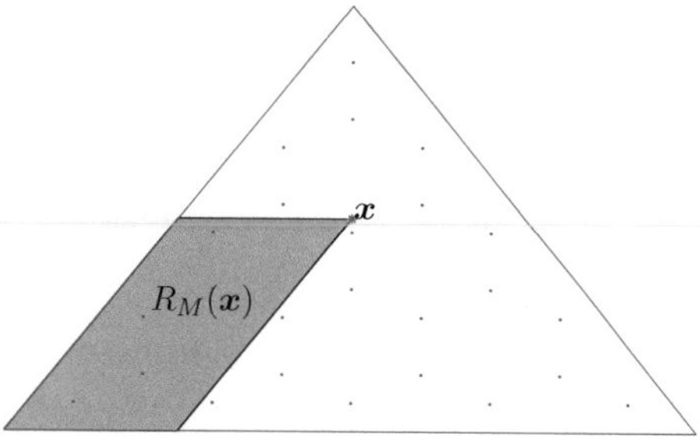

Fig. 8.4 Neighborhood $R_M(\boldsymbol{x})$ of $\boldsymbol{x}$ in space T^3

directly on the region T^s. This discrepancy can be considered as an extension of L_2-star discrepancy from the hypercube to simplex.

Let $\mathcal{P} = \{\boldsymbol{x}_1, \boldsymbol{x}_s, \ldots, \boldsymbol{x}_n\}$ be the n-point set on T^s. Denote the mixture design matrix of $\mathcal{P}$ with n runs and s factors be as follows:

$$\mathcal{P} = \begin{bmatrix} x_{11} & x_{12} & \cdots\cdots & x_{1s} \\ x_{21} & x_{22} & \cdots\cdots & x_{2s} \\ \vdots & \vdots & \vdots \quad \vdots & \vdots \\ x_{n1} & x_{n2} & \cdots\cdots & x_{ns} \end{bmatrix} = (z_1, \ldots, z_s), \tag{8.2.1}$$

where $z_i,\ i = 1, 2, \ldots, s$, is the ith column of $\mathcal{P}$. The DM_2-discrepancy for designs with mixtures is defined as following:

$$DM_2(\mathcal{P}) = \left[\int_{T^s} \left| \frac{Vol(R_M(\boldsymbol{x}))}{Vol(T^s)} - \frac{N(\mathcal{P}, R_M(\boldsymbol{x}))}{n} \right|^2 d\boldsymbol{x}\right]^{1/2}, \tag{8.2.2}$$

where $Vol(A)$ is the volume of the region A, $N(\mathcal{P}, R_M(\boldsymbol{x}))$ is the number of points of $\mathcal{P}$ falling in $R_M(\boldsymbol{x})$, and the region $R_M(\boldsymbol{x})$ for any $\boldsymbol{x} \in T^s$ is defined as follows:

$$R_M(\boldsymbol{x}) = \{\boldsymbol{t} = (t_1, \ldots, t_s) : \boldsymbol{t} \in T^s \text{and } t_i \le x_i, i = 2, 3, \ldots, s\}. \tag{8.2.3}$$

When $s = 3$, $R_M(\boldsymbol{x})$ is showed in Fig. 8.4.

For obtaining the computational formulas of DM_2-discrepancy, the tool of reproducing kernel Hilbert space defined in Chap. 2 can also be used. From the expression of kernel function (2.4.16), we have

$$\begin{aligned} \mathcal{K}(\boldsymbol{z}, \boldsymbol{t}) &= \int_{T^s} 1_{R_M(\boldsymbol{x})}(\boldsymbol{z}) 1_{R_M(\boldsymbol{x})}(\boldsymbol{t}) d\boldsymbol{x} \\ &= \frac{\sqrt{s}}{(s-1)!} \left[\max\left(1 - \sum_{i=2}^{s} \max(z_i, t_i), 0\right)\right]^{s-1}. \end{aligned} \tag{8.2.4}$$

The detailed proof of (8.2.4) can be found in the Appendix A of Ning et al. (2011b). Denote $F_u(t)$ be the uniform distribution on T^s. The density function of uniform distribution on T^s be

$$f_u(t) = F_u'(t) = \begin{cases} 1/Vol(T^s), & t \in T^s, \\ 0, & otherwise. \end{cases}$$

Substituting the density function and Eq. (8.2.4) into (2.4.15), we can get a computational formula of DM_2-discrepancy as follows:

$$\begin{aligned} DM_2(\mathcal{P}) &= \Bigg\{ \int_{T^s \times T^s} \mathcal{K}(z, t) dF_u(z) dF_u(t) \\ &\quad - \frac{2}{n} \sum_{i=1}^{n} \int_{T^s} \mathcal{K}(z_i, t) dF_u(t) + \frac{1}{n^2} \sum_{i=1,k=1}^{n} \mathcal{K}(z_i, t_k) \Bigg\}^{1/2} \\ &= \left(\frac{\sqrt{s}}{(s-1)!} \right)^{\frac{1}{2}} \Bigg\{ C_{n,s} - \frac{2(s-1)!}{n} \sum_{i=1}^{n} \sum_{(\tau_2,\ldots,\tau_s) \in \{0,1\}^{s-1}} a_\tau \cdot (x_{i1})^{\tau_1} \cdot \prod_{j=2}^{s} x_{ij}^{\tau_j} \\ &\quad + \frac{1}{n^2} \sum_{i=1,k=1}^{n} \left(\max \left(1 - \sum_{j=2}^{s} \max(x_{ij}, x_{kj}), 0 \right) \right)^{s-1} \Bigg\}^{1/2}, \end{aligned} \tag{8.2.5}$$

where $\{0, 1\}^{s-1} = \{(t_1, \ldots, t_{s-1}) : \quad t_i = 0 \text{ or } 1\}, \quad C_{n,s} = \frac{((s-1)!)^3 2^{s-1}}{(2(s-1))! \prod_{k=0}^{s-2} (2s+k-1)}$, $a_\tau = \frac{(s-1)!}{(2(s-1)-\sum \tau_i)!}$ and $\tau_1 = 2(s-1) - \sum_2^s \tau_j$.

The proof of (8.2.5) was given in Appendix B of Ning et al. (2011b). Using the formula (8.2.5), we can calculate the discrepancy value for any design with mixtures. And the explicit computational formula of the discrepancy is very useful for searching uniform mixture designs.

From the formula (8.2.5), it can be seen that the DM_2-discrepancy has the following property.

- For any mixture design matrix $\mathcal{P}$ in (8.2.1), the value of DM_2 is invariant under the row permutations.
- For any mixture design matrix $\mathcal{P}$, the value of DM_2 is almost invariant about column permutations. For any permutation $\{i_2, i_3, \ldots, i_s\}$ of $\{2, 3, \ldots, s\}$, denote the column permuted mixture design matrix $\mathcal{P}' = (z_1, z_{i_2}, \ldots, z_{i_s})$, then $DM_2(\mathcal{P}) = DM_2(\mathcal{P}')$.

The above property of DM_2-discrepancy shows that it does not lose any information for reordering the experimental points under this discrepancy. This property is useful in the practice since randomization is applied to the allocation of units to treatments.

8.2.2 Construction Methods for Uniform Mixture Design

A uniform design of experiments with s-ingredient mixtures is a set of points that are uniformly scattered on the domain T^s. The *transformation method* based on (8.1.8) for construction of such uniform designs is as following steps:

Algorithm 8.2.1 (*Transformation method for uniform mixture designs*)

Step 1. Choose a uniform design $U_n(n^{s-1})$, $\boldsymbol{U} = (u_{ki})$.

Step 2. Calculate $c_{ki} = (u_{ki} - 0.5)/n$, then

$$\boldsymbol{C} = \{\boldsymbol{c}_k = (c_{k1}, \dots, c_{k,s-1}), k = 1, \dots, n\}$$

is a UD on C^{s-1}.

Step 3. Calculate

$$\begin{cases} x_{ki} = \left(1 - c_{ki}^{\frac{1}{s-i}}\right) \prod_{j=1}^{i-1} c_{kj}^{\frac{1}{s-j}}, \ j = 1, \dots, s-1, \\ x_{ks} = \prod_{j=1}^{s-1} c_{kj}^{\frac{1}{s-j}}, \ k = 1, \dots, n. \end{cases} \tag{8.2.6}$$

Then $\mathcal{P} = \{\boldsymbol{x}_k = (x_{k1}, \dots, x_{ks}), k = 1, \dots, n\}$ is a uniform design on T^s.

In Step 1 of Algorithm 8.2.1, the uniform design on the $(s-1)$-dimensional uniform design can be obtained by the construction methods in Chaps. 3 and 4, such as the good lattice point method or threshold-accepting algorithm. The transformation method is an indirect method for construction of uniform mixture designs. Usually, it can obtain designs with good uniformity. More important, the transformation method is simple to use.

Example 8.2.1 Give a UMD for $n = 11, s = 3$.

The first two columns of Table 8.3 forms a $U_{11}(11^2)$, $\boldsymbol{U}_1$. The corresponding $\boldsymbol{C} = (c_{ki})$, where $c_{ki} = (U_{ki} - 0.5)/11$ is a UD on $[0, 1]^2$. Formula (8.2.6) for $s = 3$ has a simpler form as follows

$$\begin{cases} x_{k1} = 1 - \sqrt{c_{k1}}, \\ x_{k2} = \sqrt{c_{k1}}(1 - c_{k2}), \quad k = 1, \dots, n. \\ x_{k3} = \sqrt{c_{k1}} c_{k2}, \end{cases} \tag{8.2.7}$$

The corresponding design with mixture $\mathcal{P}_1$ on T^3 is listed in the next three columns of Table 8.3 and their plot is given by Fig. 8.5a. The material of this example is from Fang and Ma (2001).

Let $\boldsymbol{C}_1$ and $\boldsymbol{C}_2$ be two designs on C^{s-1}, and $\mathcal{P}_1$ and $\mathcal{P}_2$ be the two designs with mixture on T^s by the transformation method, respectively. Since the transformation method is an indirect method, it may occur some unreasonable phenomena. For example, although the designs $\boldsymbol{C}_1$ is more uniform than $\boldsymbol{C}_2$ on C^{s-1}, the $\mathcal{P}_2$ is more uniform than $\mathcal{P}_1$ on T^s.

Table 8.3 Two designs with mixtures for $n = 11, s = 3$

U_1		C_1		$\mathcal{P}_1$			U_2		C_2		$\mathcal{P}_2$		
1	4	0.045	0.318	0.787	0.145	0.068	1	8	0.045	0.682	0.787	0.068	0.145
2	9	0.136	0.773	0.631	0.084	0.285	2		0.136	0.409	0.631	0.218	0.151
3	7	0.227	0.591	0.523	0.195	0.282	3	2	0.227	0.136	0.523	0.412	0.065
4	1	0.318	0.045	0.436	0.538	0.026	4	10	0.318	0.864	0.436	0.077	0.487
5	11	0.409	0.955	0.360	0.029	0.611	5	7	0.409	0.591	0.360	0.262	0.378
6	3	0.500	0.227	0.293	0.546	0.161	6	4	0.500	0.318	0.293	0.482	0.225
7	6	0.591	0.500	0.231	0.384	0.384	7		0.591	0.045	0.231	0.734	0.035
8	8	0.682	0.682	0.174	0.263	0.563	8	9	0.682	0.773	0.174	0.188	0.638
9	2	0.773	0.136	0.121	0.759	0.120	9	6	0.773	0.500	0.121	0.440	0.440
10	10	0.864	0.864	0.071	0.127	0.803	10	3	0.864	0.227	0.071	0.718	0.211
11	5	0.955	0.409	0.023	0.577	0.400	11	11	0.955	0.955	0.023	0.044	0.933

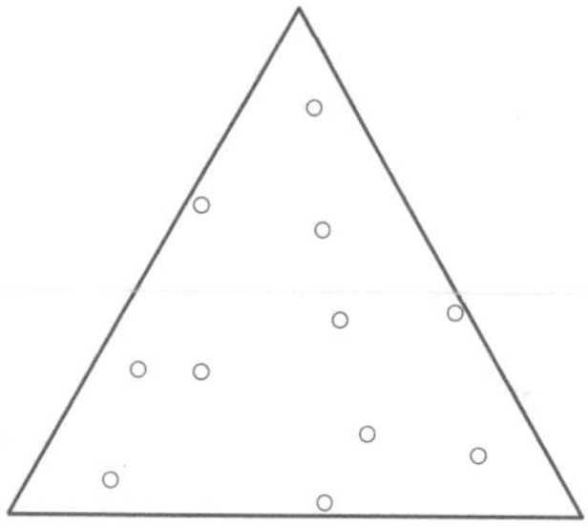
(a) design $\mathcal{P}_1$

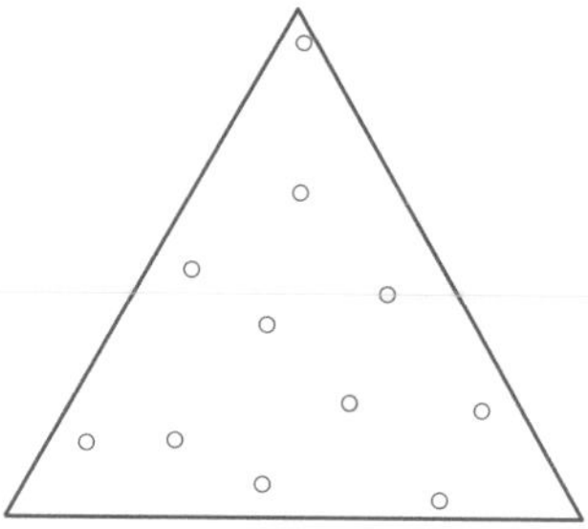
(b) design $\mathcal{P}_2$

Fig. 8.5 Two designs with mixtures on T^3

Example 8.2.2 (*Example* 8.2.1 *Continuity*) Consider the two designs $\mathcal{P}_1$ and $\mathcal{P}_2$ in Table 8.3. The corresponding $\boldsymbol{C}_1$ and $\boldsymbol{C}_2$ on C^2, $\mathcal{P}_1$ and $\mathcal{P}_2$ on T^3 are also shown. The design points of $\mathcal{P}_2$ are plotted in Fig. 8.5b. It can be calculated that the mixture discrepancies of $\boldsymbol{C}_1$ and $\boldsymbol{C}_2$ are 0.0692 and 0.0695, respectively. Then, $\boldsymbol{C}_1$ is more uniform than $\boldsymbol{C}_2$. However, the DM_2-discrepancies of $\mathcal{P}_1$ and $\mathcal{P}_2$, respectively, are 0.0486 and 0.0453, i.e., $\mathcal{P}_2$ is more uniform than $\mathcal{P}_1$ under DM_2-discrepancy. From Fig. 8.5, it also can be seen that $\mathcal{P}_2$ is more uniform in the intuitive view.

From Example 8.2.2, the transformation method may be not the best choice for searching a uniform mixture design. One can consider some direct methods to seek a uniform mixture design. Usually, we can employ some stochastic optimization methods such as the simulated annealing algorithm and the threshold-accepting algorithm under some uniformity criterion, such as the DM_2-discrepancy.

Ning et al. (2011a) considered a direct approach to construct uniform design for mixture experiments. The approach is based on the numerical method NTLBG algorithm and can be applied to search uniform designs on the simplex or other experimental region. The NTLBG algorithm was proposed by Fang et al. (1994), which combined the number-theoretical method and the LBG algorithm proposed by Linde et al. (1980) to generate the representative points for elliptically contoured distributions. The detailed procedure of the NTLBG algorithm for searching n-run uniform mixture designs on T^s is shown as follows.

Algorithm 8.2.2 (*NTLBG algorithm for uniform mixture designs*)

Step 1. For given the number of runs, n and the number of factors, s, choose a positive integer $N >> n$. Generate a N-run uniform design $\mathcal{P}_N = \{\boldsymbol{t}_1, \ldots, \boldsymbol{t}_N\}$ by the transformation method in Algorithm 8.2.1 as a training sample on T^s.

Step 2. Randomly generate n points on the experimental region T^s and take it as the initial design $\mathcal{P}_0 = \{\boldsymbol{x}_1, \ldots, \boldsymbol{x}_n\}$. Calculate the uniformity criterion $D(\mathcal{P}_0)$.

Step 3. Partition the training sample $\mathcal{P}$ into n subsets, i.e.,

$$P_{\boldsymbol{x}_i} = \left\{ \boldsymbol{t}_k : d(\boldsymbol{t}_k, \boldsymbol{x}_i) = \min_{\boldsymbol{x}_j \in \mathcal{P}_0} d(\boldsymbol{t}_k, \boldsymbol{x}_j) \right\}$$

for each $\boldsymbol{x}_i \in \mathcal{P}_0$, $i = 1, \ldots, n$.

Step 4. Calculate the sample mean of $P_{\boldsymbol{x}_i}$

$$\bar{\boldsymbol{x}}_i = \frac{1}{N_i} \sum_{\boldsymbol{t}_j \in P_{\boldsymbol{x}_i}} \boldsymbol{t}_j,$$

where N_i is the cardinal number of $P_{\boldsymbol{x}_i}$. Take $\mathcal{P}_{new} = \{\bar{\boldsymbol{x}}_1, \ldots, \bar{\boldsymbol{x}}_s\}$ as the new design and calculate the uniformity criteria $D(\mathcal{P}_{new})$.

Step 5. If $D(\mathcal{P}_0) - D(\mathcal{P}_{new}) > \alpha > 0$ (α is given small number in advance), set $\mathcal{P}_0 = \mathcal{P}_{new}$ and repeat Steps 3–5. Otherwise, stop the algorithm and take $\mathcal{P}_{new}$ as the finial design.

Some explanations of the Algorithm 8.2.2 are shown as follows. In the Step 1, the number N should be much larger than n. In the Step 2, one can also generate a n-run set by Algorithm 8.2.1 as the initial design. The uniform criterion can be chosen as the DM_2-discrepancy, mean square distance, average distance, or maximum distance. In the Step 3, the partition $\{P_{\boldsymbol{x}}, \boldsymbol{x} \in \mathcal{P}_0\}$ have two properties: (1) $\bigcup_{\boldsymbol{x} \in \mathcal{P}_0} P_{\boldsymbol{x}} = T^s$. (2) $P_{\boldsymbol{x}_i} \bigcap P_{\boldsymbol{x}_j} = \emptyset$ for any $i \neq j$. Ning et al. (2011a) showed that the NTLBG algorithm is powerful to construct uniform designs with mixture.

8.2.3 Uniform Design with Restricted Mixtures

In Example 8.1.1, there are 11 components in a coffee bread, but water and flour should be the major ingredients while sugar, salt, and others have a small percentage. More constraints are needed, for example, $0 \leqslant a_i \leqslant x_i \leqslant b_1 \leqslant 1, i = 1, \ldots, s$, or $\boldsymbol{0} \leqslant \boldsymbol{a} \leqslant \boldsymbol{x} \leqslant \boldsymbol{b} \leqslant \boldsymbol{1}$ where $\boldsymbol{a} = (a_1, \ldots, a_s), \boldsymbol{b} = (b_1, \ldots, b_s)$ and $\boldsymbol{0}$ and $\boldsymbol{1}$ are vectors of 0's and 1's, respectively. The experimental domain becomes $T^s(\boldsymbol{a}, \boldsymbol{b})$ denoted in (8.1.2). It can be easily shown that the domain $T^s(\boldsymbol{a}, \boldsymbol{b})$ is not empty if and only if

$$a \equiv \sum_{i=1}^{n} a_i < 1 < \sum_{i=1}^{n} b_i \equiv b. \tag{8.2.8}$$

The above condition may involve some superfluous constraints that can be removed by the following operation:

$$a_i := \max(a_i, b_i + 1 - b), b_i := \min(b_i, a_i + 1 - a). \tag{8.2.9}$$

Then, for the ith ingredient, its lower bound and upper bound are the updated a_i and b_i by (8.2.9), respectively.

In this subsection, we consider the construction method for uniform design with restricted mixtures on $T^s(\boldsymbol{a}, \boldsymbol{b})$. Wang and Fang (1996) applied the transformation method for the construction of uniform designs on $T^s(\boldsymbol{a}, \boldsymbol{b})$, but their method cannot obtain a good design, especially when some $d_i = b_i - a_i$ are very small. Lately, in order to overcome the disadvantage of the transformation method, Fang and Yang (2000) employed the *conditional method* to establish an alternative method. More discussions of conditional method can be seen in Johnson (1988).

Let $\boldsymbol{x} = (X_1, \ldots, X_s)$ follow the uniform distribution on $T^s(\boldsymbol{a}, \boldsymbol{b})$. The conditional method is based on the facts:

(a) The marginal distribution of X_i can be analytically expressed in a simple form;

(b) The conditional distribution of $X_1, \ldots, X_{s-1}$ for given $X_s = x_s^*$ is the uniform distribution on $T^s(\boldsymbol{a}^*, \boldsymbol{b}^*)$, where $\boldsymbol{a}^*$ and $\boldsymbol{b}^*$ can be easily calculated.

Let us introduce some key formulas for the above (a) and (b). Let $F(x_1, \ldots, x_s)$ be a multivariate distribution function of $(X_1, \ldots, X_s)$. The conditional method in Monte Carlo methods for generating a sample from this distribution is based on the following formula:

$$F(x_1, \ldots, x_s) = F_s(x_s)F_{s-1}(x_{s-1}|x_s) \cdots F_1(x_1|x_2, \ldots, x_s),$$

where $F_s(x_s)$ is the marginal distribution of X_s, $F_{s-1}(x_{s-1}|x_s)$ is the conditional distribution of X_{s-1} given $X_s = x_s$, $F_{s-2}(x_{s-2}|x_{s-1}, x_s)$ is the conditional distribution of X_{s-2} given $X_{s-1} = x_{s-1}, X_s = x_s$, and so on. The conditional method for generating a sample from $F(x_1, \ldots, x_s)$ follows the following steps:

Step 1. Generating a sample, x_s, from the population distribution $F_s(x_s)$.

Step 2. Generating a sample, x_{s-1}, from the conditional distribution $F_{s-1}(x_{s-1}|x_s)$, where x_s is the sample generated in Step 1.

Step 3. Generating a sample, x_{s-k}, from the conditional distribution $F_{s-k}(x_{s-k}|x_s, \ldots, x_{s-k+1})$, where $x_s, \ldots, x_{s-k+1}$ are obtained in the previous steps, $k = 2, \ldots, s-1$.

Step 4. Deliver $\boldsymbol{x} = (x_1, \ldots, x_s)$.

Obviously, implementing the conditional method requires

(1) There is an analytic formula of $F_1, F_2, \ldots, F_s$.

(2) It is easy to generate a sample from $F_s(x_s), F_{s-1}(x_{s-1}|x_s), \ldots, F_1(x_1|x_2, \ldots, x_s)$.

Applying the conditional method to generate a sample from the uniform distribution on the simplex T^s is based on the following results:

(A) The marginal distribution of X_s is given by

$$F_s(x) = \int_0^x (s-1)(1-y)^{s-2} dy = 1 - (1-x)^{s-1}.$$

(B) Transformation. Let $u_2, \ldots, u_s$ be $s-1$ random numbers, i.e., they are *i.i.d.* U(0,1). Set

$$x_s = 1 - (1 - u_s)^{\frac{1}{s-1}},$$

$$x_{s-i} = \{1 - (1 - u_{s-i})^{\frac{1}{s-i-1}}\}\left(1 - \sum_{j=0}^{i-1} x_{s-j}\right), i = 1, \ldots, s-2,$$

$$x_1 = 1 - \sum_{i=2}^{s} x_i.$$

Then $(x_1, \ldots, x_s)$ is a sample from $U(T^s)$ and $(x_2, \ldots, x_s)$ uniformly distributed on $S^{n-1} = \{\boldsymbol{x} : \boldsymbol{x} \in R_+^s, \sum_{i=1}^{s} x_i \leqslant 1\}$.

(C) The conditional distribution $F(x_1, \ldots, x_{s-1}|x_s)$ is the uniform distribution on the region

$$T^{s-1}(1 - x_s) = \{(x_1, \ldots, x_{s-1}) : 0 \leqslant x_j \leqslant 1 - x_s, \\ j = 1, \ldots, s-1, x_1 + \cdots + x_{s-1} = 1 - x_s\}.$$

These properties imply that we can iterate the above (A), (B), and (C) for $F(x_1, \ldots, x_{s-1}|x_s)$ to find the conditional distribution $F_{s-1}(x_{s-1}|x_s)$. Moreover, the distribution of $F(x_1, \ldots, x_{s-2}|x_s, x_{s-1})$ is also the uniform distribution on

$$T^{s-2}(1 - x_s - x_{s-1}) = \{(x_1, \ldots, x_{s-2}) : 0 \leqslant x_j \leqslant 1 - x_s - x_{s-1}, \\ j = 1, \ldots, s-2, x_1 + \cdots + x_{s-2} = 1 - x_{s-1} - x_s\}.$$

The above process can be iterated until to find $F_1(x_1|x_2, \ldots, x_s)$.

If $\boldsymbol{x}$ follows the uniform distribution on $T^s(\boldsymbol{a}, \boldsymbol{b})$, the above theory and method can also be applied. Let $\boldsymbol{b}^* = (\boldsymbol{b} - \boldsymbol{a})/(1 - a)$, and $\boldsymbol{y} = (Y_1, \ldots, Y_s) = (\boldsymbol{x} - \boldsymbol{a})/(1 - a)$ where a is defined in (8.2.8). Then $\boldsymbol{y}$ follows the uniform distribution on $T^s(\boldsymbol{0}, \boldsymbol{b}^*)$. Without loss of generality, we can assume $\boldsymbol{a} = \boldsymbol{0}$. Then we only need to focus on generation of samples from the uniform distribution on $T^s(\boldsymbol{0}, \boldsymbol{b})$. It can be found that the marginal distribution of Y_s is

$$F_{Y_s}(y) = \frac{P(d_s \leqslant X_s \leqslant y)}{P(d_s \leqslant X_s \leqslant b_s)} = \begin{cases} 1, & y \geqslant b_s, \\ \dfrac{(1 - d_s)^{s-1} - (1 - y)^{s-1}}{(1 - d_s)^{s-1} - (1 - b_s)^{s-1}}, & d_s \leqslant y \leqslant b_s, \\ 0, & y \leqslant d_s, \end{cases}$$

where $d_s = \max(0, 1 + b_s - b)$. Let u follow $U(0, 1)$. From the inverse transformation method in Monte Carlo method, it is easy to know that

$$y_s \equiv F_{Y_s}^{-1}(u) = 1 - [u(1 - b_s)^{s-1} + (1 - u)(1 - d_s)^{s-1}]^{1/(s-1)}$$

is a random sample from $F_{Y_s}(y)$. For a given $Y_s = y_s$, the conditional distribution of $(Y_1, \ldots, Y_{s-1})$ is the uniform distribution on $T^{s-1}(\boldsymbol{0}, \boldsymbol{b}_{(1)})$, where $\boldsymbol{b}_{(1)} = (b_1, \ldots, b_{s-1})/(1 - y_s)$. Then apply the above method to the uniform distribution on $T^{s-1}(\boldsymbol{0}, \boldsymbol{b}_{(1)})$.

Let

$$\Delta_s = 1,\ \Delta_k = 1 - \sum_{i=k+1}^{s} y_i,\ k = s-1, \ldots, 2,$$

$$d_k = \max\left\{a_k/\Delta_k,\ 1 - \sum_{i=1}^{k-1} b_i/\Delta_k\right\},\ k = s, s-1, \ldots, 2,$$

$$\Phi_k = \max\left\{b_k/\Delta_k,\ 1 - \sum_{i=1}^{k-1} a_i/\Delta_i\right\},\ k = s, s-1, \ldots, 2,$$

$$G(u, d, b, c, k) = c\{1 - [u(1-b)^k + (1-u)(1-d)^k]^{1/k}\}.$$

If we generate $s-1$ random numbers $u_2, \ldots, u_s$ from $U(0, 1)$ and denote

$$x_k = G(u_k, d_k, \Phi_k, \Delta_k, k-1),\ k = s, s-1, \ldots, 2,$$

$$x_1 = 1 - \sum_{k=2}^{s} x_k,$$

then $\boldsymbol{x} = (x_1, \ldots, x_s)$ is a sample from the uniform distribution on $T^s(\boldsymbol{a}, \boldsymbol{b})$. The proof can be seen in Fang and Yang (2000). Then, the construction of uniform design with restricted mixtures on $T^s(\boldsymbol{a}, \boldsymbol{b})$ can be seen as follows.

Algorithm 8.2.3 (*Conditional method for UD with restricted mixtures*)

Step 1. Choose a uniform design $U_n(n^{s-1})$ and denote it by $\boldsymbol{U} = (u_{ij})$;
Step 2. Calculate

$$t_{ij} = \frac{u_{ij} - 0.5}{n},\ i = 1, \ldots, n, j = 1, \ldots, s-1;$$

Step 3. For each i, apply the above step to find $(t_{i1}, \ldots, t_{i,s-1})$ as $u_2, \ldots, u_s$ and to calculate

$$x_{ik} = G(t_{ik}, d_k, \Phi_k, \Delta_k, k-1),\ k = s, s-2, \ldots, 2,$$

$$x_{i1} = 1 - \sum_{k=2}^{s} x_k.$$

Then $\{\boldsymbol{x}_i = (x_{i1}, \ldots, x_{is}), i = 1, \ldots, n\}$ is a UD on $T^s(\boldsymbol{a}, \boldsymbol{b})$.

Example 8.2.3 This example is from Fang (2002). In a pharmaceutical study, the task is to dissolve a slightly polar drug in a mixture of water and two cosolvents, ethanol and propylene glycol for increasing the drug's solubility. The experimenter was of interest to know whether and where a maximum exists in the solubility profile of the drug in the mixture of solvents. The response measured is the vapor pressure

Table 8.4 Design and responses in Example 8.2.3

No.	x_1	x_2	x_3	y
1	0.12296	0.11323	0.76381	31.5042
2	0.35453	0.04379	0.60168	51.5640
3	0.24176	0.36565	0.39259	30.0888
4	0.26535	0.47616	0.25849	28.9297
5	0.05914	0.41827	0.52258	21.5290
6	0.55981	0.07786	0.36233	56.6437
7	0.52303	0.22904	0.24793	44.9973
8	0.20319	0.18860	0.60821	36.6152
9	0.11576	0.54129	0.34296	20.3480
10	0.40656	0.15008	0.44336	45.6282
11	0.14105	0.27173	0.58723	30.8623
12	0.33631	0.31707	0.34662	37.0407

(y) (mm Hg). The three factors ethanol (x_1), propylene glycol (x_2), and water (x_3) were chosen on the domain

$$T^3(\boldsymbol{a}, \boldsymbol{b}) = \{\boldsymbol{x} = (x_1, x_2, x_3) : 0.0463 \leqslant x_1 \leqslant 0.7188,$$
$$0.0272 \leqslant x_2 \leqslant 0.5776, 0.2272 \leqslant x_3 \leqslant 0.9265, x_1 + x_2 + x_3 = 1\}.$$

Choose $n = 12$ for this experiment with restricted mixtures on $T^3(\boldsymbol{a}, \boldsymbol{b})$. The 12 points and the corresponding responses are shown in Table 8.4.

A real-life case study is given by Jing et al. (2007) who applied the method introduced in this section to the laccase production from trametes versicolor by solid fermentation. They chose a $U_{15}(5^4)$ table for the experiment. More details can be seen in the original paper.

8.2.4 Uniform Design on Irregular region

In some experiments with mixtures, there are more constraints among the components $x_1, \ldots, x_s$ and the corresponding experimental domain may be irregular region in R^s.

Example 8.2.4 Consider the problem in Borkowski and Piepel (2009). Three components x_1, x_2, and x_3 are needed for generating some product, and they have constraints:

$$0.1 \leqslant x_1 \leqslant 0.7,\ 0 \leqslant x_2 \leqslant 0.8,\ 0.1 \leqslant x_3 \leqslant 0.6 \text{ and } x_1 - x_2 \geqslant 0.$$

The corresponding experimental region, denoted by S, can be seen in Fig. 8.6a, i.e., the area formed by sequentially linking the points C, D, E, F, G, and the equi-

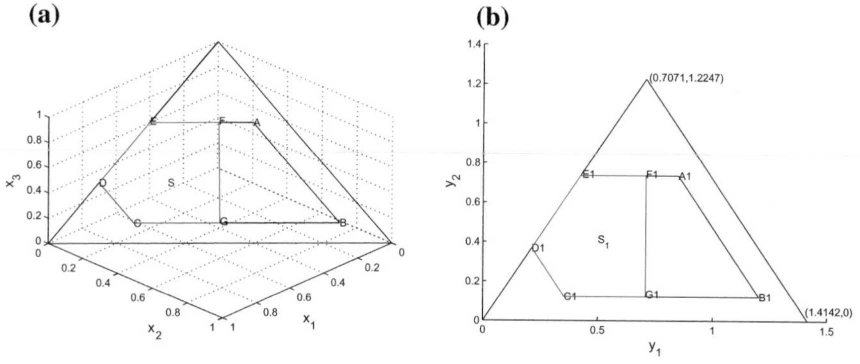

Fig. 8.6 Illustration for Example 8.2.4: **a** the experimental region S (i.e., the region enclosed by points C, D, E, F, G), **b** the image of S after the mapping (i.e., the region enclosed by points C_1, D_1, E_1, F_1, G_1)

lateral triangle with the vertices $(1, 0, 0)$, $(0, 1, 0)$, and $(0, 0, 1)$ refers to the simplex $T^3 = \{(x_1, x_2, x_3)|x_1 + x_2 + x_3 = 1, x_i \geqslant 0\}$. For the two-dimensional S, we can transform it into R^2 by using the isometric transformation, which will be shown in this subsection. Figure 8.6b shows the image of S by using the isometric transformation, denoted by S_1, i.e., the region enclosed by the points C_1, D_1, E_1, F_1, G_1, and the vertices of the simplex T^3, respectively, become $(0, 0)$, $(\sqrt{2}, 0)$ and $(\sqrt{2}/2, \sqrt{6}/2)$ through the mapping.

A. Uniformity Measure

For measuring the uniformity of the designs on the irregular region S, the DM_2-discrepancy introduced in the last subsection is not suitable, and the distance criteria such as mean square distance, average distance, and maximum distance, respectively, in (8.1.5)–(8.1.7) may be used. However, those distance criteria do not have explicit expressions. Chuang and Hung (2010) proposed the *central composite discrepancy* (CCD) to measure the uniformity of the designs on S.

For any fixed point $\boldsymbol{x} = (x_1, \ldots, x_s)$ in S, the $(s-1)$-dimensional hyperplane which is perpendicular to the ith axis can chop the ith axis into two parts, i.e., $(-\infty, x_i]$ and $(x_i, +\infty)$, through the point $\boldsymbol{x}$, where $i = 1, \ldots, s$. Then the region S can be divided into 2^s small grids by the s hyperplanes referred to above, denoted by $S_1(\boldsymbol{x}), \ldots, S_{2^s}(\boldsymbol{x})$, with point $\boldsymbol{x}$ being the center of them. Let $\mathcal{P}$ denote a n-point design on S. The CCD of $\mathcal{P}$ is defined by

$$CCD_p(n, \mathcal{P}) = \left\{ \frac{1}{V(S)} \int_S \frac{1}{2^s} \sum_{t=1}^{2^s} \left| \frac{N(S_t(\boldsymbol{x}), \mathcal{P})}{n} - \frac{V(S_t(\boldsymbol{x}))}{V(S)} \right|^p \mathrm{d}\boldsymbol{x} \right\}^{1/p}, \quad (8.2.10)$$

where $p > 0$, $N(S_t(\boldsymbol{x}), \mathcal{P})$ denotes the number of points of $\mathcal{P}$ falling into $S_t(\boldsymbol{x})$, and $V(S_t(\boldsymbol{x}))$ and $V(S)$ denote the volumes of $S_t(\boldsymbol{x})$ and S, respectively. Apparently, a small $CCD_p(n, \mathcal{P})$-value implies a relatively uniform design. Denote $Z(n)$ as the set composed of all the n-point designs on S. The uniform design on S with a given run size n, $\mathcal{P}^*$, is defined by

$$\mathcal{P}^* = \arg \min_{\mathcal{P} \in Z(n)} CCD_p(n, \mathcal{P}).$$

The positive value p in (8.2.10) is often equal to 2, as similar as that in CD, WD, and MD for the designs on hypercubes. In a practical application, the volumes $V(S_t(\boldsymbol{x}))$ and $V(S)$ may be difficult to compute, as well as the integral in the region S. Therefore, the region S is often discretized into N points ($N \gg n$), and the $CCD_p(n, \mathcal{P})$-value in (8.2.10) can be approximately calculated through the following expression:

$$CCD_2(n, \mathcal{P}) \approx \left\{ \frac{1}{N} \sum_{i=1}^{N} \frac{1}{2^s} \sum_{t=1}^{2^s} \left| \frac{N(S_t(x_i), \mathcal{P})}{n} - \frac{N(S_t(x_i))}{N} \right|^2 \right\}^{1/2}. \tag{8.2.11}$$

The N points can be chosen as the nearly N-point uniform design or the lattice points on S.

B. Construction Methods

It is not easy to construct uniform designs on the irregular region S under a given uniformity criterion such as CCD. There are some construction methods in the literature such as the switching algorithm (Chuang and Hung 2010) and the discrete particle swarm optimization (Chen et al. 2014). Moreover, Liu and Liu (2016) proposed a construction method of uniform designs for mixture experiments with complex constraints.

Chuang and Hung (2010) showed that the *switching algorithm* has less iteration times and quicker convergence speed saves time dramatically compared with the exhaustive search, and the designs obtained via such algorithm are extremely close to the uniform designs. The procedure of switching algorithm is as follows.

Algorithm 8.2.4 (*Switching Algorithm*)

Step 1. Choose N-point nearly uniform design $\mathcal{P} = \{\boldsymbol{x}_1, \ldots, \boldsymbol{x}_N\}$ on S, where $N \gg n$.

Step 2. Arbitrary choose n points in $\mathcal{P}$ as the initial current design "Cdesign," for example, choose Cdesign $= \{\boldsymbol{x}_1, \ldots, \boldsymbol{x}_n\}$; set the iteration counter $i = 0$ and the next design Ndesign $=$ Cdesign.

Step 3. While $i = 0$ or Ndesign $\neq$ Cdesign
    set $i = i + 1$, Cdesign=Ndesign;
    for $j = 1$ to n do
        let $\boldsymbol{x}^* = \arg\min_{\boldsymbol{x} \in \mathcal{P} \setminus \text{Ndesign}} CCD_p(n, \{\boldsymbol{x}\} \bigcup \text{Ndesign} \setminus \{\boldsymbol{x}_j\})$;
        if $CCD_p(n, \{\boldsymbol{x}^*\} \bigcup \text{Ndesign} \setminus \{\boldsymbol{x}_j\}) \leqslant CCD_p(n, \text{Ndesign})$
            set Ndesign $= \{\boldsymbol{x}^*\} \bigcup \text{Ndesign} \setminus \{\boldsymbol{x}_j\}$;
        end if

end for

end while.

Step 4. Export Cdesign, $CCD_p(n, \text{Cdesign})$ and i.

In Step 1 of Algorithm 8.2.4, a larger number of points N means that a more uniform final design will be obtained but under a more time-consuming process. Usually, we choose N such that $N/n > 5$. In the Step 3 of Algorithm 8.2.4, the notation Ndesign$\backslash\{\boldsymbol{x}_j\}$ means that the point $\{\boldsymbol{x}_j\}$ is deleted from the design Ndesign. The CCD-value of the designs can be calculated by the approximate expression in (8.2.11). The Algorithm 8.2.4 is a local searching algorithm and can be used for any irregular region S on R^s.

Additionally, Liu and Liu (2016) proposed a construction method when the irregular region S is a partial region of simplex, i.e.,

$$S = \left\{\boldsymbol{x} = (x_1, \ldots, x_s) | \sum_{i=1}^{s} x_i = 1, x_i \geqslant 0, f_j(\boldsymbol{x}) \leqslant 0, j = 1, \ldots, t\right\}, \quad (8.2.12)$$

where $f_j(\boldsymbol{x})$ can be linear or nonlinear function. For example, the constraints in Example 8.2.4 can be rewritten by some linear inequalities $f_j(\boldsymbol{x}) \leqslant 0, j = 1, \ldots, t$.

The construction method by Liu and Liu (2016) is based on some transformation of the simplex T^s. It is known that the simplex T^s can be transformed into the hyperplane $H = \{(z_1, \ldots, z_s)|z_s = 0\}$ with its shape and size invariant via the mapping

$$M : \boldsymbol{z} = [\boldsymbol{x} - (1, 0, \ldots, 0)]Q, \quad (8.2.13)$$

where $\boldsymbol{x} \in T^s$, $\boldsymbol{z} \in H$ and Q is the orthogonal matrix coming from the matrix QR decomposition $\begin{pmatrix} -1_{1\times(s-1)} \\ I_{s-1} \end{pmatrix} = Q \begin{pmatrix} R_{(s-1)\times(s-1)} \\ 0_{1\times(s-1)} \end{pmatrix}$, with I_{s-1} being a unity matrix of order $s-1$, $R_{(s-1)\times(s-1)}$ being an upper triangular matrix, $-1_{1\times(s-1)}$ and $0_{1\times(s-1)}$, respectively, being the $(s-1)$-dimensional row vector whose elements are -1 and 0. The geometry formed by any subset of the simplex T^s is identical with its image in H through the mapping M. The inverse mapping of M can be written as

$$M^{-1} : \boldsymbol{x} = \boldsymbol{z}Q' + (1, 0, \ldots, 0), \quad (8.2.14)$$

where $\boldsymbol{z} \in M(T^s)$ and $\boldsymbol{x} \in T^s$.

Under the CCD criterion, the construction method by Liu and Liu (2016) for nearly uniform designs (NUD) for mixture experiments on the region in (8.2.12) is as follows.

Algorithm 8.2.5 (*Constructing uniform designs on T^s*)

Step 1 Let S denote the experimental region of a mixture experiment with some constraints, as defined in (8.2.12).

step 2. Transform S into R^{s-1} through the mapping M in (8.2.13). Denote the image after the transformation as S_1.

step 3. Given the run size n, search the NUD in S_1 under the CCD criterion by the switching algorithm in Algorithm 8.2.4.

step 4. Transform the points of the NUD into the simplex T^s, by the inverse mapping M^{-1} in (8.2.14), to obtain the final NUD in S.

An advantage of the Algorithm 8.2.5 is that it can deal with mixture experiments with any complex constraints, e.g., the functions $f_i(\boldsymbol{x}), i = 1, \ldots, t$, can be nonlinear.

Example 8.2.5 (*Example* 8.2.4 *Continues*) Use the Algorithm 8.2.5 to construct the (nearly) uniform design on the irregular region S. Let $n = 21$. It needs to obtain the orthogonal matrix Q in the mapping M in (8.2.13). According to the QR decomposition, we have

$$\begin{pmatrix} -1 & -1 \\ 1 & 0 \\ 0 & 1 \end{pmatrix} = Q_{3\times 3} \begin{pmatrix} R_{2\times 2} \\ 0_{1\times 2} \end{pmatrix}$$

$$= \begin{pmatrix} -\sqrt{2}/2 & -\sqrt{6}/6 & \sqrt{3}/3 \\ \sqrt{2}/2 & -\sqrt{6}/6 & \sqrt{3}/3 \\ 0 & \sqrt{6}/3 & \sqrt{3}/3 \end{pmatrix} \begin{pmatrix} \sqrt{2} & \sqrt{2}/2 \\ 0 & \sqrt{6}/2 \\ 0 & 0 \end{pmatrix},$$

then the mapping and its inverse mapping can be expressed as

$$M: \quad \boldsymbol{z} = [\boldsymbol{x} - (1, 0, 0)] \begin{pmatrix} -\sqrt{2}/2 & -\sqrt{6}/6 & \sqrt{3}/3 \\ \sqrt{2}/2 & -\sqrt{6}/6 & \sqrt{3}/3 \\ 0 & \sqrt{6}/3 & \sqrt{3}/3 \end{pmatrix}, \text{ and}$$

$$M^{-1}: \quad \boldsymbol{x} = \boldsymbol{z} \begin{pmatrix} -\sqrt{2}/2 & \sqrt{2}/2 & 0 \\ -\sqrt{6}/6 & -\sqrt{6}/6 & \sqrt{6}/3 \\ \sqrt{3}/3 & \sqrt{3}/3 & \sqrt{3}/3 \end{pmatrix} + (1, 0, 0),$$

respectively.

By the mapping M, we transform the region S to S_1 in R^2, as shown in Fig. 8.6b. We divide the rectangle $[0, \sqrt{2}] \times [0, \sqrt{6}/2]$ into 30×30 small grids with the same size, take all the center points of these grids, and keep the ones just falling into S_1. There are 128 points in the continuous region S_1. Let $N = 128$. For the given run size $n = 21$, the NUD obtained by the switching algorithm in Algorithm 8.2.4 on S_1 is shown in Fig. 8.7a, and the corresponding NUD on S by Algorithm 8.2.5 is drawn in Fig. 8.7b. It can be seen that the design points are uniformly scattered on S.

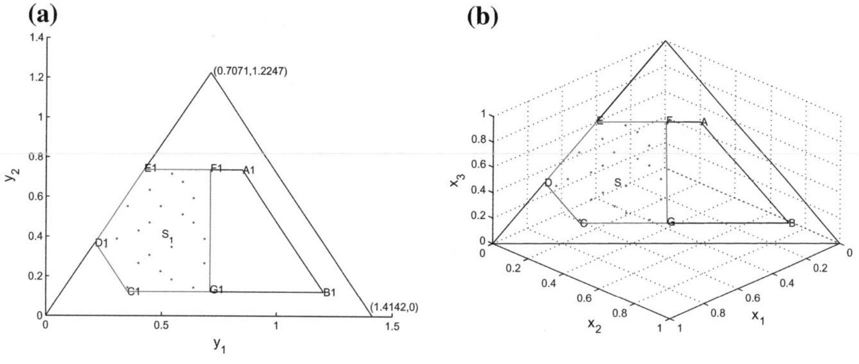

Fig. 8.7 Uniform design for Example 8.2.4: **a** the NUD with $n = 21$ in the image of S, S_1 and **b** the final NUD with $n = 21$ in S

8.3 Modeling Technique for Designs with Mixtures

The first-order model

$$E(y) = \beta_0 + \sum_{i=1}^{s} \beta_i x_i, \tag{8.3.1}$$

the second-order model

$$E(y) = \beta_0 + \sum_{i=1}^{s} \beta_i x_i + \sum_{i=1}^{s} \beta_{ii} x_i^2 + \sum_{i<j}^{s} \beta_{ij} x_i x_j, \tag{8.3.2}$$

and the centered second-order model

$$E(y) = \beta_0 + \sum_{i=1}^{s} \beta_i x_i + \sum_{i=1}^{s} \beta_{ii}(x_i - \bar{x}_i)^2 + \sum_{i<j}^{s} \beta_{ij}(x_i - \bar{x}_i)(x_j - \bar{x}_j), \tag{8.3.3}$$

are popularly used on hypercube. These polynomial regression models are also often employed as suitable response surface model for experiments with mixtures. Moreover, according to the constraint $x_1 + \cdots + x_s = 1$ for experimental design with mixtures, the first-order model (8.3.1) becomes

$$E(y) = \beta_0 \sum_{i=1}^{s} x_i + \sum_{i=1}^{s} \beta_i x_i = \sum_{i=1}^{s} \beta_i^* x_i, \tag{8.3.4}$$

where $\beta_i^* = \beta_0 + \beta_i, i = 1, \ldots, s$. For the second-order model (8.3.2), since $\sum_{i=1}^{s} x_i = 1$ and $x_i^2 = x_i(1 - \sum_{j \neq i} x_j)$, we have

$$
\begin{aligned}
E(y) &= \beta_0 \sum_{i=1}^{s} x_i + \sum_{i=1}^{s} \beta_i x_i + \sum_{i=1}^{s} \beta_{ii} x_i \left(1 - \sum_{j \neq i} x_j \right) + \sum_{i<j}^{s} \beta_{ij} x_i x_j \\
&= \sum_{i=1}^{s} (\beta_0 + \beta_i + \beta_{ii}) x_i - \sum_{i=1}^{s} \beta_{ii} x_i \sum_{j \neq i} x_j + \sum_{i<j}^{s} \beta_{ij} x_i x_j \\
&= \sum_{i=1}^{s} \beta_i^* x_i + \sum_{i<j}^{s} \beta_{ij}^* x_i x_j,
\end{aligned} \tag{8.3.5}
$$

where $\beta_i^* = \beta_0 + \beta_i + \beta_{ii}$ and $\beta_{ij}^* = \beta_{ij} - \beta_{ii} - \beta_{jj}$. Similarly, the third-order model can be rewritten as

$$
E(y) = \sum_{i=1}^{s} \beta_i^* x_i + \sum_{i<j}^{s} \beta_{ij}^* x_i x_j + \sum_{i<j}^{s} \delta_{ij} x_i x_j (x_i - x_j) + \sum_{i<j<k}^{s} \beta_{ijk}^* x_i x_j x_k. \tag{8.3.6}
$$

Note that the models (8.3.4)–(8.3.6) do not include the intercept term, quadratic term or cubic term, and these models are called as the *Scheffé's polynomial models* or *canonical form of the polynomial*. Optimal design theory can be applied to these Scheffé's polynomial models. A comprehensive review for optimal design for experiments with mixture can refer to Chan (2000).

According to the constraint $x_1 + \cdots + x_s = 1$, then $x_s = 1 - x_1 - \cdots - x_{s-1}$ and we can delete the factor x_s in the models (8.3.1)–(8.3.3), i.e., the following models can be fitted to the s-factor design with mixtures

$$
E(y) = \beta_0 + \sum_{i=1}^{s-1} \beta_i x_i, \tag{8.3.7}
$$

$$
E(y) = \beta_0 + \sum_{i=1}^{s-1} \beta_i x_i + \sum_{i=1}^{s-1} \beta_{ii} x_i^2 + \sum_{i<j}^{s-1} \beta_{ij} x_i x_j, \tag{8.3.8}
$$

$$
E(y) = \beta_0 + \sum_{i=1}^{s-1} \beta_i x_i + \sum_{i=1}^{s-1} \beta_{ii} (x_i - \bar{x}_i)^2 + \sum_{i<j}^{s-1} \beta_{ij} (x_i - \bar{x}_i)(x_j - \bar{x}_j). \tag{8.3.9}
$$

The following example exhibits new problems in modeling for data of experiments with mixtures. More detailed discussion can refer to Cornell (2002).

Example 8.3.1 Choose three metals x_1, x_2, x_3 in a new material for investigation. A UD $U_{15}(15^2)$ was employed for the design. The design and related responses are listed in Table 8.5, where there list only x_1 and x_2 as $x_1 + x_2 + x_3 = 1$. With the same reason, fitting regression models to the data involve only x_1 and x_2. By the use of selection of variables, the model

Table 8.5 Design and responses

No.	x_1	x_2	y	No.	x_1	x_2	y
1	0.81743	0.10346	8.2256	9	0.24723	0.17565	10.1362
2	0.68377	0.05271	8.7794	10	0.20418	0.76930	9.3760
3	0.59175	0.36742	9.5115	11	0.16334	0.25100	10.2772
4	0.51695	0.17712	9.5619	12	0.12440	0.55454	9.8652
5	0.45228	0.41992	9.9145	13	0.08713	0.09129	10.1022
6	0.39447	0.02018	9.5526	14	0.05132	0.79057	9.1792
7	0.34172	0.32914	9.9481	15	0.01681	0.42605	9.9565
8	0.29289	0.49497	10.1241				

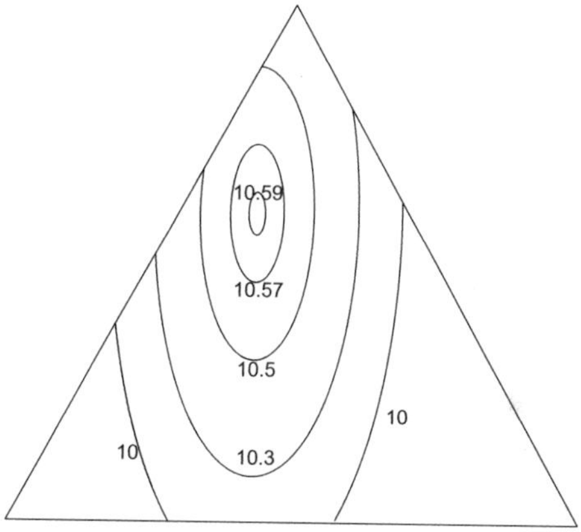

Fig. 8.8 Contour plot of (8.3.10)

$$\hat{y} = 10.472 - 1.20(x_1 - 0.3324) - 3.475(x_1 - 0.3324)^2 - 3.333(x_2 - 0.3349)^2 + 2.322(x_1 - 0.3324)(x_2 - 0.3349) \qquad (8.3.10)$$

with $R = 0.9887$, $\hat{\sigma} = 0.07$ is recommended. The model shows that there is an interaction between x_1 and x_2. Its contours are shown in Fig. 8.8. Note that due to the constraint $x_1 + x_2 + x_3 = 1$, the metal x_3 does not appear in the model (8.3.10) that is not convenience in the practice. There are a lot of discussions in Cornell (2002) to solve this problem.

Example 8.3.1 considers the modeling technique for the data of design with mixtures. Next two examples consider the modeling technique for the data of design with restricted mixtures.

Example 8.3.2 (*Example* 8.2.3 *Continuity*) For the data in Table 8.4, the underlying model between the response and factors is unknown. The major goal here is to establish a suitable model. The best result among the 12 responses is $y_6 = 56.6437$

mm Hg, which can be served as a benchmark. We wish to know whether there is any level-combination to produce a better vapor pressure.

The First-Order Linear Regression Model

The simplest model is the first-order regression. Based on the data in Table 8.4, we have model

$$\hat{y} = -2.4344 + 79.7565x_1 + 36.4051x_3,$$

with $R^2 = 96.87\%$ and $s^2 = 4.9638$. Statistical diagnostics based on the ANOVA, residual plot, normal plot, and partial residual plots indicate that this model is not satisfactory. Therefore, we consider the more complicated second-order quadratic regression model.

Quadratic Regression Model

With model selection technique, we find a metamodel

$$\hat{y} = 22.6130 - 11.3570x_2 + 112.205x_1x_3 + 38.0345x_1^2, \qquad (8.3.11)$$

with $R^2 = 99.17\%$ and $s^2 = 1.4747$. The corresponding ANOVA table is shown in Table 8.6. Statistical diagnostics, the residual plot, and normal plot are shown in Figs. 8.9 and 8.10, which indicate that the model (8.3.11) is acceptable. Maximizing y with respect to x_i, $i = 1, 2, 3$ under models (8.3.11) on the domain $T^3(\boldsymbol{a}, \boldsymbol{b})$, we find that $\max \hat{y} = 62.4414$ at $x_1 = 0.7188, x_2 = 0.0272, x_3 = 0.2540$.

Centered Quadratic Regression Model

Next, we consider the second-order centered quadratic regression model. Once again, by using model selection techniques, a metamodel is

Table 8.6 ANOVA table for metamodel (8.3.11)

Analysis of Variance					
Source	DF	Sum of Squares	Mean Square	F Stat	Prob > F
Model	3	1416.4800	472.1600	320.1736	0.0001
Error	8	11.7976	1.4747		
C Total	11	1428.2776			

Type III Tests					
Source	DF	Sum of Squares	Mean Square	F Stat	Prob > F
X2	1	9.0476	9.0476	6.1352	0.0383
X1X3	1	75.8957	75.8957	51.4652	0.0001
X1X1	1	71.9499	71.9499	48.7895	0.0001

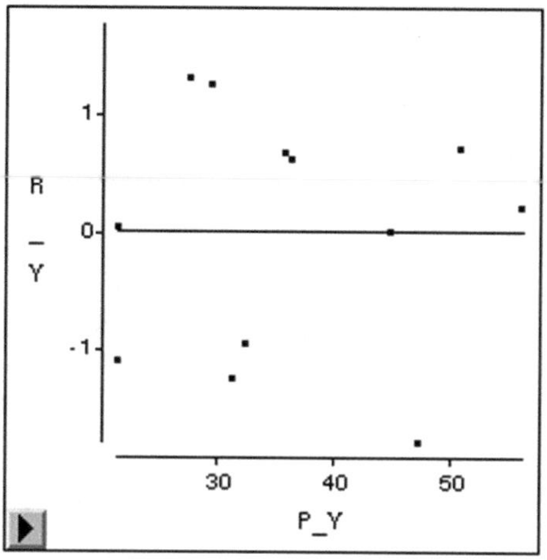

Fig. 8.9 Residual plot for metamodel (8.3.11)

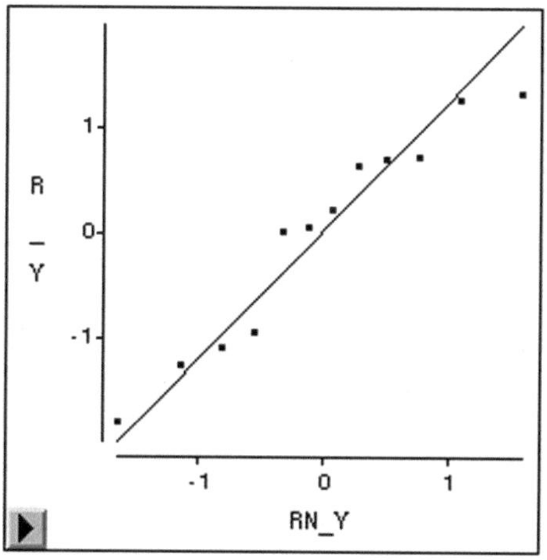

Fig. 8.10 Normal plot for metamodel (8.3.11)

$$\hat{y} = 37.3126 + 84.6285(x_1 - 0.2775) + 40.9079(x_3 - 0.4565)$$
$$+84.0939(x_1 - 0.2775)(x_3 - 0.4565), \qquad (8.3.12)$$

with $R^2 = 98.94\%$ and $s^2 = 1.8969$. The corresponding ANOVA table is shown in Table 8.7. The residual plot and normal plot are shown in Figures 8.11–8.12. Similarly, these plots indicate that the model (8.3.12) is also acceptable. Then we maximize y with respect to x_i, $i = 1, 2, 3$ under models (8.3.12) on the domain $T^3(\boldsymbol{a}, \boldsymbol{b})$ and find that $\max \hat{y} = 59.2179$ at the same point to that of model (8.3.11), i.e., $x_1 = 0.7188, x_2 = 0.0272, x_3 = 0.2540$.

By some additional experiments at the ingredient-combination $x_1 = 0.7188$, $x_2 = 0.0272$, $x_3 = 0.2540$, the average of vapor pressure is 61.75. From the viewpoint

Table 8.7 ANOVA table for metamodel (8.3.12)

Analysis of Variance					
Source	DF	Sum of Squares	Mean Square	F Stat	Prob > F
Model	3	1413.1024	471.0341	248.3185	0.0001
Error	8	15.1752	1.8969		
C Total	11	1428.2776			

Type III Tests					
Source	DF	Sum of Squares	Mean Square	F Stat	Prob > F
X1	1	1321.3692	1321.3692	696.5958	0.0001
X3	1	307.5655	307.5655	162.1415	0.0001
X1X3	1	29.4660	29.4660	15.5338	0.0043

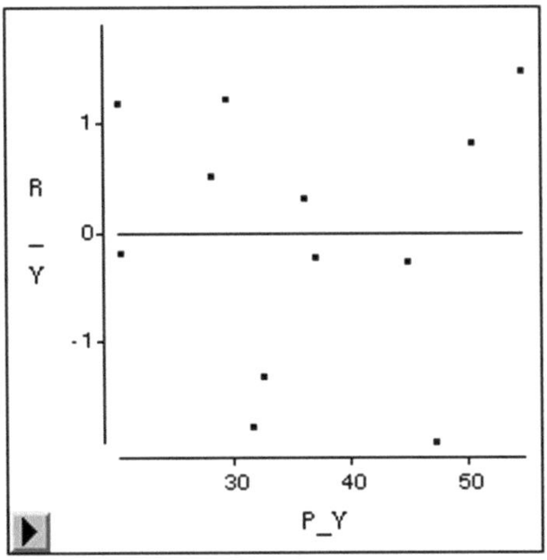

Fig. 8.11 Residual plot for metamodel (8.3.12)

of optimization and prediction, we prefer metamodel (8.3.11) as our last model. Figure 8.13 shows contours of the metamodel (8.3.11).

The following example adopted from Tang et al. (2004)) shows applications of the uniform design with mixtures in product formation in the cement manufacturing industry.

Example 8.3.3 Cement matrix grouting material has been commonly used in the construction industry, since it has high durability and high strength. The material is non-toxic, non-polluting, and relatively low in cost. However, there are some disadvantages of ordinary cement matrix grouting material, such as low stability, low workability, and low water retentivity. This is especially true when the water/cement ratio is high. For overcoming these shortcomings, there are some experience as follows: The presence of appropriate additives can improve the quality of this grouting

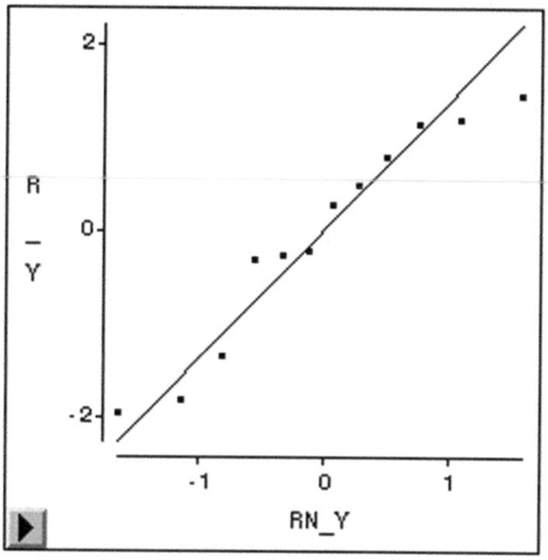

Fig. 8.12 Normal plot for metamodel (8.3.12)

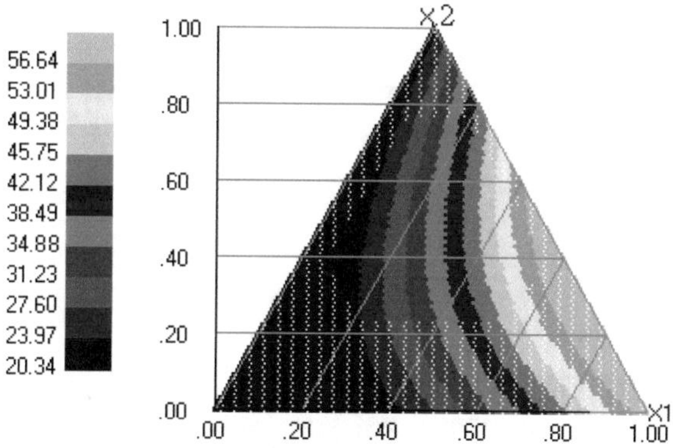

Fig. 8.13 Contours of metamodel (8.3.11)

material; inorganic materials such as silica fume will increase the strength, water retentivity, and stickiness of the mixture, and reduce segregation; organic materials such as carboxyl methyl cellulose (CMC) will increase the stickiness of the mixture and thus reduce segregation; flyash will increase the workability of the mixture. As flyash is an industrial waste from thermal power stations, making use of it will help protect the environment. The engineer has to determine how much should each of these additives be added so that a cost-effective grouting material of good quality can be formed.

Four controllable factors denoted by x_1, x_2, x_3, and x_4 were chosen: percentages of flyash, silica fume, CMC, and cement. Experience suggested to consider ranges of the three factors in percentages to be:

$$5 \leqslant x_1 \leqslant 20, \;\; 1 \leqslant x_2 \leqslant 2.4, \;\; 0.3 \leqslant x_3 \leqslant 1.0.$$

Because of the constraint $x_1 + x_2 + x_3 + x_4 = 100$, the amount of cement x_4 should lie within the range $76.6 \leqslant x_4 \leqslant 93.7$. The objectives of this experiment were to minimize the coefficient of bleeding BL (at water/cement ratio of 0.6) and maximize the compressive strength R_{28} (at water/cement ration of 0.8) which is measured twenty-eight days after the cement mixture has set.

A uniform design table $U_{16}(16 \times 8^2)$ was used for a design, and two quadratic models are separately chosen for the two responses

$$\begin{aligned} BL &= 3.337x_1 - 341.3x_3 + 0.3655x_4 + 2.998x_1x_3 - 0.04575x_1x_4 \\ &\quad +5.512x_2x_3 - 0.0561x_2x_4 + 3.323x_3x_4, \qquad (8.3.13) \\ R_{28} &= -6.694x_1 + 3.435x_2 - 4824x_3 - 0.5854x_1x_2 + 50.49x_1x_3 \\ &\quad +0.0858x_1x_4 + 61.29x_2x_3 + 48.53x_3x_4. \qquad (8.3.14) \end{aligned}$$

Tang et al. (2004) gave more discussion on behavior of BL and R_{28} and an optimal material cost was obtained. The authors concluded "Factorial designs and orthogonal arrays have been widely used in design of industrial experiments. When the number of factors is large or the numbers of levels of the factors are large, these designs require a large number of runs, which may not be possible to achieve in practice because of various constraints. In such a case, the uniform design is an excellent alternative that can be used for the experiments."

Exercises

8.1

Give experimental points of the simplex-lattice designs $\{3, 3\}$, $\{4, 3\}$ and their plots by the use of MATLAB or other software.

8.2

The domain T^3 is an equilateral triangular with side-length $\sqrt{2}$, denoted by V^2, say. Therefore, any point (z_1, z_2) on V^2 corresponds to a point (x_1, x_2, x_3) on T^3. Choose a new coordinate system on V^2 and give the mapping rule of $(x_1, x_2, x_3) \Rightarrow (z_1, z_2)$.

8.3

Suppose we choose a uniform design $U_7(7^2)$ as follows:
Construct a uniform design on $T^3 = \{(x_1, x_2, x_3) : x_i > 0, i = 1, 2, 3, x_1 + x_2 + x_3 = 1.\}$ with 7 runs by using the translation method based on the given $U_7(7^2)$.

8.4

Let $n = 17$.

(1) Randomly choose n points on $[0, 1]^2$ to form the design D_1 and calculate its mixture discrepancy.
(2) Use the translation method to obtain the design D_2 on T^3. Calculate the mean square distance, average distance, maximum distance of D_2.

No.	1	2
1	1	5
2	2	2
3	3	7
4	4	4
5	5	1
6	6	6
7	7	3

Repeat Steps (1)–(2) m times, compare their results, and give your conclusion.

8.5

Let $n = 17$. Use the NTLBG algorithm to construct the uniform mixture designs on T^3.

8.6

For the designs with restricted mixtures, prove the restriction in (8.2.8).

8.7

Consider the three factors in Example 8.2.3. Use the conditional method to construct a 17-run uniform design with restricted mixtures.

8.8

Consider the design region

$$S_2 = \{(x_1, x_2, x_3)|x_1 + x_2 + x_3 = 1, x_1^2 + x_2^2 \leqslant 0.36, x_i \geqslant 0\}.$$

Under the uniformity criterion CCD, use the switching algorithm in Algorithm 8.2.4 to construct a 15-point uniform design on S_2.

8.9

To explore the influence of component compatibility changes on antipyretic effect of Maxing Shigan decoction, the uniform design of experiments with mixtures was used. Ephedrae Herba (x_1/g), Armeniacae Semen Amarum (x_2/g), Glycyrrhizae Radix et Rhizoma Preparata Cum Melle (x_3/g), and Gypsum Fibrosum (x_4/g) were considered as 4 factors. The originally used treatment in hospitals is (6, 6, 6, 24), and the total weight is 42 g, and the response, the heat inhibition rate after 6 h, denoted by $y(\%)$, is equal to 52.19%. For investigating the reasonableness of the original treatment and finding better treatment, the researcher designed 12 different allocated proportions of Maxing Shigan decoction. The total weight of the four factors are kept to 42 g, and the corresponding design points and the response are as follows.

Analyze the data, compare the result of the original treatment, and give your conclusion.

No.	x_1(g)	x_2(g)	x_3(g)	x_4(g)	y
1	3.15	25.12	12.02	1.72	41.97
2	17.1	19.82	2.33	2.75	36.13
3	21	2.32	3.89	14.79	28.47
4	1.83	15.57	7.17	17.42	52.92
5	0.59	6.56	18.88	15.97	53.28
6	11.71	16.46	1.73	12.1	29.93
7	14.15	5.83	21.1	0.92	16.79
8	6.09	2.32	26.59	7	29.1
9	7.76	15.75	11.56	6.93	49.64
10	4.56	9.88	1.15	26.41	56.75
11	9.62	0.68	11.89	19.81	52.19
12	27.44	4.7	6.98	2.88	10.53

8.10

In an experiment for Chinese medicinal material, five components are considered and the restricted ranges of the components $x_1 \sim x_5$ are 10%~60%, 10% ~60%, 30%~60%, 10% ~12%, 10% ~ 12%, respectively. The average yield (g) and survival rate (%) are two responses and denoted by y_1 and y_2.
Analyze the data and find the optimal components.

No.	x_1(%)	x_2(%)	x_3(%)	x_4(%)	x_5(%)	y_1	y_2
1	15.66	36.69	45.31	0.98	1.36	284.5	44.44
2	33.89	16.77	41.14	1.66	6.53	356.8	44.44
3	19.77	19.03	57.39	1.69	2.11	337.9	88.89
4	36.21	13.36	32.77	8.28	9.37	463.8	100
5	47.08	16.54	33.83	1.28	1.27	326.3	66.66
6	15.57	39.57	35.26	0.65	8.95	454.3	100
7	20.95	33.3	35.23	4.42	6.1	359.1	88.89
8	38.23	14.16	45.3	1.45	0.85	381.4	55.56
9	40.21	16.78	33.03	0.68	9.3	446	55.56
10	17.32	18.97	52.68	0.97	10.05	433.3	77.78
11	18.57	16.08	54.43	9.88	1.04	342.7	66.67
12	31.03	28.74	33.45	5.66	1.12	374	55.56
13	15.96	40.91	32.56	9.72	0.85	397	44.44
14	13.05	34.97	33.66	9.27	9.05	416	88.89
15	14	14.02	52.72	9.33	9.92	475.9	100
16	14.72	33.72	42.64	7.91	1.01	290	22.22
17	32.14	32.36	33.02	0.6	1.88	317.4	88.89
18	40.5	13.49	36.32	9.03	0.66	349.8	44.44
19	26.26	20.98	38.29	8.7	5.77	474.25	44.44
20	15.57	48.33	33.9	1.14	1.06	0	0

References

Borkowski, J.J., Piepel, G.F.: Uniform designs for highly constrained mixture experiments. J. Qual. Technol. **41**, 35–47 (2009)

Chan, L.Y.: Optimal designs for experiments with mixtures: a survey. Commun. Stat. Theory Methods **29**, 2281–2312 (2000)

Chen, R.B., Hsu, Y.W., Hung, Y., Wang, W.C.: Discrete particle swarm optimization for constructing uniform design on irregular regions. Comput. Stat. Data Anal. **72**, 282–297 (2014)

Chuang, S.C., Hung, Y.C.: Uniform design over general input domains with applications to target region estimation in computer experiments. Comput. Stat. Data Anal. **54**, 219–232 (2010)

Cornell, J.A.: Experiments with Mixtures, Designs, Models and the Analysis of Mixture Data, 3rd edn. Wiley, New York (2002)

Cornell, J.A.: A Primer on Experiments with Mixtures. Wiley, New Jersey (2011)

Fang, K.T.: Experimental designs for computer experiments and for industrial experiments with model unknown. J. Korean Stat. Soc. **31**, 277–299 (2002)

Fang, K.T., Ma, C.X.: Orthogonal and Uniform Experimental Designs. Science Press, Beijing (2001)

Fang, K.T., Wang, Y.: Number-Theoretic Methods in Statistics. Chapman and Hall, London (1994)

Fang, K.T., Yang, Z.H.: On uniform design of experiments with restricted mixtures and generation of uniform distribution on some domains. Stat. Probab. Lett. **46**, 113–120 (2000)

Fang, K.T., Yuan, K.H., Bentler, P.M.: Applications of number-theoretic metods to quantizers of elliptically contoured distributions. Multivar. Anal. Appl. **24**, 237–251 (1994)

Jing, D., Li, P., Stagnitti, F., Xiong, X.: Optimization of laccase production from trametes versicolor by solid fermentation. Can. J. Microbiol. **53**, 245–251 (2007)

Johnson, M.E.: Multivariate statistical simulation. J. R. Stat. Soc. Ser. A **151**, 930–932 (1988)

Linde, Y.L., Buzo, A., Gray, R.M.: An algorithm for vector quantizer design. IEEE Trans. Commun. **28**, 84–95 (1980)

Liu, Y., Liu, M.Q.: Construction of uniform designs for mixture experiments with complex constraints. Commun. Stat. Theory Methods **45**, 2172–2180 (2016)

Ning, J.H., Fang, K.T., Zhou, Y.D.: Uniform design for experiments with mixtures. Commun. Stat. Theory Methods **40**, 1734–1742 (2011a)

Ning, J.H., Zhou, Y.D., Fang, K.T.: Discrepancy for uniform design of experiments with mixtures. J. Stat. Plan. Inference **141**, 1487–1496 (2011b)

Piepel, G., Anderson, C.M., Redgate, P.E.: Response surface designs for irregularly-shaped region - parts 1, 2, and 3. 1993 Proccedings of the Section on Physical and Engineering Sciences, pp. 169–179. American Statistical Associetion, Alexandria (1993)

Piepel, G., Cooley, S., Gan, H., Kot, W., Pegg, I.: Test matrix support TLCP model development for RPP-WTP HWL glasses, Vsl-03s3780-1, vitreous state laboratory. The Catholic University of America, Washington (2002)

Prescott, P.: Nearly uniform designs for mixture experiments. Commun. Stat. Theory Methods **37**, 2095–2115 (2008)

Scheffé, H.: Experiments with mixtures. J. R. Stat. Soc. Ser. B **20**, 344–360 (1958)

Scheffé, H.: The simplex-centroid design for experiments with mixtures. J. R. Stat. Soc. Ser. B **25**, 235–263 (1963)

Tang, M., Li, J., Chan, L.Y., Lin, D.K.J.: Application of uniform design in the formation of cement mixtures. Qual. Eng. **16**, 461–474 (2004)

Wang, Y., Fang, K.T.: Number theoretic methods in applied statistics (II). Chin. Ann. Math. Ser. B **11**, 41–55 (1990)

Wang, Y., Fang, K.T.: Uniform design of experiments with mixtures. Sci. China Ser. A **39**, 264–275 (1996)

Subject Index

K.-T. Fang et al., *Theory and Application of Uniform Experimental Designs*, Lecture Notes in Statistics 221,
https://doi.org/10.1007/978-981-13-2041-5

Printed in the United States
By Bookmasters